# The Foundations of Population Genetics

# The Foundations of Population Genetics

Daniel M. Weinreich

The MIT Press
Cambridge, Massachusetts
London, England

The MIT Press would like to thank the anonymous peer reviewers who provided comments on drafts of this book. The generous work of academic experts is essential for establishing the authority and quality of our publications. We acknowledge with gratitude the contributions of these otherwise uncredited readers.

This book was set in Times New Roman by Westchester Publishing Services. Printed and bound in the United States of America.

Library of Congress Cataloging-in-Publication Data is available.

ISBN: 978-0-262-04757-9

10   9   8   7   6   5   4   3   2   1

To Wendy Nelson, my dearest friend and biggest fan. Her indefatigable enthusiasm for this project (among many) immensely eased its completion. And to Abdi Bashir and Pat Fraze, my two favorite math teachers. And to the generations of population geneticists, now and past, who together have built a coherent and rigorous theoretical apparatus with which to study the most interesting material process I know. It has been my absolute joy to collate their work in this textbook.

# Contents

Preface    xi

## 1 Deterministic Single-Locus Population Genetics    1
1.1    Natural Selection    3
   1.1.1    An Exponentially Growing Population    3
   1.1.2    Darwin's Model of Natural Selection    7
   1.1.3    The Time for a Selective Fixation    12
   1.1.4    Fisher's Fundamental Theorem of Natural Selection    15
   1.1.5    Natural Selection in Populations with Nonoverlapping Generations    18
1.2    Mutation    21
   1.2.1    The Evolution of Allele Frequencies under Mutation Alone    21
   1.2.2    Mutation/Selection Equilibrium    23
   1.2.3    The Error Catastrophe    25
   1.2.4    Beneficial Mutations    26
   1.2.5    Soft Selection and Hard Selection    26
1.3    One-Locus Sexual Reproduction    28
   1.3.1    The Population Genetics of Diploid Reproduction    30
   1.3.2    Natural Selection in Diploids    35
1.4    Population Structure, Migration, and Nonrandom Mating    39
   1.4.1    Models of Population Structure and Migration    40
   1.4.2    Migration/Selection Equilibrium    41
   1.4.3    Allele Frequency Clines    42
   1.4.4    Nonrandom Mating in Diploids    43
      1.4.4.1    Assortative and disassortative mating    44
      1.4.4.2    Inbreeding and the probability of identity by descent    45
      1.4.4.3    Population subdivision and the Wahlund effect    47
      1.4.4.4    Nonrandom mating within a subdivided population    50
1.5    Chapter Summary    50

## 2 Stochastic Single-Locus Population Genetics    51
2.1    Stochasticity While Rare, and Establishment    53
   2.1.1    The Establishment Problem    54
   2.1.2    Establishment of a New Beneficial Allele    59
2.2    Stochasticity at All Frequencies and Random Genetic Drift    62
   2.2.1    The Chapman–Kolmogorov Equation    63
   2.2.2    The Diffusion Approximations of the Chapman–Kolmogorov Equation    67

2.2.2.1 Probability mass moves through time   68
2.2.2.2 Restricting movement of probability mass to adjacent stochastic states   69
2.2.2.3 Partitioning movement of probability mass into biased and unbiased components   69
2.2.2.4 Moving to continuous time and frequency   70
2.2.3 Representing Biology in the Chapman–Kolmogorov and Diffusion Equations   73
2.2.3.1 The Wright–Fisher model of random genetic drift   73
2.2.3.2 Incorporating biased processes into the Wright–Fisher model   76
2.2.3.3 The Moran model of random genetic drift   77
2.2.4 Using the Backward Diffusion Approximation to Study Fixation Events   79
2.2.4.1 The probability of fixation for a selectively neutral allele   79
2.2.4.2 The probability of fixation for a selected allele   82
2.2.4.3 Time to fixation for a new allele   87
2.2.5 Using the Forward Diffusion Approximation to Study Internal Equilibrium Allele Frequencies   88
2.3 Random Genetic Drift and Heterozygosity   91
2.3.1 The Rate of Decay in Heterozygosity under the Wright–Fisher Model   91
2.3.2 The Infinite Alleles Model and Heterozygosity at Mutation/Drift Equilibrium under the Wright–Fisher Model   94
2.3.3 Heterozygosity at Migration/Drift Equilibrium under the Wright–Fisher Model   97
2.4 Coalescent Theory   97
2.4.1 Modeling Genealogies under the Wright–Fisher Model   98
2.4.2 The Infinite Sites Model and the Site Frequency Spectrum   102
2.4.3 Population Subdivision and Incomplete Lineage Sorting   106
2.5 The Effective Size of a Population   108
2.6 Chapter Summary   112

3   Multilocus Population Genetics   115
3.1 Deterministic Multilocus Theory   117
3.1.1 Pairwise Linkage Disequilibrium and Genetic Recombination   117
3.1.2 The Two-Locus Wahlund Effect   121
3.1.3 Pairwise Epistasis, Natural Selection, and Recombination   123
3.1.4 Multilocus Mutation/Selection Equilibrium   128
3.1.5 The Multilocus Error Catastrophe   133
3.1.6 Sequence Space and Fitness Landscapes   136
3.1.7 Multilocus Adaptation on Fitness Landscapes under the Strong Selection/Weak Mutation Assumption   140
3.1.8 Stochastic Tunneling   144
3.2 Stochastic Multilocus Theory   147
3.2.1 Multilocus Coalescent Theory   148
3.2.1.1 The ancestral recombination graph   148
3.2.1.2 Recombination/drift equilibrium   150
3.2.2 Selective Sweeps and Genetic Draft   151
3.2.2.1 Recurrent selective sweeps without recombination   151
3.2.2.2 Coalescent time immediately after a selective sweep with recombination   154
3.2.2.3 Recurrent selective sweeps with recombination   157

3.2.3  Background Selection    159
3.2.4  The Hill–Robertson Effect    160
    3.2.4.1  The Fisher–Muller effect    162
    3.2.4.2  Muller's ratchet    166
3.3  Modifier Theory    168
3.3.1  The Evolution of Sex    170
3.3.2  The Evolution of Mutation Rate    173
3.3.3  Other Modifiers    173
3.4  Chapter Summary    174

4  **The Data    175**
4.1  Measuring Genetic Parameters    177
4.1.1  Mutation Rate    178
4.1.2  Recombination Rate    180
4.2  Measuring Effective Population Size    181
4.2.1  Direct Observation of Reproductive Variance    182
4.2.2  Equilibrium Effective Population Size Estimates    182
4.2.3  Nonequilibrium Mutation-Based Effective Population Size Estimates in Asexuals    185
    4.2.3.1  The distribution of pairwise Hamming distances    185
    4.2.3.2  Skyline plots    188
4.2.4  Nonequilibrium Mutation-Based Effective Population Size Estimates in Sexuals    190
    4.2.4.1  The site frequency spectrum    190
    4.2.4.2  Tajima's $D$ statistic    191
    4.2.4.3  The pairwise sequential Markovian coalescent    193
4.2.5  Nonequilibrium Recombination-Based Effective Population Size Estimates    195
4.3  Describing Population Structure, Migration, and Admixture    198
4.3.1  Detecting Population Structure    198
4.3.2  Estimating Population Structure with $F_{ST}$    200
4.3.3  Estimating Migration with the Joint Site Frequency Spectrum    201
4.3.4  Estimating Admixture from Haplotype Structure    203
4.4  Describing Natural Selection    205
4.4.1  Quantifying Natural Selection with the Molecular Clock    205
4.4.2  The McDonald/Kreitman Test    209
4.4.3  Scanning Whole Genome Sequence Data for Natural Selection    211
    4.4.3.1  Detecting natural selection from heterozygosity    211
    4.4.3.2  Detecting natural selection from the site frequency spectrum    212
    4.4.3.3  Detecting natural selection from deme-specific frequency differences    213
    4.4.3.4  Detecting natural selection from haplotype structure    213
4.4.4  Hard and Soft Selective Sweeps    214
4.5  Chapter Summary    217

References    219
Index    231

# Preface

The twenty-first century's genomics revolution has transformed the field of population genetics. A staggering million-fold drop in the cost of DNA sequencing since 2001 has finally given this 100-year-old field direct access to its object of study: large samples of whole genome sequence data from almost any species on Earth. This, in turn, has driven an explosion of new theory. The field's impact on the genomics revolution is similarly profound, because the analysis of genome-scale data is intrinsically a population genetic enterprise. One needs to go no further than the COVID-19 pandemic to appreciate this synergy. Our ability to understand and predict the virus' history and continuing evolution rest almost entirely on the nearly 15 million (and counting) complete viral genome sequences now on hand. Other examples of the two-way exchange between genomics and population genetics are found in current work on cancer progression and treatment, the microbiomes of plants and animals, and the study of human migration and evolution over the last half-million years.

It is thus an ideal time to introduce a book that develops the foundations of population genetics from first principles. New technologies are drawing students of biology to diverse, exciting, and important questions. These opportunities are also attracting scientists from other fields such as medicine, computer science, and physics, many of whom lack formal training in population genetics. This book is intended to serve both audiences.

Population genetics uses mathematical models to predict how frequencies of genetic variants change over time, and this book capitalizes on the pedagogical opportunities that its mathematical foundations provide. Mathematics proceeds by developing a cumulative, principled, and extensible system of knowledge from simple, explicitly stated assumptions. As a consequence, mathematics teaches us to understand, rather than to memorize. This book teaches the rich and deep structure of theoretical population genetics.

Nevertheless, the focus throughout is on biology; mathematics is simply the appropriate language. Each lesson translates a biological phenomenon into a mathematical representation, demonstrates the relevant mathematical derivation, and then translates its evolutionary implications back into English. The book employs many mathematical techniques, but their mastery is not expected nor required to succeed with the biology. Throughout, the mathematics work for the biologist rather than vice versa.

All that is expected from the reader is a curiosity about mathematics, and the book anticipates a continuum of such readers. At one end will be those who perhaps enjoyed mathematics in high school or college before finding a deeper passion for biology or one of its allied disciplines. For that reader (which described your author when he entered the field), the book provides dozens of "teachable moment" text-boxes that illustrate how general mathematical principles are applied to specific biological problems. The reader is not expected to internalize these; the point is simply to demonstrate the mathematical framing and to provide an intuition into how it works. At the other end of the continuum will be those with a strong background in calculus, linear algebra, probability theory, and perhaps stochastic processes. These readers will appreciate the book's thoroughly rigorous structure and may not mind being reminded of one or two technical points along the way.

In addition to teachable moments, the book has 50 original figures to provide complementary visual perspectives on the biology. It also contains almost 40 "empirical aside" text boxes that offer specific biological motivations for many of the questions treated in the text. Finally, the book includes nearly 80 "power user challenges." These are carefully scaffolded questions that offer ambitious readers the opportunity to amplify on, generalize, or reframe particular lessons in the book. Answers to these and other teaching ancillaries are available on the MIT Press website.

This book is designed for a one-semester graduate or upper level undergraduate course on population genetics. The first three chapters develop the theory, and the fourth provides dozens of case studies that illustrate how theory can be inverted to yield evolutionary inferences about a species' history from sequence data. In the interest of time, instructors can omit sections 2.2.1–2.2.2 (the most challenging in the book), 3.1.5–3.1.8, 3.3, and 4.1 without loss of continuity. Chapter 4 can also serve as a superb stand-alone resource for a unit on population genetics embedded in courses on genetics, evolution, or computational biology. In that setting, advanced students could optionally explore earlier chapters for a deeper understanding of the theory underlying specific applications of interest. Finally, the book will serve as an invaluable resource to newcomers from related fields who seek to capitalize on the opportunities posed by the genomic revolution but lack essential background knowledge.

### Acknowledgments

My training in population genetics began when the late Dick Lewontin agreed to take me as a graduate student in 1992, despite the fact that my last biology class had been in high school. (My undergraduate degree is in computer science.) Dick's generosity, intellectual enthusiasm, and rigor remain inspirational. Postdoctoral mentors David Rand, Lin Chao, and Dan Hartl also provided essential education in the field, and I surely learned at least as much from the many friends I made in my natal labs.

This book grew out of lecture notes developed while teaching theoretical population genetics at Brown University in 2015. To paraphrase the proverb, it takes a village to write a textbook. First, I owe a great debt to C. Scott Wylie, a former post-doc in my group, who essentially dared me to try this pedagogical approach. I also learned a great deal from several lectures Scott contributed to that first iteration of the course. The present volume benefited immensely from thoughtful comments provided by many friends and colleagues. Nick Barton, Andrew Berry, Maciej Boni, Armita Manafzadeh, Dmitri Petrov, Sawyer Smith, and three anonymous reviewers read and commented on most or all of the manuscript. Nick provided particularly thorough and trenchant feedback. Emilia Huerta-Sanchez, Sally Otto, David Rand, Sohini Ramachandran, Julian Stamp, Lindi Wahl and John Wakeley all volunteered to focus on content in areas of their specific expertise. Sally was particularly generous, even as she was more-than-fully occupied with COVID-19 epidemiology. All remaining errors and points of confusion are my responsibility, and feedback from readers will be gratefully accepted at daniel.weinreich@gmail.com.

Last but by no means least, my experience with the MIT Press could not have been better. My acquisitions editor, Anne-Marie Bono, responded enthusiastically within a day of my initial inquiry. She secured three excellent reviews from highly qualified readers, shepherded the prospectus through the publishing committee, and most importantly, told me to "Write the book you need to write," when I told her I wouldn't make our initial deadline. Assistant production manager Kate Elwell and her colleagues, and production editor Madhulika Jain at Westchester Publishing Services, answered countless questions with precision and alacrity.

Finally, I acknowledge the tremendous support that I received from two institutions of higher education, without which the book would not exist: Brown University and l'Université de Montpellier. I feel very fortunate to be spending my faculty career at an institution like Brown, which values classroom teaching as much as I do. From the day I arrived, I have been given license to develop curricula that fuel my intellectual curiosities. This book is the culmination of one such effort, and I was deeply gratified to have been named a Royce Family Professor of Teaching Excellence in the midst of its writing. I also benefited immensely from the opportunity to do much of the writing at l'Université de Montpellier, first on a sabbatical during the spring of 2021 (COVID-19 pandemic notwithstanding) and then on a return visit in June of 2022. Many rich collaborations and important friendships have developed there, and I look forward to many more trips to that intellectual community in that wonderful city.

D. M. W.
Providence, Rhode Island
Montpellier, France

# 1

## Deterministic Single-Locus Population Genetics

Theoretical population genetics builds mathematical models with which to study evolution. More precisely, it seeks to model temporal changes in the genetic composition of a population in response to five biological processes: natural selection, mutation, sexual reproduction, migration, and stochastic effects like random genetic drift. It also provides the framework for inferring a population's evolutionary history from its current genetic composition.

As noted in the preface, the twenty-first-century–omics revolution makes this the golden age of population genetics. The current flood of genome sequence data finally presents this 100-year-old field with its object of study: immense samples of whole-genome data from almost any species on Earth. This in turn has driven an explosion of new theory. Conversely, population genetics represents the best toolkit for the analysis of these data. Current examples of the two-way exchange between genomics and population genetics are found in work on tracking and predicting the evolution of the SARS-CoV-2 virus, cancer progression and treatment, the microbiomes of plants and animals, and the study of human migration and evolution over the last half-million years.

This chapter presents deterministic treatments of natural selection, mutation, genetic recombination, and migration. Chapter 2 introduces stochastic effects, and chapter 3 extends results from the first two to accommodate more than one locus in the genome. Chapter 4 illustrates how theory can be inverted to allow biological inferences from sequence data in natural populations.

But before beginning the biology, teachable moments 1.1 and 1.2 ask two even more basic questions.

---

**Teachable Moment 1.1:** What is a mathematical model?

Models are abstract representations of the world. *Mathematical models* represent some feature(s) of the world with mathematical expressions. Population genetic models almost always begin with a mathematical description of the rate at which an evolving population changes in response to biological process(es). *Solving* the model consists of using the rules of mathematics to find new expressions that teach us something quantitative about the future state of a population as a function of elapsed time, its current state, and the relevant biology.

Our mathematical expressions are comprised of three kinds of variables. *Parameters* are constants that quantify the population's state in the current moment as well as the magnitudes of all relevant biological processes. *Dependent variables* are those that describe the population at other times. They are so called because they are dependent on the *independent variable*, for us almost always time. To emphasize this relationship, dependent variables are almost always written as explicit functions of time, as for example in equation (1.1).

Mathematical models are always simplifications, and indeed, we often favor models that are as simple as possible, both because progress is easier and because results are more interpretable. Consequently, all models are fundamentally wrong. But with skill and luck, it is often possible to construct models that capture some essential aspects of the underlying biology, which, in turn, allow us to build intuition about more interesting problems that are too hard to rigorously model. These points were famously summarized by the statistician George Box (1979): "All models are wrong, but some are useful." Putting things a bit differently, the mathematician Sam Karlin (1983) said, "The purpose of models is not to fit the data but to sharpen the questions." (Indeed, for many, framing rather than answering sharp questions is the goal of theoretical science.)

---

**Teachable Moment 1.2:** What is the difference between deterministic and stochastic models?

As explained in teachable moment 1.1, population genetic models predict a population's future state as a function of parameter values, elapsed time, and the current state of the population. There are two kinds of models: deterministic and stochastic. We employ both in this textbook.

*Deterministic models* assume that the future is completely determined by parameter values and elapsed time. They disregard all random influences on the population's behavior. We develop only deterministic models in chapter 1. In contrast, *stochastic models* also capture statistical fluctuations in the population's behavior due to random effects, which are then propagated through time. Consequently, stochastic models only predict the future in a probabilistic sense. More specifically, they yield the probabilities of finding the population in each conceivable state at some point in the future. We introduce stochastic models in chapter 2.

Importantly, stochastic models do more than just add a bit of noise around deterministic predictions. As we shall see, they yield answers to quite different questions than those asked in deterministic models, thereby enriching and complementing those results. On the other hand, we will also find many deep and instructive connections between deterministic and stochastic models describing analogous problems.

---

As explained in teachable moment 1.1, model building always involves simplification. For example, we begin this book with a focus on evolution in uniparental organisms like bacteria. Unlike us, these organisms undergo *clonal reproduction*—for example, by fission in the case of single-celled microbes. Of course, many organisms of interest are biparental—that is, reproduce via *sexual reproduction*—and we will address those complications beginning in section 1.3. But studying clonal organisms already allows us to understand many important and complex population genetic concepts and phenomena, thereby giving insight into the evolution of all organisms.

---

**Empirical Aside 1.1**

Strictly biparental organisms comprise a small proportion of species on Earth (Dykhuizen 2005; "Microbiology by numbers" 2011; Sweetlove 2011) and a vanishingly small proportion of its biomass (Dance 2021). Thus, in addition to their pedagogical value, the population genetics of clonal organisms are of broad importance in their own right.

---

With these preliminaries out of the way, we begin the biology.

## 1.1 Natural Selection

One of Darwin's two great insights in *On the Origin of Species* (Darwin 1859) was that heritable differences in rates of reproduction and death among organisms would drive the evolution of the species through a process he called *natural selection*. (In point of fact, Darwin and Alfred Russel Wallace independently hit on the same idea; Darwin and Wallace 1858.) We now develop a deterministic model of that process.

---

**Power User Challenge 1.1**

What was Darwin's other main insight in *On the Origin of Species*?

---

### 1.1.1 An Exponentially Growing Population

To build intuition, we ask an even simpler question: If a population is of size $N(0)$ organisms at time $t = 0$, what is its size $N(t)$ at some other point in time $t \neq 0$? Suppose these organisms heritably have per capita birth and mortality rates $b$ and $m$, respectively. In English, that means that in time interval $\Delta t$, a fraction $b\Delta t$ of the organisms reproduce and a fraction $m\Delta t$ of them die. Because the change in population size is determined by the net effect of birth and death, we can focus on the difference in rates, called the organism's *Malthusian parameter* or *intrinsic rate of growth*, written as $r = b - m$. The population size thus changes by an amount $r\Delta t$ in time interval $\Delta t$. Mathematically, this reads

$$\Delta N(t) = N(t + \Delta t) - N(t) = rN(t)\Delta t. \tag{1.1}$$

Biology tells us that neither $b$ nor $m$ can be negative, although no such constraint exists on $r$. Rather, the sign of $r$ tells us whether population size is increasing ($r > 0$), decreasing ($r < 0$), or constant ($r = 0$).

Equation (1.1) captures some biology but doesn't yet let us answer the question posed at the start of the previous paragraph. We now use the rules of mathematics to make progress. To begin, we let $\Delta t$ become arbitrarily small, written as $\Delta t \to 0$. This doesn't mean that $\Delta t = 0$, since this would render $\Delta N(t) = N(t+0) - N(t) = N(t) - N(t) = 0$. Rather, what we are doing biologically is to make $\Delta t$ sufficiently small that we can discount the possibility of more than one birth or death occurring during the interval. This is represented mathematically by the first derivative of $N(t)$ (teachable moment 1.3), written as

$$\frac{dN(t)}{dt} = \lim_{\Delta t \to 0} \frac{N(t+\Delta t) - N(t)}{\Delta t} = \lim_{\Delta t \to 0} \frac{\Delta N(t)}{\Delta t} = rN(t). \tag{1.2}$$

---

**Teachable Moment 1.3**: Derivatives

The *first derivative* of a function $f(x)$, written as $\frac{df(x)}{dx}$, is its rate of change at $x$, or geometrically, the slope of $f(x)$ at $x$. Recalling that a function's slope is its change over an interval divided by the size of the interval draws our attention to $\frac{f(x+\Delta x) - f(x)}{\Delta x} = \frac{\Delta f(x)}{\Delta x}$ as an approximation of the first derivative. This becomes exact in the limit $\Delta x \to 0$, mathematically written as $\frac{df(x)}{dx} = \lim_{\Delta x \to 0} \frac{\Delta f(x)}{\Delta x}$. We can continue in this fashion to find the *second derivative* of $f(x)$, defined as $\frac{d^2 f(x)}{dx^2} = \lim_{\Delta x \to 0} \frac{\frac{f(x+\Delta x) - f(x)}{\Delta x} - \frac{f(x) - f(x-\Delta x)}{\Delta x}}{\Delta x}$. In English, the second derivative is the rate of change in the first derivative (or slope) of $f(x)$ at $x$.

Yet-higher derivatives are computed by iteratively applying the same framework. One technical point is important: not all functions are differentiable. For example, the slope of a function that jumps discontinuously between two values at some $x$ is undefined at that $x$.

We will also occasionally encounter functions of two independent variables, written as $f(x, y)$. We can again investigate the first and higher derivatives of such functions with respect to each of its independent variables, holding all others constant. We write such *partial derivatives* as $\frac{\partial f(x,y)}{\partial x}, \frac{\partial f(x,y)}{\partial y}, \frac{\partial^2 f(x,y)}{\partial x^2}$, and so on. Note the cursive $\partial$, which signals that these are partial derivatives, and that the denominator indicates which independent variable this derivative is relative to.

---

To solve equation (1.2), we first rearrange to find

$$\frac{dN(t)}{N(t)} = rdt,$$

which on integration yields

$$\ln[N(t)] = rt + C. \tag{1.3}$$

Here, we take advantage of a fact from calculus: the integral of $\frac{dx}{x}$ is the natural logarithm of $x$, written as $\ln(x)$. Finally, exponentiating both sides of equation (1.3) yields

$$N(t) = e^{rt+C}. \tag{1.4}$$

---

**Teachable Moment 1.4:** The constant of integration

The derivation of equation (1.3) introduced the symbol $C$, called the constant of integration. To understand $C$, think about doing this last step in reverse: take the derivative of the right side of equation (1.3) with respect to $t$ (teachable moment 1.3). The derivative is the slope; here, equal to $r$. But the slope of $rdt + 1$ is also $r$; ditto $rdt + 2$, and so on. You see that by plotting those functions, or more formally by recalling that the derivative of a sum is the sum of the derivatives, and that the derivative of any constant is 0. This is what the constant of integration does: it acknowledges the fact that there is a whole family of functions whose derivatives are $r$, namely $rdt + C$. The value of $C$ in any particular case depends on additional information in the problem, as we see next.

---

What is $C$ in terms of the biological problem at hand? So far, we've been focused on the influence of Malthusian parameter $r$, but $N(t)$ must somehow also depend on $N(0)$, the population size at time $t = 0$, which enters our solution via $C$. Specifically, setting $t = 0$ in equation (1.4) yields $N(0) = e^C$, so $C = \ln[N(0)]$. Substituting this into equation (1.4) finishes the job

$$N(t) = e^{rt + \ln[N(0)]} = e^{rt} \times e^{\ln[N(0)]} = N(0)e^{rt}. \tag{1.5}$$

(Recall that $a^{b+c} = a^b \times a^c$.) In English, we have learned that if the rate of change in population size is proportional to its current value (equation 1.2), population size changes exponentially in time. The dynamics of this process are illustrated in figure 1.1.

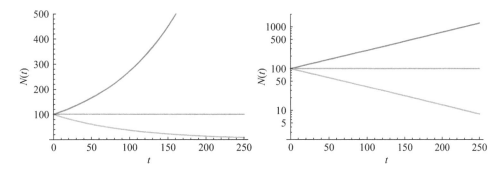

**Figure 1.1**
Exponential growth of a population. Equation (1.5), plotted for $r = 0.01$ (blue), $r = 0$ (mustard), $r = -0.01$ (green), and $N(0) = 100$. The two panels differ only in that the $y$-axis on the right is log transformed. The left panel graphically illustrates the key feature of equation (1.2): the slope is proportional to $N$. That exponential models are linear on the right (this is called a *semi-log plot*) follows immediately from their derivation. Conversely, data that appear linear on a semi-log plot immediately suggest an exponential mechanism.

**Teachable Moment 1.5:** Logarithms

The *logarithm* is the inverse of exponentiation, and vice versa. Mathematically, if $y = a^x$, it necessarily follows that $\log_a(y) = x$, and vice versa. Here, $a$ is called the *base* of the logarithm. Importantly, logarithms of different bases are simply proportional to one another. Mathematically, $\log_b(x) = \dfrac{\log_a(x)}{\log_a(b)}$, so for example $\log_{10}(x) = \dfrac{\log_2(x)}{\log_2(10)} \approx \dfrac{\log_2(x)}{3.32}$. In English, rescaling between bases only requires multiplying by a constant.

In calculus, logarithm always means natural logarithm; that is, log base e, a constant roughly equal to 2.718 that emerges in many contexts. We will write $\ln(x)$ as shorthand for $\log_e(x)$. Our preference for base e in place of base 10 reflects at least two features of the former quantity. First, as we have just seen, $y = e^x$ emerges as the solution to the simplest differential equation possible: $\dfrac{dy}{dx} = y$. Perhaps relatedly, $e^x$ has an interesting feature on differentiation: $\dfrac{de^x}{dx} = e^x$.

On the other hand, we find $\log_{10}$ an easier scale to comprehend intuitively. For this reason, when one encounters a logarithm with no base specified, it can be assumed to be base 10, and we sometimes plot $\log_{10}$-transformed data. Figure 1.1 illustrates why. By the time $t = 250$, equation (1.5) predicts that a population with $r = 0.01$ (blue) will be almost 150 times the size of one with $r = -0.01$ (green). Representing that large a dynamic range on the linear scale employed on the left would largely obscure the behavior when populations are small. (This is why the blue trace was truncated at $t \approx 160$ in that panel.) In contrast, the right panel easily represents the full dynamic range of population sizes, because its $y$-axis is log transformed. Finally, we sometimes describe a quantity's dynamic range on a $\log_{10}$ scale. For example, the $y$-axis in the right panel of figure 1.1 spans three *log orders* because $\log_{10}(2000) - \log_{10}(2) = 3$.

---

**Teachable Moment 1.6:** Our model predicts that population size will not always be an integer

The attentive reader may have noticed that equation (1.5) need not equal an integer. Despite the biological impossibility of noninteger population sizes, we often disregard this issue. First, the discrepancy becomes proportionally smaller as population size increases. And the issue disappears entirely if we regard $N$ as a density per unit space rather than the number of organisms (Felsenstein 2019). But most importantly, this defect doesn't undermine most of what the equation has to teach us.

---

**Teachable Moment 1.7:** Units of time

What are the units of time in equation (1.5)? Figure 1.1 shows that when $N(0) = 100$ and $r = 0.01$, $N(150) \approx 100e^{0.01 \times 150} \approx 448$, but is that 150 seconds, days, years, millennia? (Incidentally, the symbol "$\approx$" means "approximately equal to.") The answer lies in the math: $e^{rt}$ must be unitless, since the units of $N(0)$ and $N(t)$ are the same. Thus, the units of $t$ must be the reciprocal of the units of $r$. For example, if $r = 0.01$ per day, then in 150 days a population of 100 individuals will grow to approximately 448.

---

**Power User Challenge 1.2**

If an organism's birth and death rates are $b$ and $m$, respectively, what are its generation time and lifespan? In what units?

---

**Power User Challenge 1.3**

Laboratory cultures of the bacterium *Escherichia coli* double roughly three times per hour, giving an $r = 3 \times \ln(2) \approx 2.08$ hour$^{-1}$. (Note that hours$^{-1}$ means "per hour." This value is the solution for $r$ after setting $2N(0) = N(0)e^{\frac{r\,\text{hours}^{-1}}{3}}$, which is the mathematical representation (equation 1.5) of the fact that the population doubles three times each hour.) What is $r$ in units of minutes$^{-1}$? Next, one *E. coli* cell weighs $\approx 10^{-12}$ grams. Assuming unlimited nutrients, how many hours would it take for the descendants of one cell to weigh 1 gram? Hint: see teachable moment 1.5.

---

The foregoing allows us to illustrate many of the points made in teachable moment 1.1. Equation (1.1) captures the biology, which we then solve to find equation (1.5). That derivation is mathematics and not biology. For example, the same model can be used to compute compound interest, reading $r$ as the interest rate and $N(0)$ as the initial deposit. The model has two parameters $r$ and $N(0)$, one independent variable $t$, and one dependent variable $N(t)$. Finally, it is often wrong (teachable moment 1.6).

---

**Teachable Moment 1.8**

We have implicitly assumed that a population's rate of change ($r$) is constant in time. This has the unrealistic implication that a growing population ($r > 0$) will grow without bound. Introducing $K$, the maximum population size allowed (or *carrying capacity*) suggests the next-simplest model of population growth, called the *logistic model*, written $\frac{dN(t)}{dt} = rN(t)[K - N(t)]$. In English, the term in square brackets becomes small (thereby slowing the rate of change) as population size approaches $K$. See power user challenge 1.5 to learn more about this model.

We have also assumed that birth and death rates are independent of an organism's age. This is again unrealistic for many species whose population growth rate depends not only on total population number but also on the number of organisms at reproductive age. The mathematics of growth in so called *age-structured populations* is treated in many ecology textbooks, and the reader is directed to Charlesworth (1994) to learn more about their evolutionary implications.

---

### 1.1.2 Darwin's Model of Natural Selection

We are now in a position to write a mathematical representation of Darwin's model of natural selection. Assuming two types in the population, we are interested in what happens if they heritably reproduce at different rates. Such heritability implies a genetic basis for the difference, and we label the two types $A$ and $a$. (These will shortly be recognized as alternative alleles at a genetic locus.) We now write $r_A$ and $r_a$ for their Malthusian parameters (or *Malthusian fitnesses*), $N_A(0)$ and $N_a(0)$ for their initial counts, and introduce a new dependent variable,

$$p_A(t) = \frac{N_A(t)}{N_A(t) + N_a(t)}, \tag{1.6}$$

for the fraction of the population composed of type $A$ individuals at time $t$.

Equation (1.5) gives us $N_A(t)=N_A(0)e^{r_A t}$ and $N_a(t)=N_a(0)e^{r_a t}$. Substituting into equation (1.6) and simplifying yields

$$p_A(t)=\frac{N_A(0)e^{r_A t}}{N_A(0)e^{r_A t}+N_a(0)e^{r_a t}}=\frac{p_A(0)e^{st}}{1+p_A(0)\,(e^{st}-1)}, \tag{1.7}$$

where $s=r_A-r_a$ is the *Malthusian selection coefficient* between the two types in our population. (This derivation follows first by multiplying equation (1.6) by

$$1=\frac{\dfrac{1}{N_A(0)+N_a(0)}}{\dfrac{1}{N_A(0)+N_a(0)}}\quad\text{to find}\quad\frac{p_A(0)e^{r_A t}}{p_A(0)e^{r_A t}+[(1-p_A(0)]e^{r_a t}}\quad\text{and then multiplying by}\quad 1=\frac{\dfrac{1}{e^{r_a t}}}{\dfrac{1}{e^{r_a t}}}$$

to find $\dfrac{p_A(0)e^{st}}{p_A(0)e^{st}+[1-p_A(0)]}$, taking advantage of the facts that $\dfrac{a^x}{a^y}=a^{x-y}$ (seen also

immediately below equation [1.5]) and the definition of $s$. Finally factoring $p_A(0)$ in the denominator yields equation (1.7).) Note that we have gone from four parameters to just two: equation (1.7) depends on only the ratio of starting counts, and the difference in Malthusian parameters. Also note its dependence on the *compound parameter st*. In English, results depend only on the product of these two quantities, not on their values individually.

This expression can be used to describe the selective displacement of the resident or *wild type* by another type with higher fitness. For example, imagine a single more-fit type $A$ individual invading a population of less-fit type $a$ residents. Mathematically, that means assuming $N_A(0)=1$ and $s>0$. Figure 1.2 illustrates the time course over which the $A$ type individuals come to replace type $a$ individuals.

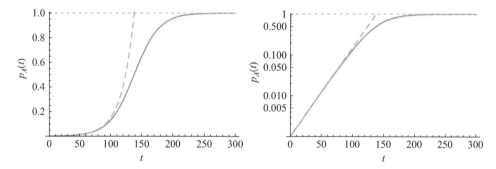

**Figure 1.2**

Time course of natural selection. Equation (1.7), where $N(0)=N_A(0)+N_a(0)=1,000$, $p_A(0)=\dfrac{1}{N(0)}=0.001$, and

$s=0.05$ (blue solid line). The fitter type ($A$) is selectively displacing the less fit type ($a$). Early in the process, this displacement is nearly exponential (mustard dashed line): mathematically, when $t$ is small, $e^{st}-1$ is nearly 0, meaning that that equation (1.7) is nearly identical to equation (1.5) after replacing $s$ for $r$. Later, $p_A(t)$ asymptotically approaches 1 (green dotted line). The two panels differ only in that the $y$-axis is log transformed on the right.

**Empirical Aside 1.2**

Data for the peppered moth *Biston betularia* provides a compelling, empirical illustration of the theory represented in figure 1.2. This organism has two common forms, white with black speckles (called typical) and dark (called *carbonaria*), and coloration is heritable. In the middle of the twentieth century, the *carbonaria* form was dominant in industrial, smoke-darkened areas of England, plausibly because the speckled form disproportionately suffered predation in those locations. After air pollution regulations were introduced in the UK, soot dissipated from around industrial centers, and the typical form returned to high frequency. Figure 1.3 presents data from West Kirby, near Liverpool, an industrial center in England.

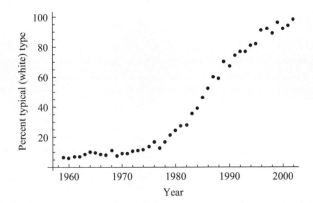

**Figure 1.3**
Frequency of white (typical) form of the peppered moth *Biston betularia* (*y*-axis) as a function of year (*x*-axis) from West Kirby, near Liverpool. (Data from Clarke et al. 1994; Cook 2003.)

Interestingly, introduction of similar air pollution regulations in the US are correlated with a parallel decline in the frequency of *carbonaria* in southeastern Michigan, also an industrial center (Grant et al. 1996). The replicated correlation between *carbonaria* frequency and air pollution on two continents represents strong evidence for the role of natural selection in this species. See power user challenge 1.8 and empirical asides 1.15 and 4.2 to learn more about the biology of this organism, and see section 4.4.3.4 for another case of replicate evolution.

**Power User Challenge 1.4**

What are the units of time $t$ in figure 1.2?

**Power User Challenge 1.5**

Reading $p_A(t)$ as $\dfrac{N(t)}{K}$, figure 1.2 also describes a population's approach to carrying capacity $K$ introduced in teachable moment 1.8. What value of $N(0)$ does the figure assume? Sketch the situation if $N(0) > K$.

This illustrates perhaps the most fundamental biological result of this book: individuals do not evolve. Rather, it is populations that evolve, by virtue of changing frequencies over the timescale of generations (see power user challenge 1.2). Also notice that the increase (or decrease) is mediated by differences in growth rates, not necessarily by direct combat. Of course, differences in fitness could be mediated by superior killing ability, but our model is agnostic on the mechanism underlying selection coefficient $s$.

---

**Empirical Aside 1.3**

What are selection coefficients in nature?

This is a fundamental and empirically challenging question for our field. The fundamental aspect is self-evident: selection coefficients are what drive natural selection. The empirical challenges are several. First, natural selection makes it difficult to sample deleterious selection coefficients because their carriers are often rapidly eliminated from the population. And beneficial selection coefficients are also hard to sample because their carriers quickly displace the wild type. Finally, while we will encounter indirect evidence for widespread, beneficial variants with very small selection coefficients (i.e., on the order of $10^{-4}$ or smaller) in chapter 4, direct measurements of such small effects are difficult. (Moreover, in chapter 2 we shall see that selection coefficients very close to zero can have important evolutionary implications.)

Experiments to sample the fitness of *de novo* variants in the laboratory have been undertaken in many organisms. Several general patterns are seen in such *distributions of fitness effects* (or *DFEs*). (See figure 1 in Firnberg et al. [2014] for typical results at atypically high resolution.) First, deleterious variants represent the largest fraction, and these can usually be partitioned into two distinct subsets: lethals and those of much more modest effect. And most beneficial variants are of modest effect, with selection coefficients less than 1%. For example, Levy et al. (2015) found that selection coefficients between approximately 5% and 10% are 50-fold less common than are those of smaller effects.

Natural populations sample vastly more mutations than is possible in the laboratory, confirming that rare, large-effect beneficial mutations exist. For example, in power user challenge 1.8, we will see that one of the *Biston betularia* forms (empirical aside 1.2) may have enjoyed a 20% fitness advantage over the other. And some mutations in *E. coli* exposed to antibacterials confer 100-fold or greater fitness advantages (Van den Bergh et al. 2016).

---

To better understand the shape of figure 1.2, we differentiate equation (1.7) with respect to time (teachable moment 1.3) to find its rate of change

$$\frac{dp_A}{dt} = sp_A(1 - p_A). \tag{1.8}$$

(We write $p_A$ rather than $p_A(t)$ for simplicity, since this result is independent of $t$. Also note that while many authors write $q$ for the frequency of the $a$ type, we instead write $1 - p_A$ to emphasize the fact that these two frequencies must always sum to 1. On the other hand, our choice obscures a symmetry in this result evident when written as $\frac{dp_A}{dt} = spq$.) Since neither $p_A$ nor $1 - p_A$ can be less than 0 or greater than 1 (they are both frequencies), equation (1.8) illustrates that $A$ type will displace type $a$ if and only if its Malthusian fitness is larger (i.e., if $s > 0$). If $s < 0$, the opposite will occur, and if $s = 0$, the two types' frequencies will remain unchanged. In addition to these rather commonsense findings, note that equation (1.8) depends on the product of $p_A$ and $1 - p_A$. This means that our model of natural selection is logistic (teachable moment 1.8): the early exponential growth of $p_A$ becomes damped as it approaches 1 by an amount that is proportional to the remaining distance to that limit.

Equation (1.8) equally describes the selective elimination of a less-fit $A$-bearing individual that appears in a population of more-fit $a$-bearing residents. In English, the selection coefficient (and thus $\frac{dp_A}{dt}$) is now negative. We often refer to beneficial types as responding to *positive selection*, while deleterious types respond to *negative* or *purifying selection*. In either case, we say that the population is *polymorphic* during the course of this process, because more than one type is present. We also sometimes describe such populations as carrying *segregating variation*, language motivated by the genetic underpinnings we take up in section 1.2. The process ends with the *fixation* of the more fit type, after which the population is again *monomorphic*: it is composed of just one type.

Importantly, we have assumed that our clonally produced offspring faithfully inherit their parent's Malthusian fitness. This is central to our model: at each moment in time, natural selection rewards the type with higher fitness by increasing its frequency in the population. But absent at least some correlation in fitness between parent and offspring, the children of high fitness parents will not themselves leave more offspring, and the process will have no long-term effect on the population's composition. Of course, no real organism is precisely identical to its parent(s), and we explore the influence of mutation on natural selection in section 1.2, and of sexual reproduction in section 1.3 and again in chapter 3. Finally, an organism's fitness can itself depend on the environment. We avoid this important complication by assuming throughout that fitness $r$ and selection coefficient $s$ are values averaged over all environments.

We conclude this section by noting that the dynamics captured by equation (1.7) are independent of population size, since we have made no assumptions about $N(t) = N_A(t) + N_a(t)$, which depends on $r_A$ and $r_a$ and not just on their difference. Thus, even in a population that is itself growing (or decreasing), $p_A(t)$ within the population is changing logistically (i.e., as in figure 1.2). This point is illustrated in figure 1.4.

---

**Power User Challenge 1.6**

Write an expression for $N(t)$ with the parameters used in each panel of figure 1.4. Compare this with $N_A(t)$ (written with equation [1.5]) to explain mathematically why the lines converge in both panels.

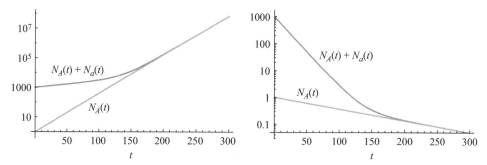

**Figure 1.4**
Time course of natural selection in populations of changing size. Mustard: more-fit type. Blue: total population size. Here $N(0) = 1{,}000$, $p_A(0) = \dfrac{1}{N(0)} = 0.001$, and $s = r_A - r_a = 0.05$, as in figure 1.2. On the left, $r_A = 0.06$ and $r_a = 0.01$. On the right, $r_A = -0.01$ and $r_a = -0.06$. Plotting $p_A(t) = \dfrac{N_A(t)}{N_A(t) + N_a(t)}$ (equation 1.6) for either panel would yield figure 1.2, even though $N(t) = N_A(t) + N_a(t)$ is not constant. Put another way, figure 1.2 is correct for all pairs of values of $r_A$ and $r_a$ that differ by $s = 0.05$.

---

**Power User Challenge 1.7**

Another way of understanding the effect illustrated in figures 1.2 and 1.4 is to focus not on frequencies, but on the ratio of frequencies. Following the strategy leading to equation (1.7), show that

$$\frac{p_A(t)}{1 - p_A(t)} = \frac{p_A(0)}{1 - p_A(0)} e^{st}$$

In English, this shows that the ratio of frequencies grows exponentially as a beneficial variant goes to fixation. This result also holds regardless of total population size.

---

### 1.1.3 The Time for a Selective Fixation

For the parameters employed in figure 1.2, the selective fixation occurs at $t \approx 275$. We now develop a mathematical treatment for this quantity, designated $t_{\text{fix}}$. Starting with a single invader of type $A$ in a population of $N$ individuals means $p_A(0) = \dfrac{1}{N}$, and inspection of the left panel in figure 1.2 demonstrates a kind of symmetry in the process: the second half of the fixation is a mirror image of the first half. The same was noted in equation (1.8) and is intrinsic to all logistic models. This implies that the time for fixation can be found by doubling the time it takes for the invading type to reach 50% of the population.

Putting these two ideas into equation (1.7) yields

$$p_A\left(\frac{t_{\text{fix}}}{2}\right) = \frac{1}{2} = \frac{\dfrac{1}{N} e^{\frac{st_{\text{fix}}}{2}}}{1 + \dfrac{1}{N}\left(e^{\frac{st_{\text{fix}}}{2}} - 1\right)}. \tag{1.9}$$

Solving for $t_{\text{fix}}$ (teachable moment 1.9) gives us

$$t_{\text{fix}} = \frac{2\ln(N-1)}{s} \approx \frac{2\ln(N)}{s}.$$  (1.10)

Substituting the parameter values used in figure 1.2 ($N=1{,}000$ and $s=0.05$) yields $t_{\text{fix}} \approx 276.31$, consistent with what is shown.

---

**Teachable Moment 1.9:** Solving equation (1.9)

Under the assumptions of our model, equation (1.9) is a true statement that connects $t_{\text{fix}}$, $s$ and $N$. But as written, it does not directly shed light on our problem, which is to compute $t_{\text{fix}}$ for given values of $N$ and $s$. To make progress, we must solve it for $t_{\text{fix}}$. More specifically, we will apply a judiciously chosen sequence of mathematical operations to both sides of the equal sign. In so doing, we preserve the truth of the original equality.

All that remains is to find the correct sequence of operations. The approach is to successively invert mathematical operations in equation (1.9) in such a way as to collect all expressions in $t_{\text{fix}}$ on one side of the equal sign. Recall from basic algebra that addition and subtraction are each the inverse of the other (e.g., if $x+y=z$ then it must be that $x=z-y$ and vice versa) and, ditto, multiplication and division $\left(\text{e.g., } xy=z \text{ if an only if } x=\frac{z}{y}\right)$. And as seen in teachable moment 1.5, logarithms and exponentiation are each other's inverses.

In the present case, we observe that $t_{\text{fix}}$ appears in both the numerator and the denominator of the right side of equation (1.9). To abolish that division, we first multiply both sides of the expression by the denominator to find

$$\frac{1}{2}\left[1+\frac{1}{N}\left(e^{\frac{st_{\text{fix}}}{2}}-1\right)\right] = \frac{e^{\frac{st_{\text{fix}}}{2}}}{N}.$$

Next, we expand the left side of the expression first by multiplying (distributing) the $\frac{1}{N}$ with (over) both terms in parentheses, and then distributing the leading $\frac{1}{2}$ over both terms in square brackets. Our rationale is to liberate $t_{\text{fix}}$ on the left so that it can be combined with $t_{\text{fix}}$ on the right. Indeed,

$$\frac{1}{2}+\frac{e^{\frac{st_{\text{fix}}}{2}}}{2N}-\frac{1}{2N} = \frac{e^{\frac{st_{\text{fix}}}{2}}}{N}$$

represents progress. Now subtracting $\dfrac{e^{\frac{st_{\text{fix}}}{2}}}{2N}$ from both sides yields an expression in which $t_{\text{fix}}$ only appears once

$$\frac{1}{2}-\frac{1}{2N} = \frac{e^{\frac{st_{\text{fix}}}{2}}}{2N}.$$

Finally, to fully isolate $t_{\text{fix}}$, we first eliminate the $2N$ in the denominator on the right by multiplying both sides of the equation by that term, reaching

$$N-1=e^{\frac{st_{\text{fix}}}{2}}$$

Taking the natural logarithm of both sides eliminates the base-e exponentiation, giving us

$$\ln(N-1) = \frac{st_{\text{fix}}}{2}.$$

Lastly, inverting the division by 2 and the multiplication by $s$ on the right yields equation (1.10).

**Teachable Moment 1.10:** Approximations are often quite informative

Equation (1.10) makes an approximation (mathematically, $\ln(N-1) \approx \ln(N)$). This is motivated by the greater simplicity of the latter, and warranted by the fact that the two quantities quickly converge as $N$ grows. For example, the error is less than 0.25% when $N = 100$ and less than 0.015% when $N = 1,000$. Recalling George Box, all models are wrong, but this approximation scarcely undermines the general conclusions developed unless $N$ is very small. Indeed, errors in our empirical estimates of $N$ and $s$ in any particular biological context are likely to introduce much greater discrepancies in our estimation of $t_{\text{fix}}$ than will the difference between $\ln(N-1)$ and $\ln(N)$.

Also note that the symmetry underlying equation (1.9) is not always preserved in more complicated models (e.g., section 1.3.2). Nevertheless, equation (1.10) is an excellent approximation that we will return to more than once.

**Power User Challenge 1.8**

The *carbonaria* form of *Biston betularia* (empirical aside 1.2) spread very rapidly with English industrialization in the nineteenth century. For example, the first specimens were reported near Manchester in approximately 1850, and the white, typical form was nearly absent by 1900 (Clarke et al. 2008). Assuming one generation per year and a population size of $e^{10} \approx 22,000$, use equation (1.10) to show that this variant enjoyed a selection coefficient $s \approx 0.2$.

**Power User Challenge 1.9**

Solve equation (1.10) for $N$. Use your answer to show that in a population of approximately 270,000 individuals, a single invading type with advantage $s = 0.025$ will displace a less-fit resident in approximately 1,000 generations.

Beyond the immediate utility of being able to compute $t_{\text{fix}}$ for specific values of $N$ and $s$, equation (1.10) yields two biological insights. First, we see that $t_{\text{fix}}$ depends only weakly on population size. For example, increasing the population size by a factor 100 only increases the time for fixation by

$$\frac{2\ln(100N)}{s} - \frac{2\ln(N)}{s} = \frac{2[\ln(100N) - \ln(N)]}{s} = \frac{2\ln\left(\frac{100N}{N}\right)}{s} = \frac{2\ln(100)}{s} \approx \frac{9.2}{s}.$$

generations. (The second step takes advantage of another interesting fact about logarithms: $\ln(x) + \ln(y) = \ln(x \times y)$, or equivalently, $\ln(x) - \ln(y) = \ln\left(\dfrac{x}{y}\right)$.) In contrast, $t_{\text{fix}}$ is inversely proportional to the selection coefficient $s$: doubling this quantity halves the time. (Pick some pairs of values of $N$ and of $s$ to convince yourself of both claims.) These conclusions illustrate that mathematical models often yield results that are far from clear on verbal considerations.

---

**Teachable Moment 1.11**

The attentive reader may wonder why we set $p_A(t_{\text{fix}}) = \dfrac{1}{2}$ and then doubled the result. Why not set $p_A(t_{\text{fix}}) = 1$ and solve directly for $t_{\text{fix}}$? If you try this approach, you will find $t_{\text{fix}} = \infty$ (after taking the logarithm of zero). Indeed, figure 1.2 also illustrates that $p_A(t)$ only *asymptotically approaches* 1, meaning that as $t$ grows, $p_A(t)$ becomes ever closer to 1 without ever quite equaling it. Yet on biological considerations, we reasonably expect that at some (finite) time, the last individual of type $a$ will be displaced by the $N$th individual of type $A$.

This discrepancy between model and reality reflects the fact that the mathematics allows the frequency $p_A$ to be any real value, whereas biologically, frequencies can only assume some integer multiple of $\dfrac{1}{N}$. (This shortcoming was already raised in teachable moment 1.6.)

Since half of $t_{\text{fix}}$ is the time for the type $A$ to go from frequency $\dfrac{1}{N}$ to $\dfrac{1}{2}$, strictly speaking $t_{\text{fix}}$ is the time it takes to increase in frequency from $\dfrac{1}{N}$ to $1 - \dfrac{1}{N}$. We ignore the time for the frequency to make the last jump to 1. As in teachable moment 1.10, that is in general good enough for our purposes.

---

### 1.1.4 Fisher's Fundamental Theorem of Natural Selection

As Darwin emphasized in *On the Origin of Species*, natural selection requires variation in fitness; it cannot change a homogeneous population (see also equation [1.8]). R. A. Fisher (one of the pioneers of our discipline) demonstrated a remarkable result: the instantaneous rate of fitness increase in a population during the selective substitution of a fitter type is numerically equal to the fitness variance in the population at that moment. He called this his fundamental theorem of natural selection. Maintaining our assumption of just two types in the population, we compute a population's mean fitness at time $t$ as

$$\bar{r}(t) = p_A(t) r_A + [1 - p_A(t)] r_a \tag{1.11a}$$

and its fitness variance as

$$\text{Var}[r(t)] = p_A(t) r_A^2 + [1 - p_A(t)] r_a^2 - \bar{r}(t)^2 \tag{1.11b}$$

(teachable moment 1.12). We can now write Fisher's fundamental theorem mathematically as

$$\frac{d\bar{r}(t)}{dt} = \text{Var}[r(t)].$$

Note its focus on the temporal change in mean fitness, rather than in variant frequency as in equation (1.8). Not surprisingly, the two effects are intimately connected: mean fitness is changing precisely because natural selection is changing variant frequency. This is the basis of the proof for the theorem given in teachable moment 1.13.

---

**Teachable Moment 1.12:** Means, variances, and discrete probability mass functions

The fundamental theorem of natural selection is written in terms of means and variances, two *summary statistics* commonly used to distill some quantitative feature of a set of observations into single numbers. Indeed, mean and variance can be computed for any trait seen in any group of objects, biological or otherwise. This book focuses primarily on organismal fitness, although occasionally other organismal traits (or *phenotypes*) will be examined.

Given some trait $x$ that assumes one of $k$ possible values $x_i$ at frequency $p_i$ in a population ($1 \leq i < k$), its mean is

$$\bar{x} = p_1 x_1 + p_2 x_2 + \cdots + p_k x_k = \sum_{i=1}^{k} p_i x_i. \tag{1.12}$$

The overbar on $x$ signifies that this is its mean and $\Sigma$ represents the summation of the product $p_i x_i$ over all possible observations. (We note in passing that $\sum_{i=1}^{k} p_i = 1$, since $x$ must assume one of the $k$ possibilities.) The mean is a measure of the central tendency of the trait in the population.

The variance of $x$ is defined as

$$\text{Var}(x) = \sum_{i=1}^{k} p_i (x_i - \bar{x})^2. \tag{1.13a}$$

It is a measure of the dispersion of values around the mean. Two additional, equivalent expressions of the variance are sometimes of use. First, expanding the quadratic term in equation (1.13a), we find $\sum_{i=1}^{k} p_i (x_i^2 - 2x_i\bar{x} + \bar{x}^2)$. We then expand the summation as $\sum_{i=1}^{k} p_i x_i^2 - \sum_{i=1}^{k} 2p_i x_i \bar{x} + \sum_{i=1}^{k} p_i \bar{x}^2$. Next, we factor 2, $\bar{x}$, and $\bar{x}^2$ (these are independent of index $i$) to find $\sum_{i=1}^{k} p_i x_i^2 - 2\bar{x}\sum_{i=1}^{k} p_i x_i + \bar{x}^2 \sum_{i=1}^{k} p_i$. Finally, recognizing that $\sum p_i x_i = \bar{x}$ and recalling $\sum p_i = 1$, gives $\sum_{i=1}^{k} p_i x_i^2 - 2\bar{x}^2 + \bar{x}^2$ or

$$\text{Var}(x) = \sum_{i=1}^{k} p_i x_i^2 - \bar{x}^2. \tag{1.13b}$$

Additionally, referring to the definition of the mean, we also have $\sum_{i=1}^{k} p_i x_i^2 = \overline{x^2}$, which lets us rewrite equation (1.13b) as

$$\text{Var}(x) = \overline{x^2} - \bar{x}^2. \tag{1.13c}$$

We will employ all three formulations in this book.

Note that we motivated these expressions with reference to a sample taken from a population. In this case, the $p_i$ are the observed frequencies at which we sample trait value $x_i$. Importantly, working with samples leaves an uncertainty about the underlying population that is the central problem of statistical inference, a topic beyond the scope of this book.

In contrast, our mathematical representation of Fisher's fundamental theorem used equations (1.12) and (1.13b) to compute the mean (equation 1.11a) and variance (equation 1.11b) of a trait (fitness) whose $p_i$ are given by a model (equation 1.8), called a *probability mass function*. Similarly, in chapter 2 we will develop stochastic models that capture random variation in a process (see teachable moment 1.2). In that context, we will be interested in the mean and variance in the stochastic model's behavior. In both cases, because we find our $p_i$ from a model rather than by sampling, we have no statistical uncertainty in results.

Finally, note that equations (1.12) and (1.13) assume a finite number ($k$) of possible trait values of $x$, and the corresponding probability mass function is thus discrete. Means and variances of continuous traits can similarly be computed, as described in teachable moment 2.2.

---

**Teachable Moment 1.13:** Proof of Fisher's fundamental theorem

Fisher's fundamental theorem equates the rate at which a population's mean fitness changes over time to the population's current variance in fitness. Its proof simply invokes the definitions of each of those terms. First, equation (1.11a) lets us write the time derivative of mean fitness as

$$\frac{d\bar{r}(t)}{dt} = \frac{dp_A(t)}{dt}r_A + \frac{d[1-p_A(t)]}{dt}r_a.$$

(We are entitled to factor $r_A$ and $r_a$ out of the derivatives on the right because they are independent of time.) This seemingly obscure expression is actually very useful. As noted, natural selection is why mean fitness is changing, via its influence on the frequency of types in the population. This suggests that our proof should somehow employ equation (1.8), which quantifies that effect. And indeed, equation (1.8) can equally be written as $\frac{dp_A(t)}{dt} = p_A(t)$ $[r_A - \bar{r}(t)]$. (Subtract $r_A$ from both sides of equation [1.11a], factor $1 - p_A(t)$ and recall that $s = r_A - r_a$ to see why $r_A - \bar{r}(t) = s(1 - p_A)$.) And by the symmetry mentioned immediately below equation (1.8), we also have $\frac{d[1-p_A(t)]}{dt} = [1 - p_A(t)][r_a - \bar{r}(t)]$. Substituting these into the previous expression yields

$$\frac{d\bar{r}(t)}{dt} = p_A(t)[r_A - \bar{r}(t)]r_A + [1 - p_A(t)][r_a - \bar{r}(t)]r_a$$
$$= p_A(t)r_A^2 + [1 - p_A(t)]r_a^2 - \{p_A(t)r_A + [1 - p_A(t)]r_a\}\bar{r}(t)$$
$$= p_A(t)r_A^2 + [1 - p_A(t)]r_a^2 - \bar{r}(t)^2.$$

after recognizing that the quantity in curly brackets on the second line is $\bar{r}(t)$. Finally, Equation (1.11b) completes the job.
Q.E.D.

---

**Teachable Moment 1.14**

The end of a mathematical proof is often marked by the phrase "Q.E.D." for the Latin *quod erat demonstrandum*, or literally, "that which was to be proved." In the present case, we have shown that both sides of the theorem equal $p_A(t)r_A^2 + [1 - p_A(t)]r_a^2 - \bar{r}(t)^2$, proving that they are equal to each other.

Since variances are always nonnegative, Fisher's fundamental theorem gives the result that population mean fitness can never decline by the action of natural selection. This (perhaps unsurprising) interpretation of the fundamental theorem is unambiguous for our model: constant selection acting on a single variant. But many complications are known that undermine this conclusion, including *frequency-* and *density-dependent selection* (meaning that the *r*'s are dependent on $p_A$ or on $N$, respectively) and selection acting on more than one locus. (Indeed, numerous interpretations of the theorem in more complex settings have been offered; see, for example, Frank and Slatkin [1992]; Lessard [1997]; and Okasha [2008].) Nevertheless, the quantitative connection between fitness variance in the population and the rate of fitness increase offered by Fisher's fundamental theorem remains an important benchmark in many settings, and one we return to more than once.

---

**Teachable Moment 1.15**

That the rate of fitness increase should be proportional to the variance in fitness was already anticipated by our observations in equation (1.8) that $\frac{dp_A}{dt}=sp_A(1-p_A)$. To see this connection, we first define an *indicator function* for individual $x$ as

$$I(x)=\begin{cases}1 \text{ if } x \text{ is of type } A \\ 0 \text{ if } x \text{ is of type } a\end{cases}.$$

In English, the indicator function maps $A$-type individuals to 1 and $a$-type individuals to zero. Application of equations (1.12) and (1.13a) to a population gives us

$$\bar{I}=p_A\times 1+(1-p_A)\times 0=p_A \text{ and}$$
$$\text{Var}(I)=1^2\times p_A+0^2\times(1-p_A)-p_A^2=p_A(1-p_A).$$

In English, $\bar{I}$ is the mean type in the population and $\text{Var}(I)$ is the variance in type. The selection coefficient $s$ is precisely the amount by which variance in type is manifest as variance in fitness. For example, if $s=0.2$, then the population's variance in fitness equals 20% of its variance in type at any time during the variant's fixation.

---

### 1.1.5 Natural Selection in Populations with Nonoverlapping Generations

To this point, we have assumed that an organism's reproduction and death are decoupled. Consequently, ours have been models of *overlapping generations*: organisms can undergo multiple reproductive events during their lifetime and parents may continue to reproduce even after some of their offspring have begun reproducing. A great many species of life on Earth (including our own) exhibit overlapping generations.

In contrast, some organisms (e.g., annual plants and annual insects) synchronously reproduce and then synchronously die; they thus exhibit *nonoverlapping generations*. The mathematics of natural selection in populations of nonoverlapping generations differ subtly and importantly from that developed above. Fundamentally whereas the Malthusian parameter can be written in any unit of time (see teachable moment 1.7), in models of nonoverlapping generations, growth rates are always in units of generations. Why? Because from the perspective of a population's size and composition, nothing happens between the synchronous reproductive and death events that define generations.

Thus, we introduce a new parameter $W$, an organism's *absolute Wrightian fitness*, the number of offspring it produces in one generation. This is named for Sewall Wright, another of the field's pioneers. Thus, a population of $N$ individuals in the present generation will increase by that proportion each generation. Mathematically, we write

$$N' = NW, \tag{1.14}$$

where $N$ and $N'$ are the population sizes in the current and next generations, respectively. The biology demands that absolute fitness $W \geq 0$. The population is growing when $W > 1$, of constant size when $W = 1$, and shrinking when $0 \leq W < 1$. (Because $W$ must be an integer, nonoverlapping generation models avoid the difficulties noted in teachable moment 1.6.)

Now consider a population comprised of $N_A$ organisms of type $A$ and $N_a$ organisms of type $a$, with Wrightian fitness values $W_A$ and $W_a$, respectively. Substituting equation (1.14) into equation (1.6) yields

$$p'_A = \frac{N'_A}{N'_A + N'_a} = \frac{N_A W_A}{N_A W_A + N_a W_a} = \frac{p_A W_A}{p_A W_A + (1 - p_A) W_a}.$$

for the frequency of type $A$ in the next generation. Now writing $\Delta p_A$ for the per generation change in frequency of type $A$, we find

$$\Delta p_A = p'_A - p_A = \frac{p_A W_A - p_A [p_A W_A + (1 - p_A) W_a]}{p_A W_A + (1 - p_A) W_a} = \frac{p_A (1 - p_A)(W_A - W_a)}{\overline{W}}, \tag{1.15}$$

where $\overline{W} = p_A W_A + (1 - p_A) W_a$ is the population's mean absolute fitness.

Equation (1.15) illustrates that the pace at which one type will displace another in a model of nonoverlapping generations depends only on the ratio of fitness values, and not their absolute values. To see this, note that our expression for $\Delta p_A$ remains unchanged if we divide fitness values $W_A$ and $W_a$ by any nonzero constant. To simplify matters, we often normalize fitness values by $W_a$ to reach *relative Wrightian fitnesses* $w_A = \frac{W_A}{W_a}$ and $w_a = \frac{W_a}{W_a} = 1$. (Note the distinction between absolute and relative fitness values, represented by $W$ and $w$, respectively.) Substituting relative fitness values into equation (1.15) yields

$$\Delta p_A = \frac{p_A (1 - p_A)(w_A - 1)}{p_A w_A + (1 - p_A)} = \frac{p_A (1 - p_A) s}{\overline{w}}. \tag{1.16}$$

Here, we introduce $s$, the *Wrightian selection coefficient* between the two types in our population. In English, the Wrightian selection coefficient is the normalized difference in absolute fitnesses, mathematically, written $s = \dfrac{W_A - W_a}{W_a} = \dfrac{W_A}{W_a} - 1 = w_A - 1$. (Equivalently, $\dfrac{W_A}{W_a} = w_A = 1 + s$.) Since the biology requires that $w_A \geq 0$, it follows that $s \geq -1$. Finally, note that the denominator in equation (1.16) has also been normalized by $W_a$, giving mean relative fitness $\bar{w} = \dfrac{\bar{W}}{W_a} = p_A w_A + (1 - p_A) = 1 + p_A s$.

Compare equation (1.16) to (1.8), its overlapping generation analog. As before, the sign of $s$ tells us whether type $A$ increases or decreases over time, since $\dfrac{p_A(1 - p_A)}{\bar{w}}$ can never be negative. Also, as before, the rate of change in frequency is modest early and late, and more rapid during the middle of the process, since $\Delta p_A$ is approximately proportional to $p_A(1 - p_A)$.

---

**Power User Challenge 1.10**

Writing $s_W$ and $s_M$ for Wrightian and Malthusian selection coefficients respectively, derive $s_M = \ln(s_W + 1)$. Hint: first show that $W_x = e^{r_x}$, where $x$ is either $A$ or $a$. Now reconcile the units of the two selection coefficients. Finally, use your result to find the overlapping-generation model analog to the observation for the present model that the rate of increase by a beneficial variant depends only on the ratio of fitness values.

---

The difference between the overlapping and nonoverlapping generation results reflects the difference in the way we've framed the problem. While equation (1.8) gives the instantaneous rate of change in $p_A$, equation (1.16) is the amount by which it changes over one full generation. This distinction is responsible for the denominator of equation (1.16): $\bar{w} = 1 + p_A s$, which vanishes as $s$ becomes small. (Mathematically, if $s \ll 1$ then $1 + p_A s \approx 1$. Here we introduce the symbol $\ll$ to mean "much smaller than.")

---

**Teachable Moment 1.16:** Overlapping and nonoverlapping generation models

That our nonoverlapping generation model of natural selection (equation 1.16) converges to its overlapping generation analog (equation 1.8) as the selection coefficient gets small is a mathematical fact, and thinking about the biology helps us to see why. From the point of view of the dynamics of this process, making the per generation fitness difference $s$ small while holding $\Delta t$ fixed at one generation is formally the same as making $\Delta t$ small for fixed $s$, the latter being exactly how we solved equation (1.1). On the premise that beneficial selection coefficients are quite small in nature (although see empirical aside 1.3: exceptions are not unknown), we will often choose one framing or the other on the basis of expediency.

More specifically, overlapping generation models often yield mathematical results not available otherwise. For example, no *closed-form expression* analogous to equation (1.7) is possible assuming nonoverlapping generations, meaning that the only way to compute the frequency of type *A* after *t* nonoverlapping generations is to iteratively apply equation (1.7) *t* times (e.g., figures 1.9, 2.5, 3.2, and 3.4). On the other hand, models of nonoverlapping generations more easily accommodate sexual reproduction (e.g., section 1.3.1) and stochastic treatments (chapter 2), making them the far more common choice in much of the field.

## 1.2   Mutation

Thus far, we have assumed that offspring are identical to parents. However, this is rarely true, a fact with obviously deep evolutionary implications. We begin by addressing the fact that even among uniparental or clonally reproducing organisms, *mutation* can cause offspring to differ at least slightly from their parent. The additional complexity introduced with biparental reproduction will be explored in section 1.3 and chapter 3.

Correspondingly, we now recognize that the two organismal types *A* and *a* that we modeled beginning in section 1.1.2 differ from one another in their genetic composition, or equivalently, their fitness values are determined by their genetics. (This was already implicitly assumed in empirical asides 1.2 and 1.3). In order to develop intuition, the remainder of this chapter and all the next assume that an organism's fitness is determined by exactly one genetic factor, which resides at a genetic *locus* (plural *loci*). We call the genetic factor in any particular organism its *allele* (e.g., *A* and *a*, or alternate trait values in Mendel's pea plants). Of course, in truth an organism's fitness is determined by a great many loci, each with many alternative alleles, and in chapter 3 we take up some of those complications.

### 1.2.1   The Evolution of Allele Frequencies under Mutation Alone

Imagine a population of organisms defined by a single locus with two possible alleles, called *A* and *a*. (Of course, many loci have more than two alleles, and we will relax this assumption in chapter 2.) Returning to an overlapping generation model, we write $\mu$ and $v$ for the per capita rates at which mutation converts *A* into *a* and *a* into *A*, respectively. In other words, a fraction $\mu \Delta t$ of the offspring produced by each *A*-bearing organisms in interval $\Delta t$ will be *a* bearing, while a proportion $v \Delta t$ of the offspring produced by each *a*-bearing organism in that same interval will be *A* bearing. (Biologically, we have $0 \le \mu, v \le 1$.) Again, writing $p_A(t)$ for the fraction of the population that is *A* bearing at time *t* and letting $\Delta t \to 0$ gives us

$$\frac{dp_A}{dt} = -\mu p_A + v(1 - p_A). \tag{1.17}$$

We note first that, unlike the models of selection developed in section 1.1, mutation will drive genetic polymorphism into a population. In other words, mutation creates genetic variation, whereas natural selection eliminates it (although section 1.3.2

presents one exception). To see this mathematically, observe first that equations (1.8) and (1.16) both equal 0 when $p_A$ is 0 or 1. In contrast, equation (1.17) equals $v$ when $p_A$ is 0 and $-\mu$ when it is 1.

To find the equilibrium level of genetic polymorphism predicted by equation (1.17), we set the equation to zero (see teachable moment 1.17), which yields

$$\tilde{p}_A = \frac{v}{\mu + v}. \tag{1.18}$$

(The tilde signals that this is the equilibrium value of $p_A$.) Not surprisingly, if $v = \mu$ then $\tilde{p}_A = \frac{1}{2}$. In this case we say that the locus has a *symmetric mutation rate*. Mutation rates at the nucleotide level are typically roughly symmetric (see empirical aside 1.4). But at the level of a *gene* (a locus that encodes a specific product; that is, a protein or an RNA), it's generally much easier for mutation to disrupt function than it is to restore it. Writing $A$ for the wild type, we thus often assume $\mu \gg v$.

---

**Teachable Moment 1.17:** What is an equilibrium, and is it stable or unstable?

*Equilibrium* means that the system stops changing over time; or mathematically, that the time derivative of its dependent variable is zero. Equilibria are locally stable if, after a very small perturbation, the system returns to its former state. In contrast, if small perturbations to the system cause it to diverge from the equilibrium, it is described as unstable. For example, the equilibrium allele frequency in equation (1.18) is stable: perturbations toward larger values of $p_A$ render $\frac{dp_A}{dt}$ in equation (1.17) negative, while those toward smaller values render it positive. This is illustrated graphically in figure 1.5 for $\mu = 0.01$, $v = 0.005$, and thus $\tilde{p}_A = \frac{1}{3}$.

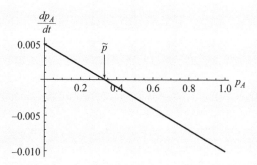

**Figure 1.5**
Rate of change of allele frequency $\left(\frac{dp_A}{dt}\right)$ as a function of allele frequency ($p_A$) under our model of two-way mutation (equation 1.17) for $\mu = 0.01$, $v = 0.005$; thus $\tilde{p}_A = \frac{1}{3}$ (equation 1.18). This equilibrium is stable, because the sign of allele frequency change is positive for allele frequencies below $\frac{1}{3}$ and negative above it. In other words, pertubations in allele frequency away from $\frac{1}{3}$ in either direction will be corrected by mutation.

---

**Power User Challenge 1.11**

Draw an analog of figure 1.5 for the case of an unstable equilibrium.

---

**Power User Challenge 1.12**

What are the equilibrium allele frequencies under equation (1.8)? Hint: there are three. Are they each stable or unstable? How do you know?

---

**Empirical Aside 1.4:** What are mutation rates in nature?

The base-substitution (or *point mutation*) rates in cellular organisms are between $10^{-8}$ and $10^{-10}$ per nucleotide per generation, while for *eukaryotes* (cellular organisms with subcellular structures) this range is narrower: $10^{-8} - 10^{-9}$ (Lynch et al. 2016). Unlike cellular organisms, the genomes of many pathogenic viruses are encoded by RNA, whose replication enzymes have much higher error rates. For example, the point mutation rate in HIV is roughly $10^{-3}$ per generation.

Interestingly, the mutation rates between different pairs of nucleotides need not be equal, a phenomenon called *mutational bias*. Among the four nucleotides in DNA (or RNA), there are $4 \times 3 = 12$ mutations possible: each nucleotide can be changed into one of the other three. Mutational bias means that the probabilities of these 12 mutations are unequal. For example, the mutation rates among the purines (guanine and adenine) and among the pyrimidines (cytosine and thymine) are generally lower than they are between groups. On the other hand, these rates rarely vary by more than a factor of 10.

Less commonly, mutation can cause insertions, deletions, and translocations of genetic material, but their rarity in nature makes measuring their rates more difficult.

---

### 1.2.2   Mutation/Selection Equilibrium

We now consider a population in which mutation and selection act simultaneously. If $A$ is the wild type allele with Malthusian fitness advantage $s > 0$ relative to mutant $a$, selection alone will drive the $A$ allele to fixation. What is the consequence of recurrent mutation between alleles? The answer is found mathematically by adding the two processes' effects on $\dfrac{dp_A}{dt}$ (equations 1.17 and 1.8) to find

$$\frac{dp_A}{dt} = sp_A(1 - p_A) - \mu p_A + v(1 - p_A) \approx sp_A(1 - p_A) - \mu p_A. \tag{1.19}$$

(The approximation is justified if we assume that selection is holding $p_A$ very close to unity, and again assuming $\mu \gg v$. Equivalently, we are disregarding *back mutations* (or *reversion*), which occur at rate $v(1 - p_A)$. We relax this assumption in teachable moment 1.20.)

Setting equation (1.19) to zero and solving yields two solutions (or *roots*)

$$\tilde{p}_A = \begin{cases} 1 - \dfrac{\mu}{s}, & \text{if } \mu < s \\ 0 & \text{otherwise} \end{cases}. \tag{1.20}$$

The first root is called the population's *mutation/selection equilibrium* and represents the balance between the recurrent introduction of deleterious mutants and their elimination by purifying selection. Like equation (1.7), this model also depends only on a compound parameter, here $\dfrac{\mu}{s}$. In this case, $\tilde{p}_A$ depends only on the ratio of the mutation rate and selection coefficient, and not on their values individually. (See teachable moment 3.6 for another derivation of this result.)

---

**Teachable Moment 1.18**

Two solutions exist in equation (1.20) because equation (1.19) is quadratic in $p_A$, meaning that it depends on $p_A^2$. This is analogous to the fact that the equation $x^2 = 1$ has two solutions: $x = 1$ and $x = -1$.

---

**Power User Challenge 1.13**

If $\mu < s$, the first root of 1.19 is stable (and thus the biologically relevant one). Use the technique in teachable moment 1.17 to show why. Explore what happens if the inequality is reversed.

---

Maintaining our assumption that selection is stronger than mutation (i.e., $s > \mu$), the population's mean fitness at equilibrium (written as $\bar{r}_{eq}$) is easily found. Substituting equation (1.20) into equation (1.11a) and simplifying gives

$$\bar{r}_{eq} = r_A - \mu. \tag{1.21}$$

In English, mean fitness in populations at mutation/selection equilibrium is depressed by the mutation rate. This result formalizes another very important point for us: natural selection cannot always maximize population mean fitness, and the fitness reduction due to nonselective processes in a population is called its *genetic load*. In the present example, recurrent deleterious mutation reduces mean fitness in spite of the action of natural selection. This form of genetic load is called the *mutation load*.

**Teachable Moment 1.19:** Genetic load

Only natural selection systematically increases fitness, and genetic load is the reduction in equilibrium mean fitness due to all nonselective processes. Besides mutation load, other forms include the *lag load*, experienced during a population's evolutionary response to environmental change; *substitution load*, experienced during the course of a selective fixation by a beneficial mutation; *segregation* or *recombination load*, caused when genetic recombination disrupts high-fitness allelic combinations at or across loci, respectively; *inbreeding load*, caused by population subdivision; *drift load*, caused by random genetic drift; and *migration load*, caused by migration of organisms into an environment to which they are not fully adapted.

Equation (1.21) also illustrates that mean fitness at mutation/selection equilibrium is independent of the selection coefficient. This remarkable point was first recognized by J. B. S. Haldane (another of our pioneers). The math tells us that as we increase selection coefficient against the mutant, its equilibrium frequency is reduced. So long as $\mu < s$, the net effect on equilibrium fitness exactly balances. This is often called the *Haldane–Muller principle*.

### 1.2.3 The Error Catastrophe

Thus far, we have only focused on the first root of equation (1.20), which is biologically relevant when $\mu < s$. When $\mu \geq s$, the second root ($\tilde{p}_A = 0$) becomes stable and biologically relevant. This transition when $\mu = s$ is called the *error threshold*, and when $\mu > s$ the population is said to be in an *error catastrophe*. Now substituting $\tilde{p}_A = 0$ into equation (1.11a) tells us that in this regime, equilibrium mean fitness $\bar{r}_{eq}$ equals $r_a$. In English, the mutation load shifts from $\mu$ to $s$ at the error threshold, and the Haldane–Muller principle fails when $\mu > s$. The population genetics on both sides of the error threshold are illustrated in figure 1.6.

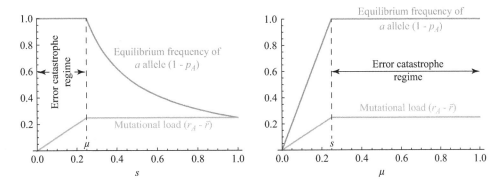

**Figure 1.6**

Allele frequency (blue) and mutation load (mustard) as a function of selection coefficient (left, fixing mutation rate $\mu = 0.25$) or mutation rate (right, fixing selection coefficient $s = 0.25$). Error threshold (vertical dashed line) occurs when $\mu = s$. When $\mu < s$, selection maintains deleterious allele frequency $(1 - \tilde{p}_A)$ below 1 and mutation load is equal to mutation rate. When $\mu > s$, mutation overwhelms selection and the population enters error the catastrophe. Here, the deleterious allele frequency is 1 and mutation load is equal to the selection coefficient.

**Teachable Moment 1.20:** What about back mutations?

Recall that in deriving our expression for allele frequency at mutation/selection equilibrium, we disregarded mutation from the deleterious $a$ allele back to the wild type $A$ allele. We justified this approximation by assuming that allele $a$ was very rare, but as mutation rate approaches or exceeds the strength of selection, this is no longer true. In this case back mutation will begin to contribute appreciably to the production of beneficial $A$ alleles, increasing their frequency above model prediction (and reducing mutation load). Setting the exact expression for $\frac{dp_A}{dt}$ in equation (1.19) to zero yields an analytic solution for $\tilde{p}_A$ that incorporates back mutation via application of the quadratic formula. Importantly, however, the larger qualitative conclusion, that mutation/selection equilibrium breaks down at $\mu = s$, is robust to this omission. On the other hand, back mutations will be of greater biological interest when we return to these questions in multilocus models (chapter 3).

**Teachable Moment 1.21:** The quasispecies

The error threshold was first described not by a population geneticist but by a theoretical chemist called Manfred Eigen. He explored the deterministic dynamics of populations of self-reproducing, error-prone molecules (although his was a multilocus model, which we will develop in chapter 3). Chemists describe structurally different molecules as different species, and consequently, these models were said to describe *quasispecies*. Readers should recognize the distinction with the biologist's use of the word species, which explicitly assumes variation.

Virologists have been particularly interested in this concept because of virus's high mutation rates and the implications of the error threshold for infectious disease biology (see empirical aside 1.5). However, the chemists' quasispecies behaves no differently than the population described by equation (1.20) (Wilke 2005).

### 1.2.4  Beneficial Mutations

Without a doubt there are vastly more deleterious mutations than beneficial ones (see empirical aside 1.3), motivating our initial attention to the interplay between natural selection and recurrent deleterious mutations. In contrast, recurrent beneficial mutations are impossible in the one-locus, two-allele models employed in this chapter. For the moment, we thus leave the matter as developed in section 1.1, but return to the question when we take up multilocus models in chapter 3.

### 1.2.5  Soft Selection and Hard Selection

We now ask whether the error catastrophe is actually catastrophic for the population. Extinction is the only possible evolutionary catastrophe, so does extinction necessarily follow when $\mu > s$? The answer draws our focus back to the very first result in this chapter: that the fate of a population depends on its Malthusian fitness, $r$. Specifically, extinction occurs if and only if this quantity is negative. Thus, when $\mu > s$ and the population crosses the error threshold, its extinction depends entirely on the sign of $r_a$, the Malthusian fitness of the lower-fitness allele.

---

**Empirical Aside 1.5**

The distinction between the error catastrophe and extinction may have clinical implications. Mutagenic drugs are clinical compounds that increase their target's mutation rate. These have been proposed as antiviral therapies, in the hopes of inducing viral extinction within patients via the error catastrophe. But we now see that even if mutagenic drugs drive an infectious viral population across the error threshold, it will only go extinct if the Malthusian parameter of the surviving viruses is negative. It could be so, but it needn't be. (And as we shall see in chapter 3, there are several theoretical challenges to the idea that populations of multilocus organisms ever cross the error threshold.)

---

The distinction between the error threshold and extinction reveals an important gap in our understanding: little contact exists between our picture of natural selection and the question of population survival. The logistic dynamics of a selective fixation are independent of any assumptions about the total population size (figure 1.4), so a beneficial mutation can fix even while the total population goes extinct. Adaptation without risk of extinction is called *soft selection*. Mathematically, this corresponds to assuming nonnegative values of Malthusian fitness values. Similarly, our treatment of natural selection with nonoverlapping generations implicitly implements soft selection by virtue of its construction in terms of frequencies rather than counts.

In contrast, we are often interested in how a population responds to environmental change. Think for example of the appearance or improved efficiency of a predator, an infectious microbe suddenly facing a drug challenge, or individuals migrating into a less hospitable location. In cases such as these, we might not be surprised to find that the wild type's Malthusian fitness has become negative, after which the population's survival depends on new beneficial mutations appearing before extinction (or on preexisting, already segregating, now beneficial mutations; see also section 4.4.4). This is called *hard selection*: without one or more genetic innovations, the population is now doomed to extinction. In this case, a population's persistence is not assured, natural selection notwithstanding.

---

**Empirical Aside 1.6**

As we shall see in chapter 4, the analysis of gene and genome sequence data has revealed widespread evidence of repeated selective fixations in every species examined. Unfortunately, most of those methods do not provide insights into the underlying biological mechanisms of fitness differences among individuals or on the ultimate, ecological pressures being addressed by natural selection. (See also empirical aside 4.2.) Thus, the relative frequencies of episodes of hard and soft selection in nature largely remain unclear.

---

Hard and soft selection are illustrated schematically in figure 1.7. The recovery in population size after a dip shown on the right is sometimes referred to as

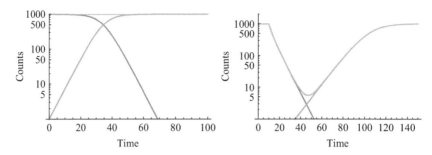

**Figure 1.7**
Soft (left) and hard (right) selection. The number of wild type (blue), beneficial mutants (mustard) and the total population size (green) are shown as a function of time. In both panels, the maximum population size is 1,000, and the beneficial mutant enjoys selection coefficient $s = 0.2$. Note that in the case of hard selection only, population survival is critically dependent on the appearance of the mustard mutant (here at $t = 35$), owing for example to an extrinsic deterioration in the environment (here at $t = 10$) that will otherwise drive the blue wild type to extinction. No such hazard is present in models of soft selection.

*evolutionary rescue.* The last 15 years have seen a growing interest in the linkage between ecological influences on population size and population genetics. For example, when the population size declines because of environmental deterioration, so too does its ability to generate new beneficial alleles by mutation, putting it at even greater risk of extinction. (See also the related idea of mutational meltdown introduced in chapter 3.) The interested reader is directed to Bell (2017) and citations therein to learn more.

---

**Power User Challenge 1.14**

Would the two panels in figure 1.7 differ if we plotted frequencies instead of counts? Why or why not?

---

### 1.3   One-Locus Sexual Reproduction

To this point, we have assumed clonal or uniparental reproduction. We now relax that assumption. Most eukaryotes inherit genetic material from each of two parents in a process called sexual reproduction. These two life cycles are illustrated in figure 1.8. (For convenience, we sometimes refer to clonally reproducing organisms as *asexuals* and sexually reproducing ones as *sexuals*.) For the moment, we continue to restrict attention to one-locus models of sexual reproduction; we will explore multilocus models in chapter 3.

Sexually reproducing organisms experience a *diploid* and *haploid* phase in each generation. Diploids carry two alleles at every locus while haploids carry just one. The diploid phase begins at *syngamy*, when two single-celled haploid cells called *gametes* fuse to form the single-celled diploid *zygote*. The reverse transition occurs at *meiosis*, when a diploid cell divides to form haploid gametes, which go on to form

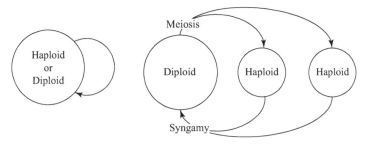

**Figure 1.8**
Uniparental and biparental life cycles. Arrows represent the passage of genetic material. Left: uniparental organisms reproduce asexually. They can be haploid or diploid. Right: biparental organisms reproduce sexually. These are almost always diploid (though see empirical aside 1.7). In each generation, diploid organisms produce haploids by meiosis, which then fuse by syngamy to form the next generation of diploids. Sexual offspring thus inherit genetic material from both parents.

diploid offspring in the next generation. We describe the pairing of two diploids whose gametes fuse in syngamy as a *mating*. In this section, we focus almost entirely on the diploid phase of sexual organisms. Consequently, we will sometimes refer to sexually reproducing organisms as diploids, but see empirical asides 1.7 and 1.8.

**Empirical Aside 1.7:** Not all diploids are entirely sexual

First, asexual, clonally reproducing diploid organisms exist. And some sexual diploids are actually partially haploid throughout their life cycle. For example, most male mammals and most female birds are functionally haploid for most of one chromosome; in those organisms, those chromosomes are transmitted clonally. And male hymenopteran insects are fully haploid.

**Empirical Aside 1.8:** Meiosis, mitosis, and multicellular development

Meiosis is cell division with reduction from two alleles at each locus (in the diploid) to just one (in the haploid); *mitosis* is cell division without genetic reduction. Mitosis is the mechanism of clonal reproduction of both haploid and diploid cells and also of development in multicellular organisms. In animals, mitosis only occurs during the diploid phase; their haploid phase is comprised of just a single cell generation, that of the gamete. However, as noted in empirical aside 1.9, many other sexual diploids also undergo mitosis and multicellular development in the haploid phase.

**Empirical Aside 1.9:** Selection can act in haploid and diploid phases of sexual organisms

In most animals, selection in the haploid phase is comparatively modest, since it is restricted to single cells that do not undergo mitosis (although not unknown, see for example, sperm competition, reviewed in Civetta and Ranz [2019]). Many plants have multicellular haploid phases, such as the pollen tubes in angiosperms, which can experience natural selection. And haploidy is the dominant life cycle phase in some algae, protists, and many fungi. Interestingly, the fraction of time a sexual organism spends in its haploid and diploid phases can itself respond to natural selection, a point to which we return briefly in section 3.3. In this chapter, we will assume selection acts only in the diploid phase of sexual organisms, but in chapter 3, we will assume the opposite for reasons articulated in teachable moment 3.1.

---

**Empirical Aside 1.10:** Dioecious vs. monoecious species, anisogamy, and sexual selection

*Dioecious* species are sexually reproducing diploids such as ourselves, in which there are two mating types, called *sexes*. In these species, syngamy is only possible between gametes produced by organisms of opposite sex. Some but not all dioecious species (including humans) are also *anisogamous*, meaning that the two mating types produce gametes of different sizes. In anisogamous species, the mating type that produces the larger gamete is called the *female*, and the one that produces the smaller gamete is called the *male*. In contrast, *monoecious* species are diploids with just one mating type. In these species, any two organisms can mate. Some monoecious species are also anisogamous. (Incidentally, botanists also apply the words dioecious and monoecious to individual flowers; the former carry both female and male sex organs, while the latter carry just one or the other.)

Anisogamy is a form of *sexual dimorphisms* (i.e., sex-based differences in organismal form) and is thus the basis of *sexual selection*. Unlike natural selection, which responds to differences in organisms' fit to their environment, sexual selection focuses exclusively on differences in sexual organisms' success in competition for mates. First recognized by Darwin (1859, 1871) to explain the evolution of costly ornamentation in males, this fascinating and complex topic is unfortunately beyond the scope of this book. The interested reader is directed to Alonzo and Servedio (2019) and citations therein to learn more.

---

There are two fundamental consequences of sexual reproduction in one-locus models. First, the process of reproduction *per se* can cause offspring to differ genetically from their parents, even in the absence of mutation. Second, the fitness effect of an allele can depend also on the identity of the allele with which it is paired in the organism. We shall address each of these in turn. Analogous complexities of sexual reproduction emerge in multilocus models, as we shall see in chapter 3.

### 1.3.1  The Population Genetics of Diploid Reproduction

The two alleles carried at a diploid locus are called *homologs*, and the genetic composition of a diploid organism is its *genotype* Assuming two segregating alleles $A$ and $a$ at a locus, three one-locus genotypes are possible: $AA$, $Aa$, and $aa$. The first and third are called *homozygotes* because they carry two copies of the same allele, while the second is called the *heterozygote*. (Note that we could as well have written the heterozygote $aA$: its essential attribute is that its two alleles differ.) During meiosis, diploid organisms produce haploid gametes carrying just one of the two homologs. Syngamy between gametes produced by two $AA$ (or $aa$) diploids yields zygotes that are genetically identical to both parents. However, all other matings yield at least some zygotes that differ from at least one of the parents. We shall assume that heterozygotes transmit their two homologs with equal frequency, as they did in Mendel's original experiments (Mendel 1866). This assumption is called *Mendelian segregation*. Cases of non-Mendelian segregation are known (empirical aside 1.11), and the corresponding theory is developed in power user challenges 1.16 and 1.17. (Genetic recombination also happens during meiosis, but we postpone exploration of that multilocus process to chapter 3.)

---

**Empirical Aside 1.11:** Mendelian segregation and meiotic drive

*Mendel's first law of segregation* asserts that heterozygotes transmit their two homologs with equal frequency. However, alleles that are transmitted into more than half of a diploid's gametes are known. Because this unequal transmission occurs during meiosis, this process is called *meiotic drive*. Interestingly, in organisms carrying a meiotic drive allele at one locus, one often also finds second loci carrying alleles that suppress meiotic drive. This suggests a picture of genomes undergoing repeated cycles of an evolutionary arms race in order to maintain Mendelian segregation ratios. (In section 3.3, we will touch on the influence of natural selection on alleles that modify segregation ratio.) Indeed, meiotic drive was first discovered in hybrids between populations or species in which the parents did not carry matched pairs of driving and suppressing alleles.

---

Now writing $N_{AA}$, $N_{Aa}$, $N_{aa}$, $p_{AA}$, $p_{Aa}$, and $p_{aa}$ for the counts and frequencies of the three one-locus genotypes defined for two alleles, we can trivially compute allele frequency as

$$p_A = \frac{N_{AA} + \frac{1}{2} N_{Aa}}{N_{AA} + N_{Aa} + N_{aa}} = p_{AA} + \frac{1}{2} p_{Aa}. \tag{1.22}$$

(The leading $\frac{1}{2}$ corresponds to the fact that heterozygotes are genetically only 50% $A$.) In general, however, the reverse calculation is underspecified. For example, allele frequency $p_A = \frac{1}{2}$ is consistent with genotype frequencies $p_{AA} = p_{aa} = \frac{1}{2}$ and $p_{Aa} = 0$, but also with genotype frequencies $p_{AA} = p_{aa} = 0$ and $p_{Aa} = 1$.

Importantly, assuming random mating, Mendelian segregation, and the absence of mutation, selection, migration and stochastic effects, a one-locus population's genotype frequencies will equilibrate to the following values:

$$p_{AA} = p_A^2 \tag{1.23a}$$

$$p_{Aa} = 2 p_A (1 - p_A) \tag{1.23b}$$

$$p_{aa} = (1 - p_A)^2. \tag{1.23c}$$

(Note that diploid reproduction only influences genotype frequencies. In the absence of mutation, sexual reproduction leaves allele frequencies unchanged.) These are called the *Hardy–Weinberg frequencies*, in honor of H. G. Hardy and Wilhelm Weinberg, who independently discovered them in 1908.

It will often prove convenient to examine the frequency of heterozygotes predicted under Hardy–Weinberg assumptions. This is called the population's *heterozygosity*, represented by the symbol $H$ and calculated as

$$H = 2p_A(1 - p_A) \tag{1.23d}$$

Critically, $H$ need not equal the frequency of heterozygotes ($p_{Aa}$) observed in a diploid population (see empirical aside 1.12). Moreover, heterozygosity can also be informative in haploid populations.

The derivation of equation (1.23) is simplified by two more assumptions: nonoverlapping generations and monoecy (or equivalently, that in dioecy, the two sexes are in 50:50 ratio). In this case, populations equilibrate to Hardy–Weinberg frequencies in one generation. (Relaxing these assumptions leaves the outcome unchanged, but the equilibration is slower. See, for example, Crow and Kimura [1970].) First, assuming nonoverlapping generations lets us model syngamy and meiosis as each occurring synchronously in the population (teachable moment 1.16). Assuming monoecy allows us to capture dynamics using population-wide allele and genotype frequencies. (In the alternative, one would have to follow sex-specific allelic and genotypic frequencies.)

The derivation of equation (1.23) is as follows. First, assuming *random mating* lets us model parents as being paired without respect to genotype. (We relax this assumption in section 1.4.4.) Mathematically, this means that the frequency of each mating is given by the product of the frequencies of each parent. These probabilities and the resulting proportions of offspring produced (assuming Mendelian segregation and nonoverlapping generations) are shown in table 1.1.

**Table 1.1**
Probabilities of one locus diploid matings and offspring produced

| Mating pair | Mating pair probability[a] | Proportion of given genotype in the next generation[b] | | |
|---|---|---|---|---|
| | | $AA$ | $Aa$ | $aa$ |
| $AA \times AA$ | $p_{AA}^2$ | $p_{AA}^2$ | $0$ | $0$ |
| $AA \times Aa$ | $2p_{AA}p_{Aa}$ | $p_{AA}p_{Aa}$ | $p_{AA}p_{Aa}$ | $0$ |
| $AA \times aa$ | $2p_{AA}p_{aa}$ | $0$ | $2p_{AA}p_{aa}$ | $0$ |
| $Aa \times Aa$ | $p_{Aa}^2$ | $\frac{1}{4}p_{Aa}^2$ | $\frac{1}{2}p_{Aa}^2$ | $\frac{1}{4}p_{Aa}^2$ |
| $Aa \times aa$ | $2p_{Aa}p_{aa}$ | $0$ | $p_{Aa}p_{aa}$ | $p_{Aa}p_{aa}$ |
| $aa \times aa$ | $p_{aa}^2$ | $0$ | $0$ | $p_{aa}^2$ |
| Sums | $1$ | $\left(p_{AA} + \frac{1}{2}p_{Aa}\right)^2$ | $2\left(p_{AA} + \frac{1}{2}p_{Aa}\right)\left(\frac{1}{2}p_{Aa} + p_{aa}\right)$ | $\left(\frac{1}{2}p_{Aa} + p_{aa}\right)^2$ |

[a] Assuming random mating. The leading 2 on the second, third, and fifth lines reflects the fact that two matings are represented: one in which the mother's genotype is on the left and one in which hers is on the right.
[b] Values for matings involving at least one heterozygous parent assume Mendelian segregation. See empirical aside 1.11 and power user challenge 1.16 for more.

---

**Power User Challenge 1.15**

Use the sums in table 1.1 to show that the frequency of allele $A$ after diploid reproduction is still $p_A$. Hint: apply equation (1.22). Now show that the frequency of the $A$ allele in the offspring is $p_A$ even in the absence of random mating and Mendel's first law of segregation. This must be so: as noted, only mutation changes allele frequency during reproduction.

---

**Power User Challenge 1.16**

Show that the right-most three sums at the bottom of table 1.1 add to 1.0, as they must since each mating must produce some genotype. Now rewrite table 1.1 on the premise of meiotic drive. Parameterize this effect by assuming that heterozygotes produce a proportion $x$ of $A$ alleles ($0 \le x \le 1$) and proportion $1-x$ of $a$ alleles. Show that the sums of each of the three columns at the right of your new table still add to 1. Hint: what value of $x$ was used in constructing table 1.1?

---

Next, the sums in the last three columns in table 1.1 give us genotype frequencies at the start of a generation. More specifically, given arbitrary genotype frequencies $p_{AA}$, $p_{Aa}$, and $p_{aa}$ among parents, zygotes in the next generation will be at genotype frequencies

$$p_{AA}^{zygote} = \left( p_{AA} + \frac{1}{2} p_{Aa} \right)^2 = p_A^2,$$

$$p_{Aa}^{zygote} = 2 \left( p_{AA} + \frac{1}{2} p_{Aa} \right) \left( \frac{1}{2} p_{Aa} + p_{aa} \right) = 2 p_A (1 - p_A), \text{ and} \qquad (1.24)$$

$$p_{aa}^{zygote} = \left( \frac{1}{2} p_{Aa} + p_{aa} \right)^2 = (1 - p_A)^2.$$

(The last step in each expression relies on equation [1.22].) Finally, in the absence of mutation, natural selection, and migration, genotype frequencies will remain unchanged from syngamy to meiosis. Thus, parental genotype frequencies in this next generation will be

$$p'_{AA} = p_{AA}^{zygote} = p_A^2,$$

$$p'_{Aa} = p_{Aa}^{zygote} = 2 p_A (1 - p_A), \text{ and}$$

$$p'_{aa} = p_{aa}^{zygote} = (1 - p_A)^2.$$

This demonstrates that under our assumptions, genotype frequencies move to Hardy–Weinberg frequencies in one generation. Repeating the above reasoning shows that

$$p''_{AA} = p_A^2,$$

$$p''_{Aa} = 2 p_A (1 - p_A), \text{ and}$$

$$p''_{aa} = (1 - p_A)^2$$

for genotype frequencies at the end of the second generation. This demonstrates that a population at Hardy–Weinberg frequencies remains at Hardy–Weinberg frequencies.

---

**Empirical Aside 1.12**

Visible phenotypes determined by single loci have long provided empirical access to both allelic and genotypic frequencies in natural populations. Consequently, comparisons between the prediction of equation (1.23d) and empirical frequencies of heterozygotes represented one of the first experimental tests of population genetic theory. For example, there are approximately 30 distinct blood groups in humans, reflecting the many antigens on the surface of human red blood cells. Among these, the ABO and Rh groups are the best known. (The successful characterization of human blood groups with respect to their antigen reactivity by Karl Landsteiner at the turn of the twentieth century for the first time allowed safe blood transfusions. This technological breakthrough has saved $\approx 1$ billion lives (Woodward et al. 2023) and earned Landsteiner a Nobel Prize in 1930.)

Hartl and Clark (2007) illustrate the application of the Hardy–Weinberg result to MN blood group frequencies in human populations. The MN blood group is particularly useful for our present purposes, both because it is determined by a single biallelic locus (alleles designated $M$ and $N$) and because all three genotypes have antigenically distinguishable phenotypes. In a 1976 study of 6,129 individuals, 1,787 had the MM blood type, 3,037 were MN, and 1,305 were NN. Equation (1.22) tells us that the frequency of the $M$ allele in this population is $p_M = \dfrac{1{,}787 + \frac{1}{2} \times 3{,}037}{6{,}129} = 0.539$, which in turn predicts heterozygosity $H = 2 \times 0.539 \times (1 - 0.539) \approx 0.497$ via equation (1.23d). That this nearly matches the observed frequency of heterozygotes: $\dfrac{3{,}037}{6{,}129} \approx 0.496$ suggests that all Hardy–Weinberg assumptions are likely met in this system.

---

**Power User Challenge 1.17**

Write an expression for the rate at which genotype frequencies change if meiotic drive (see empirical aside 1.11) causes heterozygotes to transmit the $A$ allele to 60% of its offspring. You should assume monoecy, nonoverlapping generations and all other Hardy–Weinberg conditions.

---

**Power User Challenge 1.18**

Imagine a diploid population of organisms under Hardy–Weinberg assumptions, but now allowing mutation. Write the $A \to a$ and $a \to A$ mutation rates as $\mu$ and $\nu$, respectively. Thus, in each generation, a fraction $\mu$ of $A$ alleles become $a$ alleles, and a fraction $\nu$ of $a$ alleles become $A$ alleles. Find the analog to equation (1.24) for this model. Hint: equation (1.23) still applies, because this is a model of nonoverlapping generations.

### 1.3.2 Natural Selection in Diploids

We now ask how selection on diploid genotypes affects allele frequencies. We restrict attention to *viability selection*, defined as the proportion of organisms surviving from syngamy to meiosis (see teachable moment 1.22). In other words, we assume that selection only acts in the diploid phase of the life cycle. We also assume monoecy, or equivalently, that viability selection acts the same in the two sexes. We maintain all other Hardy–Weinberg assumptions (nonoverlapping generations, random mating, Mendel's law of segregation, and the absence of mutation and population structure). Thus, genotypes will always be at Hardy–Weinberg frequencies at syngamy.

---

**Teachable Moment 1.22:** Viability vs. fertility selection

In sexual diploids, selection can act on viability (differences in individual survival) as well as on fertility (differences in the reproductive output of mating pairs). We shall disregard *fertility selection*, which requires six parameters, one corresponding to each of the possible matings shown in table 1.1 (or even nine if there are fecundity differences when the genotypes of the two parents are reversed). The interested reader is referred to Crow and Kimura (1970) to see how this is done.

---

If allele $A$ is present at frequency $p_A$ at the start of one generation, we seek $p_A'$, its frequency at the start of the next. We first write $p_{AA}$, $p_{Aa}$, and $p_{aa}$ for the frequencies of each genotype at syngamy and correspondingly, relative Wrightian fitness values $w_{AA}$, $w_{Aa}$, and $w_{aa} = 1$. Equation (1.22) suggests

$$p_A' = \frac{p_{AA} w_{AA} + \frac{1}{2} p_{Aa} w_{Aa}}{p_{AA} w_{AA} + p_{Aa} w_{Aa} + p_{aa}}$$

after at the next meiosis. (We omit the details; one begins by defining absolute fitness values, then substitutes equation [1.14] into equation [1.22], and finally normalizes by $W_{aa}$ to reach relative fitness values.) Our assumption of Hardy–Weinberg conditions (including no selection during the haploid phase) means that genotype frequencies at the next syngamy are given by equation (1.23), letting us write

$$p_A' = \frac{p_A^2 w_{AA} + p_A(1-p_A) w_{Aa}}{\overline{w}},$$

where $\overline{w} = p_A^2 w_{AA} + 2 p_A(1-p_A) w_{Aa} + (1-p_A)^2$ is the population's mean relative fitness. Finally,

$$\Delta p_A = p_A' - p_A = \frac{p_A(1-p_A)[p_A(w_{AA} - w_{Aa}) + (1-p_A)(w_{Aa} - 1)]}{\overline{w}}, \tag{1.25a}$$

found by expanding and factoring out the leading $p_A(1-p_A)$.

Comparing this result to that in its haploid analog (equation 1.16), we see that the difference in relative fitness between haploid individuals carrying alternate alleles (written $s = w_A - 1$) has been replaced by $p_A(w_{AA} - w_{Aa}) + (1 - p_A)(w_{Aa} - 1)$. This new, seemingly cryptic expression reflects the fact that in diploids, the fitness difference between the $A$ and $a$ alleles is manifest in two genetic contexts: $AA$ vs. $Aa$ and $Aa$ vs. $aa$. Moreover, at Hardy–Weinberg frequencies, the fraction of $A$ alleles found in the homozygous context is $\dfrac{2N_{AA}}{2N_{AA} + N_{Aa}} = \dfrac{p_{AA}}{p_{AA} + \frac{1}{2}p_{Aa}} = \dfrac{p_A^2}{p_A^2 + p_A(1 - p_A)} = p_A$, which leaves a fraction $1 - p_A$ of the $A$ alleles in the heterozygotes. In English, our new expression can thus be read as the average fitness difference between alleles, weighted by the proportion of the two contexts in which $A$ alleles are found under Hardy–Weinberg assumptions.

Following our haploid framing, we write the diploid Wrightian selection coefficient $s = \dfrac{w_{AA} - 1}{2}$. (Because $w_{AA} - 1$ is the fitness difference when genotypes differ by two copies of the $A$ allele, we divide by 2 to get a per copy selection coefficient.) Of course, the real novelty of diploids is that the fitness effect of the $A$ allele can differ in its two genetic contexts. Thus, we introduce a Wrightian *dominance coefficient* $h = \dfrac{w_{Aa} - 1}{2s}$, which quantifies this effect. Substituting these definitions into equation (1.25a) yields

$$\Delta p_A = \frac{2p_A(1 - p_A)[p_A(1 - h)s + (1 - p_A)hs]}{\overline{w}}. \tag{1.25b}$$

(Now, the change compared to our haploid treatment reads $2[p_A(1 - h)s + (1 - p_A)hs]$, which again quantifies the mean fitness difference between $A$ and $a$ alleles.)

---

**Empirical Aside 1.13:** What are dominance coefficients in nature?

Surveys in diverse organisms have demonstrated that dominance coefficients for deleterious alleles are often in the range of 0–0.3. Interestingly, there appears to be a positive correlation between the magnitude of effect and dominance coefficient. In other words, more severely deleterious alleles tend to be more recessive. Owing to their rarity, much less is known about dominance coefficients for beneficial alleles.

---

Although there are no formal bounds on $h$, there are several biologically motivated benchmark values that we examine now. First, setting $h = \dfrac{1}{2}$ means that heterozygote fitness is exactly halfway between that of the homozygotes. In this case, selection influences allele frequencies independent of genetic context, and we describe it as *genic selection* (in contrast to genotypic selection). Substituting into equation (1.25b) yields

$$\Delta p_A = \frac{p_A(1-p_A)s}{\overline{w}},$$

which is exactly the value we found for the haploid case (equation 1.16).

Next, setting $h=1$ means that the heterozygote fitness is equal to that of the $AA$ homozygote. We say that $A$ is *fully dominant* over $a$ (or equivalently, that $a$ is *fully recessive* to $A$), and equation (1.25b) now reads

$$\Delta p_A = \frac{p_A(1-p_A)^2 s}{\overline{w}}.$$

The novelty here in comparison to the genic and haploid cases is that $\Delta p_A$ is proportional to $p_A(1-p_A)^2$ rather than $p_A(1-p_A)$. Biologically, this means that the selective fixation of a fully dominant, beneficial allele is slower toward the end; that is, when $1-p_A$ becomes very small. We shouldn't be surprised. By definition, when fitness is genic, natural selection eliminates the less-fit $a$ allele equally efficiently in heterozygotes and homozygotes. In contrast, when $A$ is fully dominant, selection can only eliminate $a$ alleles in the homozygote; copies of that allele in heterozygotes are unseen by selection. Under Hardy–Weinberg assumptions, homozygote $aa$ genotypes become very rare toward the end of the process.

Conversely, setting $h=0$ means that the heterozygote fitness is equal to the $aa$ homozygote. We say that $A$ is *fully recessive* to $a$ (or equivalently, that $a$ is fully dominant over $A$) and equation (1.25b) now reads

$$\Delta p_A = \frac{p_A^2(1-p_A)s}{\overline{w}}.$$

In this case, $\Delta p_A$ is proportional to $p_A^2(1-p_A)$, telling us that the selective fixation of a recessive beneficial allele starts more slowly than in the genic case, for reasons analogous to those developed in the previous paragraph.

Of course, $h$ can also assume values between $\frac{1}{2}$ and 1, in which case the $A$ allele is said to be *partially dominant*, or between 0 and $\frac{1}{2}$, in which case it is *partially recessive*. Figure 1.9 illustrates selective fixations of a beneficial allele under genic selection and when it is partially dominant and partially recessive.

Thus far, we have restricted attention to cases where heterozygote fitness lies between that of the two homozygotes; mathematically, $0 \le h \le 1$ in equation (1.25b). A heterozygote whose fitness is higher than either homozygote is said to exhibit *overdominance* (mathematically, $h > 1$), while one whose fitness is lower than either shows *underdominance* ($h < 0$). See empirical aside 1.14 for evidence of overdominance in nature.

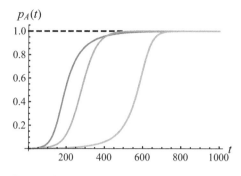
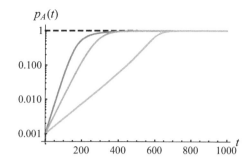

**Figure 1.9**
Time course of natural selection in diploids.
Here, $p_A(0) = 0.001$, $s = 0.05$, and $h = 0.2$ (green), 0.5 (mustard), and 0.8 (blue). The fitter type ($A$) is selectively displacing the less fit type ($a$). The two panels differ only in that the $y$-axis on the right is log transformed. Values were computed by the iterative application of equation (1.25b); see teachable moment 1.16. In all cases, $p_A(t)$ asymptotically approaches 1 (black, dashed line). In both panels, results for $h = 0.5$ are identical to those shown for haploids in figure 1.2.

---

**Empirical Aside 1.14**

Before the advent of molecular methods, considerable genetic variation in natural populations was already inferred on the basis of observed phenotypic variation. Such phenotypic variability was commonly interpreted as reflecting overdominance, the best example being the case of the *HbS* allele at the $\beta$-globin locus of humans. Here, *HbA/HbA* homozygotes are fully malaria-sensitive and *HbS/HbS* homozygotes are susceptible to sickle-cell anemia (a debilitating illness), but *HbA/HbS* heterozygotes enjoy reduced malarial sensitivity with almost no adverse consequences. The MHC locus in mammals also likely represents a case of overdominance (empirical aside 1.16). However, overdominance as a general explanation for genetic variability in natural populations raises technical concerns, which directly motivated the neutral theory of molecular evolution (empirical aside 2.2).

---

From the theoretical point of view, the most interesting feature of over- and underdominance is that natural selection can now cause populations to remain polymorphic at equilibrium. This stands in contrast to results in haploids and in diploids when $0 \leq h \leq 1$, where natural selection drives allele frequency to 0 or 1. Mathematically, this *internal equilibrium* happens when the average fitness difference between alleles $(2[p_A(1-h)s + (1-p_A)hs]$ in equation [1.25b]) is zero, thereby causing allele frequency to cease changing. To find this equilibrium, we set the average fitness difference between alleles to zero and solve for $p_A$ to find

$$\tilde{p}_A = \frac{h}{2h-1}.$$

---

**Power User Challenge 1.19**

One of the polymorphic equilibria associated with over- and under-dominance is stable and one is unstable. Which is which, and why? (See teachable moment 1.17.)

---

**Power User Challenge 1.20**

We have just seen that population polymorphism can persist indefinitely when the heterozygote is over- or underdominant. Compute the variance for fitness in populations at this equilibrium. You should assume that genotypes are at Hardy–Weinberg frequencies. Reconcile your result with Fisher's fundamental theorem of natural selection.

---

**Teachable Moment 1.23:** Balancing selection

*Balancing selection* refers to cases in which natural selection maintains polymorphism at a locus. Overdominance is one mechanism, and so is temporal or spatial variation in environment (see section 1.4). Indeed, varying environment can even cause balancing selection in haploid organisms. For example, imagine a situation where one allele is favored in one season (or portion of the species' range) and the other is favored in another. So long as the time spent in each environment is much less than the time to selective fixation of either allele, the population will remain polymorphic. Concerns raised in empirical aside 1.14 notwithstanding, data from humans and chimpanzees revealed evidence consistent with a modest but still important role for balancing selection in nature (Leffler et al. 2013).

---

**Teachable Moment 1.24:** Population genetic models can be parameterized in many ways

These choices reflect differences in authors' motivation, tradition, or mathematical expedience, and may not deeply change the behavior of the underlying model. For example, one often sees diploid selection coefficients defined as $s = w_{AA} - 1$ instead of our $s = \dfrac{w_{AA} - 1}{2}$. And models of overdominance often normalize fitness by $W_{Aa}$ instead of $W_{aa}$. Importantly, these superficial choices have no impact on the underlying theory. It is therefore important to read and think carefully when comparing models, in order to be able to distinguish between trivial and structural differences.

---

## 1.4   Population Structure, Migration, and Nonrandom Mating

So far, we have assumed *well-mixed populations*, meaning that all organisms with identical genetics are equivalent in our models. For example, our treatments of a selective fixation in haploids assumed that every $A$-bearing individual in the population has identical fitness, which allowed us to predict their rate of increase or decrease only as a function of their fitness effect and frequency. Similarly, the frequencies of diploid matings in table 1.1 assumed that mating probabilities are determined only by genotype frequencies.

We now relax the assumption of well-mixed populations. For example, imagine a population of organisms distributed across an island archipelago. If environmental conditions vary among islands, different alleles might be selectively favored in different locations. In the total absence of migration, each subpopulation will evolve in isolation according to the theory developed above. But novel behavior emerges when migration rates between islands are low enough to preserve some genetic differentiation but high enough to allow some mixing.

And even in the absence of such *diversifying selection* (i.e., selection favoring different alleles in different locations), spatial structure in diploids can perturb populations from Hardy–Weinberg frequencies, even if each subpopulation fully honors all relevant assumptions. More broadly, all forms of nonrandom mating (be they due to spatial population structure, behavior, or any other mechanism) have conceptually similar effects. These deviations are often accessible to the experimentalist, yielding important opportunities for inference.

### 1.4.1 Models of Population Structure and Migration

The study of migration is intimately connected to specific models of population structure. For example, *island models* consider spatially discrete subpopulations (sometimes called *demes*) that are assumed to be well mixed, and study the evolution of allele frequencies across the entire population (or *metapopulation*) in the presence of migration between demes. Demes can be equal sized but need not be. Some island models have explicit spatial structure, in which case migration rates are higher between spatially nearby demes and lower between more distant ones. *Stepping stone* models are a special case of spatially structured island models, in which migration is restricted to spatially adjacent demes. Other island models assume equal migration rates between all pairs of demes; these lack all spatial structure. Yet other models of spatial structure consider populations living in a continuous space in which individual organisms migrate some (small, possibly random) distance during their lifetime. And models of spatial structure (be they discrete or continuous) can incorporate local fitness differences to oppose the homogenizing effect of migration on allele frequencies. As we shall see in chapter 2, stochastic effects also cause local differences in allele frequency.

---

**Teachable Moment 1.25**

Most island models assume that relative deme size within the metapopulation remains constant. Relaxing this assumption—as, for example, if within-deme allele frequency influences local population density—can allow competition (and thus natural selection) among demes, above and beyond that among individuals. This example illustrates the fact that natural selection can act simultaneously at more than one level of biological organization. A thorough treatment of this possibility is beyond our scope, but the interested reader is directed to Chuang et al. (2009) for an illuminating example.

---

**Teachable Moment 1.26**

Reproductive isolation, as for example between subpopulations, underlies the biological species concept. Speciation is beyond the scope of this book, but the interested reader is directed to a seminal synthesis by Coyne and Orr (2004).

### 1.4.2  Migration/Selection Equilibrium

To begin, imagine a haploid, biallelic population living on some mainland and an adjacent island. (Or equivalently, imagine a diploid population under genic selection.) Let the environments differ, such that allele $A$ has a fitness advantage over allele $a$ on the island but is at fitness disadvantage on the mainland. In the absence of migration, common sense (and section 1.1.2) tells us that diversifying selection will cause allele $a$ to fix on the mainland and allele $A$ to fix on the island.

We now ask whether migration from the mainland can overwhelm selection on the island, thereby driving the $A$ allele to extinction despite its local fitness advantage. And if not, to what frequency will the $A$ allele equilibrate on the island? To simplify, we disregard migration from the island to the mainland, either because it doesn't happen or because the vastly larger mainland population means that allele frequency there is unaffected by migration. We shall also ignore mutation.

Assuming overlapping generations, write $m$ for the rate of immigration to the island. Thus, $m\Delta t$ is the fraction of organisms on the island that are replaced by immigrants during time interval $\Delta t$. (Please note that, although represented by the same symbol as mortality rate in section 1.1.1, the two are entirely unrelated.)

Writing $p_A$ for the frequency of the $A$ allele on the island, $s$ for its haploid or genic fitness advantage there, and letting $\Delta t \rightarrow 0$, we find

$$\frac{dp_A}{dt} = sp_A(1-p_A) - mp_A \qquad (1.26)$$

for the rate of change in frequency of allele $A$ on the island. The first term gives the instantaneous rate at which the more-fit $A$ allele is enriched by natural selection on the island; it is identical to equation (1.8). The second term is new: it is the rate at which immigration depletes $A$ alleles on the island. $m$ is the fraction of the population that is displaced by immigrants; by definition, these are 100% $a$ bearing. However, only a fraction $p_A$ of immigrants displace $A$-bearing individuals, since fraction $1-p_A$ of the island population is already $a$ bearing.

Repeating the approach introduced in section 1.2.2, we set equation (1.26) to zero and again find two roots,

$$\tilde{p}_A \approx \begin{cases} 1 - \dfrac{m}{s}, \\ 0 \end{cases} \qquad (1.27)$$

for the frequency of the $A$ allele on the island at *migration/selection equilibrium*. The stability approach introduced in teachable moment 1.17 tells us that the first root is appropriate when $m < s$. Thus, under this condition, allele frequency on the island will equilibrate to $1 - \frac{m}{s}$ under the joint effect of migration and selection. However, populations risk a "migration threshold" when $m = s$, and when $m \geq s$, migration overwhelms selection and will drive the fitter $A$ allele extinct on the island. (Equation [1.27] is formally identical to equation [1.20], only substituting the compound parameter $\frac{m}{s}$ for $\frac{\mu}{s}$. This is the first of several formal analogies between migration and mutation we will encounter.)

### 1.4.3   Allele Frequency Clines

Instead of assuming a population subdivided into well-mixed demes, now imagine that each organism migrates some small distance in an unbiased direction during its lifetime. Suppose this population is distributed across two spatially abutting environments, where allele $A$ is favored in one with selection coefficient $s$ and disfavored in the other by an amount $-s$. We might imagine that at any position along a one-dimensional transect across the boundary between environments, allele frequency will reach a particular equilibrium between selection and migration.

Intuitively we can already draw a few conclusions. (These ideas go back to Wright [1943] and Malécot [1948]; the interested reader is directed to the chapter 11 appendix in Coop [2020] for a quantitative treatment of this situation.) Deep within one environment, the population will be fixed for the locally favored allele. This follows from the fact that migration is finite, and indeed, the quantitative specification of "deep" in the previous sentence must somehow depend on the relative strengths of migration (rates and distances) and selection. As we then move along the transect toward the other environment, we should expect to find the disfavored allele at increasing frequency, given precisely by the balance between migrational input and selective elimination at that position. As we cross the boundary between

**Empirical Aside 1.15:** Allele frequency clines and diversifying selection

One of the first- and best-studied genetic polymorphisms is in the alcohol dehydrogenase (*Adh*) gene of the fruit fly *Drosophila melanogaster*. Adh is an enzyme that catalyzes the first step of alcohol metabolism in animals, and the frequency of the slow allele (written as *Adh-S*, so-called because of its reduced electrophoretic mobility) in eastern North America is shown as a function of latitude in figure 1.10.

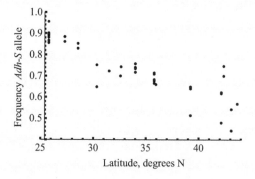

**Figure 1.10**
Frequency of the *Adh-S* allele of *Drosophila melanogaster* (*x*-axis) in 42 samples from 11 locations taken in the eastern United States as a function of latitude (*y*-axis). (Data from Vigue and Johnson 1973.)

At least two lines of evidence support the hypothesis that this gradient in allele frequency is maintained by diversifying selection. First, it is replicated on the east coast of Australia (Oakeshott et al. 1982), where its frequency again declines with distance from the equator, and thus temperature. This replicated, positive correlation with an important feature of the environment is suggestive of diversifying selection. Second, the cline is restricted to the *Adh* locus, making it unlikely that it is an artifact of recent migration (see also section 4.4.3).

Another compelling example of a cline maintained by diversifying selection comes from the pepper moth *Biston betularia* (see empirical aside 1.2). Using data collected before English air pollution regulations began to have an effect, Bishop (1972) reported that the frequency of the dark *carbonaria* type dropped from over 95% in Liverpool to near 0% at a rural location 75 km away.

environments, we will find the two alleles at 50:50 ratio, because migration is unbiased (and thus symmetric) and the selection coefficients are equal and opposite on either side of the boundary. As we now move more deeply into the other environment, we will see a similar, continued increase of the now-favored allele, until it eventually is locally fixed. Such a spatial gradient in allele frequencies is called a *cline*, one example of which is illustrated in empirical aside 1.15.

### 1.4.4 Nonrandom Mating in Diploids

The preceding models of migration and population structure apply equally to haploid and diploid organisms. However, because migration and population structure explicitly undermine the assumption of random mating, they will both also perturb diploid genotype frequencies from Hardy–Weinberg frequencies. We now explore these implications.

To begin, we introduce the *fixation index,*

$$F = 1 - \frac{p_{Aa}}{2p_A(1-p_A)}, \tag{1.28}$$

to quantify deviations from Hardy–Weinberg genotype frequencies. Thus, $F=0$ when heterozygotes are present at exactly the frequency expected under Hardy–Weinberg assumptions, positive when heterozygotes are underrepresented, and negative when they are overrepresented. Solving equation (1.28) for $p_{Aa}$ yields

$$p_{Aa} = 2p_A(1-p_A)(1-F). \tag{1.29a}$$

Substituting this into equation (1.22) and solving for $p_{AA}$ yields

$$p_{AA} = p_A^2(1-F) + p_A F, \tag{1.29b}$$

which means

$$p_{aa} = (1-p_a)^2(1-F) + (1-p_A)F \tag{1.29c}$$

(since $p_{AA} + p_{Aa} + p_{aa} = 1$).

Thus, knowing the value of $F$ in a population resolves the issue of computing genotype frequencies from allele frequencies introduced below equation (1.22). Returning to those two examples, if $p_A = \frac{1}{2}$ but now $F = 1$, we have $p_{AA} = p_{aa} = \frac{1}{2}$ and $p_{Aa} = 0$, while if $p_A = \frac{1}{2}$ and $F = -1$ we find $p_{AA} = p_{aa} = 0$ and $p_{Aa} = 1$. See also teachable moment 3.2.

---

**Teachable Moment 1.27a:** $F$ is the first of several fixation indices

Introduced by Sewall Wright (1949), each of these indices quantify the proportional reduction (or excess) in heterozygotes relative to some benchmark. In the present case, the benchmark is the Hardy–Weinberg prediction computed with the population-wide allele frequency. More broadly, fixation indices will quantify deviations from Hardy–Weinberg frequencies due to any process, a point we take up next and again in section 2.3.

---

**1.4.4.1  Assortative and disassortative mating**  We begin with the possibility that mates are chosen not at random, but with reference to their genotypes. For example, if like genotypes tend to mate, then the frequencies on the first, fourth and sixth lines of table 1.1 will be inflated at the expense of the frequencies on the second, third, and fifth lines. This is called *assortative mating*. *Disassortative mating* is the opposite: overrepresented mating frequencies among unlike genotypes (second, third, and fifth lines of table 1.1). Of course, one can imagine a great many other nonrandom mating scenarios (e.g., an enrichment of $AA \times Aa$ and $aa \times aa$ matings at the expense of the others).

---

**Empirical Aside 1.16:** Disassortative mating and the major histocompatibility complex (MHC)

In many animals, mating choice is not random but is rather made with reference to physical or behavioral characteristics of the other parent. If that process has a genetic component, it violates our assumption of random mating, and such organisms instead exhibit assortative or disassortative mating. For example, the MHC loci encode cell surface proteins essential for mammalian immune response, and good evidence exists for disassortative mating at these loci. This is likely driven by a fitness advantage associated with greater diversity in an individual's immunological repertoire. See Penn and Potts (1999) and Meyer and Thomson (2001) to learn more.

---

For example, table 1.2 presents the proportion of offspring of each genotype under perfect assortative mating. Because like mate only with like under this model, the proportions of each mating is equal to the frequency of the corresponding parent genotype. Consequently, in each generation the fraction of $Aa$ heterozygotes is halved, since heterozygotes are now only produced by $Aa \times Aa$ matings, and only half of those offspring are heterozygotes. Thus, under perfect assortative

**Table 1.2**
Probabilities of one locus diploid mating under perfect assortative mating, and offspring produced

| Mating | Mating pair probability[a] | Proportion of given genotype in the next generation[b] | | |
| --- | --- | --- | --- | --- |
| | | $AA$ | $Aa$ | $aa$ |
| $AA \times AA$ | $p_{AA}$ | $p_{AA}$ | 0 | 0 |
| $AA \times Aa$ | 0 | 0 | 0 | 0 |
| $AA \times aa$ | 0 | 0 | 0 | 0 |
| $Aa \times Aa$ | $p_{Aa}$ | $\frac{1}{4}p_{Aa}$ | $\frac{1}{2}p_{Aa}$ | $\frac{1}{4}p_{Aa}$ |
| $Aa \times aa$ | 0 | 0 | 0 | 0 |
| $aa \times aa$ | $p_{aa}$ | 0 | 0 | $p_{aa}$ |
| Sums | 1 | $p_{AA}+\frac{1}{4}p_{Aa}$ | $\frac{1}{2}p_{Aa}$ | $\frac{1}{4}p_{Aa}+p_{aa}$ |

[a] Assuming perfect assortative mating.
[b] Assuming Mendelian segregation.

mating (and retaining all other Hardy–Weinberg assumptions), meiotic segregation means the population will eventually lose all of its heterozygotes.

---

**Power User Challenge 1.21**

Use table 1.2 to show that the $AA$, $Aa$, and $aa$ genotypes will equilibrate respectively to frequencies $p_A$, 0, and $1 - p_A$ under perfect assortative mating, maintaining all other Hardy–Weinberg assumptions. Does your answer depend on assuming Mendelian segregation? Does it depend on perfect assortative mating?

---

**Teachable Moment 1.27b:** Why is $F$ called a "fixation" index?

As just seen, homozygotes become fixed under perfect assortative mating. Wright proposed conceptualizing any observed deviation from random mating as a chimeric population in which a proportion $1 - F$ of the individuals undergo random mating, and a proportion $F$ of which mate assortatively. This interpretation provides one explicit interpretation of equation (1.29) (although note that it doesn't capture the consequences of disassortative mating, under which $F$ is negative).

---

**1.4.4.2 Inbreeding and the probability of identity by descent** The fixation index defined by equation (1.28) captures the deviation in heterozygote frequency from that predicted under Hardy–Weinberg assumptions but is agnostic on mechanism. Genetically determined (dis)assortative mating is one mechanism, and *inbreeding* (i.e., matings among genetically related individuals) is another. More specifically, inbreeding can cause individuals to carry two homologs of an allele derived from a single copy found in some ancestor. Such individuals are said to be *identical by descent* (or *IBD*), and IBD individuals represent another source of homozygotes in the population not anticipated in the Hardy–Weinberg framing. Patterns of inbreeding can be represented in a *pedigree* or family tree, and we introduce an individual's *inbreeding coefficient* $F_I$, as the probability that it is IBD. It remains only to compute $F_I$ from a given pedigree.

Before starting, we make several technical points. First, in answer to power user challenge 1.1, Darwin's other major insight in *On the Origin of Species* was that all organisms on Earth are descended from a common ancestor. It thus follows that ultimately, the two homologs in any individual must descend from some single ancestral copy. (We substantially sharpen this point in chapter 2.) Consequently, IBD can only sensibly be defined over some time interval. Equivalently, $F_I$ is computed on the assumption that it was zero in some particular generation in the past (see also teachable moment 1.27a). Second, we define $F_I$ as the probability of IBD at one locus, but in chapter 4 we will regard it as the fraction of loci that are IBD. Finally, we sometimes use the terms *allozygote* and *autozygote* for diploids formed by random mating and by inbreeding, respectively. *F* is equivalently the proportion of autozygous individuals in the population, which offers another way of reading equation (1.29). All heterozygotes are allozygotes, which is why their frequency is reduced by the fraction $1 - F$. The proportions of allo- and autozygote homozygotes correspond to the two terms on the right side of equations (1.29b) and (1.29c). This restates the point made in teachable moment 1.27b.

The method for computing $F_I$ in any given pedigree follows directly from Mendel's law of segregation, which states that the probability that a focal homolog will survive meiosis and be transmitted to an offspring is $\frac{1}{2}$. To illustrate, consider the most extreme case of inbreeding: *selfing* (or self-fertilization), in which a single individual contributes both gametes to an offspring. The probability that both homologs in the offspring are derived from the same homolog in the parent (and thus, the offspring's $F_I$) is $\frac{1}{2}$. The situation for a child of half siblings (organisms sharing exactly one parent) is only slightly more complex. In the first generation on the pedigree, the probability that both half-siblings inherit the same homolog from their shared parent is $\frac{1}{2}$. In the second generation, the probability that each half-sibling transmits the homolog inherited from the common parent to the focal child is also $\frac{1}{2}$. Thus, in total, $F_I$, the probability that their child is IBD, is $\frac{1}{2^3} = \frac{1}{8}$. Put another way, for the child of half siblings to be IBD, the homolog must be transmitted through three meioses. $F_I$ for a child of full siblings is twice this value, since the foregoing logic applies equally to the homologs inherited by each sibling from either parent.

In general, we compute the probability of IBD for an individual as

$$F_I = \frac{1}{2^{k-1}}, \tag{1.30}$$

where $k$ is the number of meioses separating the two homologs in the individual. The exponent is one less than the number of meioses $(k-1)$ because the individual will be IBD if it inherits either of the two homologs in the ancestor. As illustrated by the case of the offspring of full siblings, if there are multiple ancestors by which the individual can be IBD, its $F_I$ is the sum of equation (1.30) computed through each such ancestor.

### 1.4.4.3 Population subdivision and the Wahlund effect

As noted above, subdivision can also cause diploid metapopulation genotype frequencies to deviate from Hardy–Weinberg frequencies, even if all assumptions are honored within demes. This is called the *Wahlund effect*, and its mechanism is conceptually simple. Namely, if allele frequencies differ among demes, then random mating within demes will result in a metapopulation-wide excess of matings between homozygotes carrying the locally more common allele. Put another way, population structure imposes yet another form of nonrandom mating.

To most easily model this quantitatively, imagine a population divided into two equal-sized, fully isolated demes, each honoring all Hardy–Weinberg assumptions. Equation (1.23d) tells us that in this case, heterozygosity in each deme is

$$H_1 = 2p_1(1-p_1) \text{ and}$$

$$H_2 = 2p_2(1-p_2),$$

where subscripts 1 and 2 to refer to the two subpopulations. The mean heterozygosity in this subdivided population will then be

$$H_S = \frac{1}{2}(H_1 + H_2) = p_1(1-p_1) + p_2(1-p_2). \tag{1.31}$$

The subscript on $H_S$ tells us that this is the heterozygosity for the subdivided population, and the leading $\frac{1}{2}$ reflects the fact that the two demes are assumed to be equal size. To quantify the effect of subdivision, we need to compare this result with what the heterozygosity would be in the absence of subdivision. Using subscript T to refer to the total metapopulation, allele frequency is $p_T = \frac{1}{2}(p_1 + p_2)$, which gives

$$H_T = 2p_T(1-p_T) = 2\frac{(p_1+p_2)}{2}\left[1 - \frac{(p_1+p_2)}{2}\right]$$

$$= p_1 + p_2 - \frac{1}{2}(p_1+p_2). \tag{1.32}$$

Somewhat remarkably, the reduction in heterozygosity due to subdivision is twice the variance in allele frequencies among demes. Since variances can never be negative, population subdivision inevitably reduces metapopulation heterozygosity unless demic allele frequencies are equal. Mathematically, we have

$$H_T - H_S = 2\mathrm{Var}(p) \geq 0, \tag{1.33}$$

where the $p$ are the frequencies of the $A$ allele in each deme. The proof of equation (1.33) is shown in teachable moment 1.28.

---

**Teachable Moment 1.28:** Proof of the Wahlund effect

We begin by expanding its left side of equation (1.33), finding

$$H_T - H_S = p_1 + p_2 - \frac{1}{2}(p_1 + p_2)^2 - [p_1(1-p_1) + p_2(1-p_2)]$$

$$= p_1^2 + p_2^2 - \frac{1}{2}(p_1 + p_2)^2,$$

via equations (1.31) and (1.32). And working from the right of the same equation, we apply equation (1.13b) to find

$$\mathrm{Var}(p) = \frac{1}{2}p_1^2 + \frac{1}{2}p_2^2 - \left(\frac{p_1 + p_2}{2}\right)^2 = \frac{p_1^2 + p_2^2}{2} - \frac{1}{4}(p_1 + p_2)^2,$$

or

$$2\mathrm{Var}(p) = p_1^2 + p_2^2 - \frac{1}{2}(p_1 + p_2)^2$$

Q.E.D.

---

Although teachable moment 1.28 assumes two equal-sized demes, equation (1.33) applies equally to a fully subdivided population of any number of demes of any size, now taking the weights used to compute means and variances to be the relative sizes of each deme. Figure 1.11 presents a graphical proof of the Wahlund effect for two demes of arbitrary sizes.

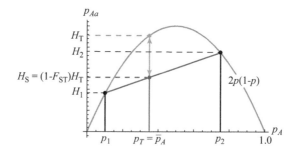

**Figure 1.11**
Graphical proof of the Wahlund effect for two demes of arbitrary size. The frequency of heterozygotes predicted under Hardy–Weinberg assumptions is $2p_A(1-p_A)$, shown in blue. Given demic allele frequencies $p_1$ and $p_2$ in demes 1 and 2, respectively, the within-deme heterozygosities under Hardy–Weinberg assumptions ($H_1$ and $H_2$) lie on this curve. The population-wide weighted allele frequency $p_T = \bar{p}_A$ must lie between $p_1$ and $p_2$, and the corresponding heterozygosity ($H_T$) also lies on the blue curve. In contrast, the true frequency of heterozygotes in the subdivided population ($H_S$) is the weighted average of $H_1$ and $H_2$, which lies on the purple line. The length of the vertical orange line represents the magnitude of the Wahlund effect; that is, the amount by which subdivision causes a deficiency of heterozygotes. Unless $p_1$ equals $p_2$, the purple line will lie below the blue curve. Q.E.D.

The amount by which mean heterozygosity is reduced by subdivision depends both on the variance in allele frequencies among demes and on allele frequency in the metapopulation, the latter through its influence on total heterozygosity. (Mathematically, recall that $H_T = 2p_T(1 - p_T)$.) To control for the effect of allele frequency, we divide equation (1.33) by $H_T$ to find a new $F$ statistic, written as

$$F_{ST} = \frac{H_T - H_S}{H_T} = 1 - \frac{H_S}{H_T}. \tag{1.34a}$$

In English, $F_{ST}$ is the proportional reduction in heterozygosity due to subdivision; the subscript ST refers to the fact that this is the deviation in mean heterozygosity among subpopulations, relative to expectation in the total population. Appealing to equations (1.23d) and (1.33), we also have

$$F_{ST} = \frac{H_T - H_S}{H_T} = \frac{\mathrm{Var}(p)}{p_T(1 - p_T)}. \tag{1.34b}$$

And coming at the same thing from another perspective, recall from teachable moment 1.15 that the metapopulation variance in allelic identity is $\mathrm{Var}(I) = p_A(1 - p_A)$. Thus $H_T = 2\mathrm{Var}(I)$ and

$$F_{ST} = \frac{H_T - H_S}{H_T} = \frac{\mathrm{Var}(p)}{\mathrm{Var}(I)}. \tag{1.34c}$$

We revisit this third understanding of $F_{ST}$ as a partitioning of variances in teachable moment 2.20.

---

**Power User Challenge 1.22**

Find expressions for population-wide genotype frequencies $p_{AA}$, $p_{Aa}$, and $p_{aa}$ in a subdivided population in terms of $p_T$ and $F_{ST}$. You can assume that each deme honors all Hardy–Weinberg assumptions. Your answer should be independent of number and sizes of demes.

---

**Empirical Aside 1.17:** The apportionment of human diversity

As first noted by Richard Lewontin (1972), the genetic diversity within individual human populations is nearly as great as the total diversity of our species. Lewontin's observation can be framed in terms of equation (1.34): average $F_{ST}$ between racially or ethnically defined groups are in the vicinity of 0.05–0.15 (see Jorde and Wooding [2004]; Mountain and Ramakrishnan [2005] for a summary of this and related statistics). In English, despite physically evident differences between races, genetically these groups are scarcely differentiated. The culturally fraught concept of race thus has very little basis in population genetics. See Novembre (2022) for a detailed description of the development and legacy of this very important finding.

Of course, since many physical differences are heritable, some loci must be differentiated between races. Lewontin's observation then implies that these most likely reflect the influence of diversifying selection rather than population divergence. For example, Field et al. (2016) treat the case of selection acting on loci responsible for blond hair and blue eyes over the last 2,000 years in Britain. Nor does Lewontin's observation contradict the observation that genetic ancestry is correlated with disease susceptibility (e.g., Lord et al. 2022), a critical avenue of research for improving human health.

**1.4.4.4 Nonrandom mating within a subdivided population**   Our treatment of the Wahlund effect assumes random mating within demes. We conclude this chapter by relaxing that assumption. This possibility motivates yet another fixation index, defined as

$$F_{IS} = 1 - \frac{H_I}{H_S},\qquad(1.35)$$

where $H_I$ is the population-wide average demic heterozygosity weighted by deme size. In English, $F_{IS}$ captures the reduction in heterozygosity due to nonrandom, within-deme mating, averaged across demes. (In contrast, $F_{ST}$ is the mean reduction in heterozygosity due to variance in allele frequencies among randomly mating demes.) Finally, we introduce

$$F_{IT} = 1 - \frac{H_I}{H_T}\qquad(1.36)$$

for the reduction in individual heterozygosity due to both nonrandom, within-deme mating and variation in allele frequencies among demes.

Subtracting each of equations (1.34a), (1.35), and (1.36) from 1 and then multiplying the first two together yields the following, rather remarkable result:

$$(1 - F_{IS})(1 - F_{ST}) = \frac{H_I}{H_S}\frac{H_S}{H_T} = \frac{H_I}{H_T} = (1 - F_{IT}).\qquad(1.37)$$

In English, the effects of subdivision and nonrandom mating on the population-wide fraction of heterozygotes can be partitioned into independent, multiplicative components. We might not be entirely surprised: as noted, population subdivision is itself a form of nonrandom mating. What we now see is that these two effects act independently.

**1.5   Chapter Summary**

- Mathematical models seek to represent real-world phenomena quantitatively. The manipulations of those mathematical expressions can yield powerful predictions.

- Natural selection, mutation, sexual reproduction, and migration can be modeled mathematically.

- Only natural selection increases fitness, and we can find the equilibrium balance between it and other processes.

- Population structure can influence both allele frequencies and diploid genotype frequencies.

# 2

## Stochastic Single-Locus Population Genetics

This book is concerned with the five processes that influence the genetic composition of an evolving population: natural selection, mutation, sexual reproduction, migration, and stochastic effects. In chapter 1, we developed one-locus deterministic models to study the effects of the first four of these. We now turn to the fifth, while maintaining our focus on single-locus models. This represents a critical advance on our understanding: as we shall see, there is overwhelming empirical support for the importance of stochastic effects in the evolution of natural populations. In chapter 3, we will introduce multilocus extensions for both deterministic and stochastic theory.

Recall that deterministic models overlook statistical fluctuations in the biology (see teachable moment 1.2), instead assuming that the future is completely determined by parameter values and time elapsed. For example, our deterministic treatment of natural selection gives allele frequency at time $t$ as a function of its current frequency and selection coefficient $s$ (equation 1.7).

However, biology is filled with random fluctuations of all sorts. This chapter develops methods to characterize the evolutionary consequences of one of these: random fluctuations in the number of offspring born to an organism. Of course, any particular organism will have left behind some particular number of offspring by the time of its death. But deterministic models assume that this number is known at the moment of its birth (given by its absolute Wrightian fitness, $W$). In contrast, stochastic models allow for the fact that random events during an organism's life can introduce uncertainty in its lifetime reproductive output. There will still be an average, but we will now allow variation in the realized number of offspring among genetically identical organisms.

Fortunately, we needn't specify the biology responsible for the randomness in an organism's lifetime reproductive output. Instead, we will write mathematical expressions for the uncertainty in offspring counts, without modeling its underlying causes. By propagating this uncertainty across individuals and through time, we will find probabilistic expressions for the future state of the population (teachable moment 2.1). Thus, for example, rather than predicting a beneficial allele's frequency at time $t$, we will find the probability with which each conceivable allele frequency will be observed at that time.

**Teachable Moment 2.1:** Stochastic probability distributions

Variability is absent in the deterministic models developed in chapter 1, which instead predict a single outcome for given parameter values and elapsed time. For example, equation (1.18) gives the deterministic equilibrium allele frequency under two-way mutation. Recalling the language of teachable moment 1.1, allele frequency is that model's dependent variable.

In contrast, stochastic models also capture uncertainty in the underlying biology. To do this, stochastic models first define a *state space*, the set of all possible configurations its system (e.g., an organism or population) can assume. The corresponding *state variable* can assume any value in the state space, and the dependent variable for stochastic model are probability distributions of state variables over state spaces. For example, figure 2.10 presents the stochastic analog to equation (1.18): the probability distribution of allele frequencies (the state variable) over the interval between 0 and 1 (the state space) in a population at equilibrium under two-way mutation.

One way of interpreting a stochastic probability distribution is to imagine a vast number of replicate systems (here, populations) initialized with identical configuration and evolving under identical parameter values. The fraction of replicates that realize each conceivable value in the model's state space is equivalent to the model's probability distribution in the limit of an infinite number of replicates. (This perspective, in turn, motivates the widespread use of computer simulations of vast numbers of replicates to characterize the underlying distribution of outcomes for some model. Indeed, we will see an example of this method shortly.) Finally, a state variable's value observed in any one replicate is its *realized value*, and its *expected value* is its mean across its probability distribution. We will sometimes write the expectation of state variable $x$ as E[$x$].

**Teachable Moment 2.2:** Means, variances, and continuous probability density functions

The mean and variance of trait $x$ were introduced in teachable moment 1.12 on the assumption that $x$ can only assume a finite number of distinct values. In that case, the probability distribution $p_i$ is discrete, and is called a probabilty mass function. That approach is applicable to stochastic models with discrete state spaces.

But we will also encounter continuous state spaces (e.g., allele frequency). In this case, the discrete probability mass function is replaced with a continuous function $p(x)$ called a *probability density function*. Here, $p(x)dx$ is the probability of observing a value in the interval $(x, x+dx)$. (This seemingly more cumbersome definition is necessary because, if $x$ is continuous, the probability that it will exactly equal any particular value is zero.) The mean and variance are now computed as

$$\overline{x} = \int_a^b p(x)x\,dx$$

and

$$\text{Var}(x) = \int_a^b p(x)(x-\overline{x})^2\,dx,$$

where $a$ and $b$ are the minimum and maximum possible values of $x$. As a reminder, the integral operator $\int$ is the continuous analog of summation $\Sigma$. And $\int_a^b p(x)dx = 1$, which captures the fact that $p(x)$ is still a probability distribution; in English, $x$ falls somewhere between $a$ and $b$ with probability 1.

As noted in teachable moment 1.2, stochastic considerations do more than just add a bit of noise around deterministic results. Rather, stochastic models allow us to ask and answer new, complementary questions that deepen our understanding of how populations evolve. For example, we will soon see that a new beneficial allele whose first few carriers happen to have the bad luck to have lower-than-expected offspring counts may fail to fix in the population, despite the allele's positive selection coefficient. This possibility is entirely outside the deterministic treatment. Moreover, as a consequence of this effect, beneficial alleles that do reach fixation will tend to have passed through carriers that enjoyed anomalously good reproductive luck, meaning that their time to fixation will usually be less than predicted by the deterministic approach.

---

**Teachable Moment 2.3:** Deterministic population genetics and infinite-sized populations

Some authors colloquially equate deterministic models with infinite sized populations and, as we shall see, the strength of one stochastic effect (random genetic drift; section 2.2) does scale inversely with population size, so this framing has some intuitive appeal. But it turns out that even in arbitrarily large populations, stochastic effects can be quite important. Indeed, this is where we begin.

---

## 2.1   Stochasticity While Rare, and Establishment

Imagine that mutation has just introduced a new beneficial allele into a large population with nonoverlapping generations. To be concrete, suppose the allele confers a 10% fitness advantage to its carrier relative to the resident wild type. How many copies will there be in the next generation? The most likely answer is one: even a jump to two copies would represent a 100% increase in frequency, far greater than the fitness advantage conferred by the allele. Moreover, as there is some chance of a jump to two copies, there must also be some probability of loss of the allele in order to achieve the necessary average (over many replicate realizations) of 1.1 copies in the next generation.

Nevertheless, beneficial alleles tend (on average) to increase in frequency. Suppose that there are now 10 copies of the allele in the population. While the foregoing considerations continue to apply to each carrier individually, the most likely number of copies in the next generation will be proportionately closer to the deterministic prediction of 11. While we might not be terribly surprised to find, say, 9 or 13 copies, the probability of loss in the next generation must now be very small, since this would require that all 10 carriers suffered the bad luck of having no offspring.

Thus, as a consequence of allowing random fluctuations in offspring counts, beneficial alleles have a nonzero probability of stochastic loss. But they suffer this risk only while rare. If a beneficial allele instead has the good luck to avoid loss early on, stochastic fluctuations are averaged out in the population. This is the law of large numbers at work, and we describe this transition as the *establishment* of a beneficial allele. It hasn't yet fixed in the population, but it will now almost surely do so.

### 2.1.1 The Establishment Problem

To build intuition into the mathematics of establishment, we first develop a stochastic analog to the first model in chapter 1: the growth of a genetically homogeneous population. Specifically, we ask: What is the probability of establishment of a population founded with exactly one individual? To look at things stochastically means focusing on individual birth and death events rather than on their rates. As we will see, there can be a nonzero probability that the founding individual dies before reproducing. Even if that first individual does reproduce, there can be a nonzero probability that all of its offspring will die before any of them reproduce, and so on. Bad luck for the population, but not impossible. On the other hand, once there are "enough" individuals in the population, the law of large numbers all but assures its continued, exponential expansion.

Recall from chapter 1 that the size of a genetically homogeneous population at time $t$ is related to its initial size according to $N(t) = N(0)e^{(b-m)t}$ (this is equation 1.5, remembering that $b - m = r$), where $b$ and $m$ are respectively the per capita rates of birth and mortality. We first recast $bdt$ and $mdt$ respectively as the probabilities per time interval $dt$ that an individual reproduces and dies. (As in chapter 1, $dt$ is shorthand for the limit of $\Delta t \rightarrow 0$. As before, we don't actually mean $\Delta t = 0$, but only that $\Delta t$ should be short enough that the probability of more than one birth or death event in a single interval is essentially zero.) We now introduce $P_i$ for the probability of eventual extinction in a population of $i$ individuals. Writing $P_{\text{est}}$ for the probability of establishment from a single individual, we have $P_{\text{est}} = 1 - P_{i=1}$.

---

**Teachable Moment 2.4:** Establishment is a subtle concept

How large do we require the population to grow before we deem it established? By contrast, extinction is clearly defined, and its logical complement thus yields a simple definition of an established populations as those that never went extinct. This reasoning is critical to our ability to quantitatively model the establishment process.

---

We can write $P_1$, the probability of eventual extinction from a single individual, as the sum of three terms. With probability $mdt$, the population of one individual goes extinct in interval $dt$. With probability $bdt$, the one individual reproduces in interval $dt$, after which the new population of two individuals nevertheless goes extinct eventually (with probability $P_2$). With probability $1 - bdt - mdt$, nothing happens in interval $dt$, after which the population of one eventually goes extinct (still with probability $P_1$). Mathematically, we have

$$P_1 = mdt + bdtP_2 + (1 - bdt - mdt)P_1 \qquad (2.1)$$

Equation (2.1) is called a *recurrence equation* because it tells us about some quantity (here, $P_1$) in terms of similar quantities (here $P_1$ and $P_2$). We will encounter many recurrence equations in this book.

---

**Power User Challenge 2.1**

What is the state space for the establishment problem?

---

To solve equation (2.1), we first subtract $P_1$ from both sides and divide through by $dt$ to find

$$0 = m + bP_2 - (b+m)P_1. \tag{2.2}$$

To go further, we assume that each individual gives birth to a number of offspring that is independent of the number of individuals already in the population. This assumption makes this a *branching process*. Biologically, the branching process assumption implies that lineages are not competing for resources, which may be reasonable while the number of lineages is small. Moreover, the population may become established before this assumption becomes untenable (see figure 2.2).

This means that the probability of $i$ extinctions is simply the product of the probabilities that each of $i$ lineages independently goes extinct. Mathematically, this is written as $P_i = P_1^i$. Substituting this into equation 2.2 lets us write

$$0 = m + bP_1^2 - (b+m)P_1,$$

which factors as

$$0 = (m - bP_1)(1 - P_1),$$

yielding

$$P_{est} = 1 - P_1 = \begin{cases} 0 \\ 1 - \dfrac{m}{b} \end{cases}, \tag{2.3}$$

where $P_{est}$ is the probability of establishment from one individual. Equation (2.3) is shown in figure 2.1.

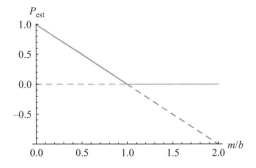

**Figure 2.1**
Probability of establishment from one individual as a function of $\dfrac{m}{b}$. Equation (2.3) has two roots (teachable moment 1.18), shown in blue and orange. Each has a biologically relevant region (solid lines) and a biologically irrelevant region (dashed lines). See text for more.

Note that if the probability of death is zero ($m = 0$) then the probability of establishment $P_{est} = 1 - \dfrac{0}{b} = 1$, as a minute's reflection suggests it must (blue $y$-intercept in figure 2.1). In this case, our deterministic result for $N(t)$ will apply from $t = 0$. (Although only approximately: early in any one realization, some variation in $N(t)$ will emerge from stochasticity in timing of births.) And if the probability of death is greater than of birth ($0 < b \leq m$), then $P_{est} = 0$; again, consistent with intuition and given by the other root (solid orange in figure 2.1). Far more interesting is the case $0 < m < b$. Because each individual now has a nonzero probability of dying before reproducing, the population can go extinct, but with probability $1 - \dfrac{m}{b}$ it will not (solid blue in figure 2.1).

Stochastic computer simulations of $N(t)$ with $m = 0.75$ and $b = 1$ shown in figure 2.2 provide a deeper understanding. Several points in particular are worth noting. First, the analytic expression for $P_{est}$ in equation (2.3) is good. It predicts that the population should become established in $1 - \dfrac{m}{b} = 25\%$ of replicates, whereas in this example, establishment occurred in 26 out of 100 of them. Next, most replicates in which the population fails to establish go extinct very quickly. Why? Because of the upward bias in population size (mathematically, $m < b$), replicates that fail to go extinct quickly tend to grow, and once there is an appreciable number of individuals in the population, the law of large numbers means that the probability of extinction collapses. For example, the largest replicate that failed to establish in the right panel of figure 2.2 only grew to nine individuals. But even at this small size, the probability of immediate extinction was already down to $P^9 = \left(\dfrac{0.25}{1}\right)^9 \approx 3.8 \times 10^{-6}$. This point explains why our branching treatment is so accurate in spite of the fact that the population size was capped at 1,000 in the simulations.

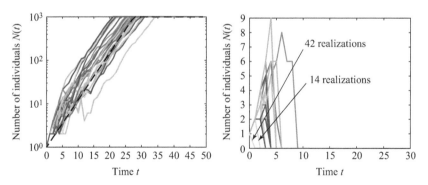

**Figure 2.2**
One hundred replicate realizations of population establishment founded with a single individual. Here, $b = 1$ and $m = 0.75$, and total population size was capped at 1,000. Twenty-six realizations yielded establishment (left panel; the deterministic result given by equation 1.5 is shown by the black, dashed line). Seventy-four realizations ended in extinction (right). Note that 42 of these 74 went extinct in the first time-step and 14 more in the second.

---

**Teachable Moment 2.5:** Absorbing barriers

The establishment problem is called a *biased random walk* with an *absorbing barrier* at zero. Although on average the population may be biased to grow (if $0 < m < b$), realizations end if the population ever reaches zero. It is this absorbing barrier that is responsible for the novelties seen in this process, and indeed, for many other important results in this chapter.

---

The third point of note in figure 2.2 is that among lineages that do become established, population growth is very nearly exponential (i.e., linear on this semilog plot; teachable moment 1.5), with rate parameter (slope) given by our deterministic treatment (black dashed line). Again, as populations grow, the law of large numbers comes to mask variability in individual reproductive output, in this case resulting in an average growth rate equal to that defined in chapter 1. Note, finally, that although their rate of growth matches the deterministic prediction, established populations are almost always larger than predicted, as explained in teachable moment 2.6.

---

**Teachable Moment 2.6**

The expected value of a stochastic state variable often equals predictions from the analogous, deterministic model. This is true in the establishment problem, explaining why established populations tend to be larger than predicted in our deterministic model (figure 2.2). Since the number of individuals in a fraction $P_1 = \dfrac{m}{b}$ of realizations becomes zero very quickly, the number of individuals in the remaining fraction must be larger than the deterministic prediction in order to maintain the average over all realizations. Importantly, deterministic predictions do not always match stochastic expectations (we will encounter a counterexample in section 2.3.1), but they often represent an instructive benchmark.

---

We can go further with the establishment problem. Again, using computer simulations, figure 2.3 presents the frequency with which each conceivable population size was observed in 1,000 replicate realizations as a function of time. At $t = 0$, every realization is in the same state ($N(0) = 1$) but thereafter, realizations begin to diverge. At $t = 1$ almost half have already gone extinct (black), and this quantity increases over the next few time steps. Over the same interval, the distribution of population sizes among realizations not yet extinct simultaneously moves to the right and spreads out. (Both of these effects are actually evident in figure 2.2; see also teachable moment 2.7.) Finally, by $t = 50$, roughly 25% of replicates have become established, and the remaining 75% have gone extinct.

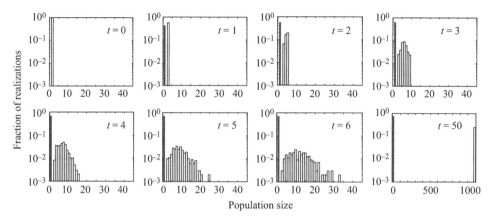

**Figure 2.3**
Frequency distributions of population sizes over time in simulations of the establishment problem. All parameters as in figure 2.2. Results are shown for 1,000 replicate realizations (in which 248 established and 752 went extinct). Black: replicates in which the population has been lost; Yellow: those in which the population has not yet been lost. Note that in all panels the $y$-axis is log transformed, and that the scale of the $x$-axis changes for $t = 50$. (By this point, all realizations have either established or gone extinct.)

---

**Teachable Moment 2.7:** Diffusion and translation of stochastic probability distributions

Figure 2.3 illustrates the simultaneous broadening and rightward movement of a stochastic frequency distribution over time. These two characteristics are called *diffusion* and *translation*, respectively. Mathematically, diffusion increases the distribution's variance while translation changes its mean. Or equivalently, diffusion has an unbiased impact on the distribution while translation is biased. In most of the models examined in this book (including the establishment problem), stochasticity exerts an unbiased effect on the state variable. In those cases, its impact is entirely manifest as diffusion of the corresponding distribution, meaning that translation is entirely driven by deterministic processes. This connection explains why the expected value of a state variable often matches the analogous, deterministic prediction (teachable moment 2.6): both are responding to quantitatively identical pressures.

---

The establishment problem illustrates many of the characteristics of stochastic population genetic models introduced in teachable moments 1.2 and 2.1. To begin, recall that the analogous deterministic model developed in chapter 1 yielded a dependent variable that predicted the precise number of individuals in a population as a function of time. But now allowing stochastic effects, we have lost predictive access to this single quantity, since by its very nature stochasticity means variation among replicate realizations of the process. Instead, we have propagated a stochastic model of individual reproductive output across individuals and through time, yielding a distribution of populations size as a function of time. Moreover, we see that stochasticity doesn't only introduce a bit of noise around our deterministic prediction. Rather the outcome of every realization is now also in doubt. Stochastic models ask and answer questions that are outside of the scope of the deterministic framings employed in chapter 1.

### 2.1.2 Establishment of a New Beneficial Allele

We continue as we did in chapter 1, and turn to the problem of natural selection. To do this we ask a biological question analogous to that in the previous section: what is the probability that a single copy of a beneficial allele will reach fixation? Loss of beneficial alleles from the population is impossible in the deterministic treatment, but now there is again a nonzero probability that the first individual endowed with the benefit will by bad luck fail to reproduce. And even if that individual does reproduce, there remains a nonzero probability that all of its offspring will die before any of them reproduce, or that some do reproduce, but that all the grandchildren fail to, and so on. This is precisely the biological consequence of stochasticity: again, bad luck for the population, but not impossible. On the other hand, once there are "enough" copies of a beneficial allele in the population, its continued increase and eventual fixation are all but assured.

First presented by Haldane (1927), our approach mirrors that of the previous section: write and solve a recurrence equation for $P_1$, the probability of eventual loss of a haploid lineage founded with a single copy of a beneficial allele with fitness advantage $s$. As before, $P_{est}(s) = 1 - P_1$ will give the probability of establishment (teachable moment 2.4). And we again assume branching process conditions, namely, that the fates of all copies of the beneficial allele in the population are independent. We further assume nonoverlapping generations, so that each organism dies after having some number of offspring, and introduce $p_i$ for the probability that an organism's offspring number is exactly $i$. (This is where the stochasticity enters the model.) Thus, our recurrence equation reads

$$P_1 = p_0 P_1^0 + p_1 P_1^1 + p_2 P_1^2 + p_3 P_1^3 + \cdots = \sum_{i=0}^{\infty} p_i P_1^i. \qquad (2.4)$$

This expression is directly analogous to equation (2.1). It is the sum of the probabilities that the first carrier will have each conceivable number of offspring $i$, written as $p_i$, multiplied by the probability that all $i$ resulting lineages are eventually lost, equal to $P_1^i$ by the branching process assumption. (By taking the sum to infinity, this expression makes no assumptions about how common the allele might become before being lost. Note too that whereas $p$ is a frequency in the deterministic notation of chapter 1, it is now a probability, and we will shortly write $x$ for frequencies. We regret the possibility of confusion; this notational shift reflects common usages in the field.)

It remains only to specify the probability distribution for offspring numbers ($p_i$) for individuals carrying the beneficial allele. The Poisson distribution is a biologically reasonable choice: it describes the number of events (here, births) in an interval (here, one generation), on the assumption that events occur independently at some rate that is constant over the interval. Algebraically, the probability of $i$

Poisson-distributed events in an interval reads $\dfrac{\lambda^i e^{-\lambda}}{i!}$, where $\lambda$ is the expected number of events per interval and $i!$ is the factorial of $i$, defined in teachable moment 2.8. (The interested reader is directed to any probability textbook to see how to derive the Poisson distribution.) Substituting relative Wrightian fitness (section 1.1.5) of carriers of the beneficial allele $w = 1 + s$ for $\lambda$, the Poisson probability of exactly $i$ offspring reads $p_i = \dfrac{(1+s)^i e^{-(1+s)}}{i!}$. And substituting this $p_i$ into equation (2.4) yields

$$P_1 = \sum_{i=0}^{\infty} P_1^i \frac{(1+s)^i e^{-(1+s)}}{i!} = e^{-(1+s)} \sum_{i=0}^{\infty} \frac{\left[ P_1(1+s) \right]^i}{i!}$$

$$= e^{-(1+s)} e^{P_1(1+s)} = e^{-(1+s)+P_1(1+s)} = e^{-(1+s)(1-P_1)}.$$

(2.5)

(In the first step, we factor $e^{-(1+s)}$ from the sum, permitted because it is independent of $i$, and in the second, we take advantage of the Taylor expansion of $e^x = \sum_{i=0}^{\infty} \frac{x^i}{i!}$; see teachable moment 2.8. We then combine exponents on e in the third step, just as we did immediately below equation [1.5]. Lastly, we factor the exponent. This derivation also illustrates that our choice to assume offspring counts are Poisson-distributed is mathematically convenient, reminding us of teachable moment 1.1.)

---

**Teachable Moment 2.8:** The Taylor expansion and approximation of a function

A theorem from calculus states that any infinitely differentiable function $f(x)$ can be expressed in terms of its derivatives as follows:

$$f(x) = f(a) + f^{(1)}(a)\frac{(x-a)}{1} + f^{(2)}(a)\frac{(x-a)^2}{2} + f^{(3)}(a)\frac{(x-a)^3}{6} + \cdots = \sum_{i=1}^{\infty} f^{(i)}(a)\frac{(x-a)^i}{i!}$$

where $a$ is any constant, $f^{(i)}$ is the $i$th derivative of $f$, and $i! = i(i-1)(i-2) \times \cdots \times 2 \times 1$ is the factorial function. This is called the Taylor expansion of $f(x)$. For example, the Taylor expansion of $e^x$ around $a = 0$ is $1 + x + \frac{x^2}{2} + \frac{x^3}{6} + \cdots = \sum_{i=0}^{\infty} \frac{x^i}{i!}$. (Recall that $\frac{de^x}{dx} = e^x$, so $f^{(i)}(a=0) = 1$ for all $i$.) We used this result in deriving equation (2.5).

The mechanism behind this more complex representation of $f(x)$ is illustrated geometrically in figure 2.4 for the function $f(x) = \ln(1+x)$ around $a = 0$. This illustrates that the first few terms of the Taylor expansion of a function can be a good approximation. However, its accuracy is dependent on the fact that $(x-a)^i$ vanishes as $i$ grows. It is thus most useful when $|x-a| \ll 1$, and the inset illustrates that the approximations for $\ln(1+x)$ disintegrate as $x$ grows beyond $a + 1 = 1$, since now ever-larger terms become ever-more important. This highlights the critical importance of one's choice of $a$: it should be as close as possible to the values of $x$ of interest, so that only a few terms of the expansion already give a useful approximation.

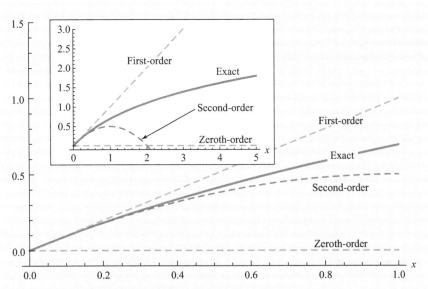

**Figure 2.4**
The first three terms of the Taylor expansion of the function $\ln(1+x)$. The exact value is shown by the solid blue line. The first term of the expansion is $f(a)+O(x) \approx \ln(1+0)=0$, a constant (dashed orange line). (The notation $O(\cdot)$ is read "of order" and here tells us that the error introduced by the approximation is proportional to terms no larger than $x$.) But if $x$ is very close to zero, then $f(a)$ is already pretty close to $f(x)$. This is called the zeroth-order approximation because the exponent on $x$ is zero. The first-order approximation (dashed green) is $f(a)+f^{(1)}(a) \times (x-a)+O(x^2) \approx 0+\dfrac{1}{1+0}(x-0)=x$. (Recall that $\dfrac{d\ln(1+x)}{dx}=\dfrac{1}{1+x}$.). This first-order approximation includes a linear correction: the slope of the function at $a$, multiplied by the offset between $x$ and $a$. The second-order approximation (dashed red) is

$$f(a)+f^{(1)}(a) \times (x-a)+\frac{f^{(2)}(a) \times (x-a)^2}{2}+O(x^3) \approx 0+\frac{1}{1+0}(x-0)-\left(\frac{1}{1+0}\right)^{-2}\frac{(x-0)^2}{2}=x-\frac{x^2}{2}.$$ which

further corrects the approximation by adding a quadratic term. We will use this result in deriving equation (2.6). Inset illustrates that the Taylor approximation disintegrates if $|x-a| \geq 1$.

Now substituting $P_{est}(s)=1-P_1$ for the probability of establishment, we have

$$1-P_{est}(s)=e^{-(1+s)P_{est}(s)}.$$

Taking the logarithm of both sides and solving for $s$ gives us

$$s=\frac{-\ln\left[1-P_{est}(s)\right]}{P_{est}(s)}-1 \approx \frac{-\left[-P_{est}(s)-\dfrac{P_{est}^2(s)}{2}-O(P_{est}^3(s))\right]}{P_{est}}-1$$

$$=\frac{P_{est}(s)}{2}+O(P_{est}^2(s))\,. \tag{2.6}$$

Here, we use the second-order Taylor approximation of $\ln(1-x)=x-\dfrac{x^2}{2}-O(x^3)$ (teachable moment 2.8). The error introduced by the approximation is proportional to terms no larger than $P_{\text{est}}^2(s)$. For selection coefficients much less than 1, the error introduced by this approximation will be modest, yielding

$$P_{\text{est}}(s) \approx 2s. \tag{2.7}$$

In English, the probability that an allele with selective advantage much smaller than one establishes is approximately twice its selection coefficient. Again, this result is at sharp odds with our deterministic treatment, in which beneficial alleles always fix.

---

**Power User Challenge 2.2**

Because $P_{\text{est}}(s)$ is a probability, it can never be larger than unity, which means that equation (2.7) must eventually break down as $s$ increases. Does this failure lie with the branching process assumption, the Taylor expansion of $\ln[1-P_{\text{est}}(s)]$, or some third feature of its derivation?

---

**Power User Challenge 2.3**

Sketch analogs to both panels of figure 2.2 for populations starting with a single copy of a beneficial allele.

---

## 2.2 Stochasticity at All Frequencies and Random Genetic Drift

Stochastic effects can cause a new beneficial allele to be lost from the population by bad luck. Importantly, stochasticity can also allow the occasional fixation of selectively neutral and even deleterious alleles. (Indeed, assuming just two alleles, the fixation of a deleterious allele is precisely what happens when a beneficial allele is lost.) However, there is an important distinction. Beneficial alleles escape the risk of stochastic loss once established; that is, once there are enough copies that the law of large numbers renders stochastic fluctuation in individual carriers' offspring counts irrelevant. Establishment depends on the upward, deterministic bias of natural selection. No such transition exists for selectively neutral and deleterious alleles destined to fix. Since they enjoy no deterministic upward bias, they remain susceptible to stochastic loss until the very moment of their fixation.

*Random genetic drift* (or simply *drift*) is a form of stochastic fluctuation in reproductive output that influences an allele's fate at all frequencies. Correspondingly, modeling drift requires a more complex mathematical framework than did modeling establishment. Indeed, this section employs the most sophisticated mathematics encountered in this book. In section 2.2.1, we derive the *Chapman–Kolmogorov equation*, a powerful, quantitatively exact framework for the problem. In section 2.2.2, we derive its diffusion approximations, partial differential equations that often allow greater analytic progress. These two sections use methods from stochastic process theory and calculus, and are almost entirely divorced from the underlying biology. Section 2.2.3 connects this apparatus to biology, and relies heavily on results from probability theory. Finally, in sections 2.2.4 and 2.2.5, we solve some modestly complex differential equations to quantitatively understand many evolutionary implications of random genetic drift.

Consequently, these technical details are often omitted from this curriculum, and their mastery is not necessary to understand what comes later. But, as noted in the preface, one of our key pedagogical objectives is to provide the reader with an appreciation of the structure of the mathematics underpinning population genetics. As elsewhere in the book, care has been taken to complement the mathematics with intuitive explanations, and we hope that the reader perseveres. We believe they will be amply rewarded.

---

**Teachable Moment 2.9:** Beneficial alleles can also experience random genetic drift

Once established, the frequency of beneficial alleles essentially ceases to be subject to stochastic effects; that is, once there are enough copies in the population that fluctuations in individual carriers' reproductive output are swamped by their average reproductive advantage. Thereafter, the allele's fate is all but certain. We will shortly develop a quantitative sense of the number of copies above which a beneficial allele is established but, intuitively, we can see that this threshold will vary inversely with selection coefficient. Consequently, the fixation of beneficial alleles with larger selection coefficients will become assured while there are fewer copies in the population than it will for those with smaller selection coefficients.

This reasoning raises the possibility that a beneficial allele might reach fixation without ever having become established. This is the situation if the number of copies required for establishment is larger than the population size; that is, if either selection coefficient or population size are sufficiently small. Under these circumstances, random genetic drift determines the fate of even beneficial alleles.

---

### 2.2.1 The Chapman–Kolmogorov Equation

In answer to power user challenge 2.1, the state space for our treatment of the establishment problem has just two elements: extinction and establishment. The same is true for our branching process treatment for the fate of a beneficial allele. Assuming two genetically distinct alleles at a locus labeled $A$ and $a$, we now need a stochastic model whose state space consists of all possible (integer) counts of the $A$ allele in the population. How large is that space? In a diploid population of $N$ individuals, the answer is $2N+1$: the $A$ allele can be present in $0, 1, 2, \ldots, 2N-1$ or $2N$ copies. (In a haploid population of $N$ individuals, the corresponding space has $N+1$ states.)

Correspondingly, we seek an expression for the probability that the population will be in each such state at time $t$ in the future as a function of its current state and all the relevant biology. To simplify, we assume nonoverlapping generations (teachable moment 1.16); thus, writing time in generations. We will first define transition probabilities between all conceivable pairs of allele counts over the course of just one generation. Mathematically iterating these per generation probabilities through time will yield the required expression, called the Chapman–Kolmogorov equation.

To begin, we introduce $t(i|k)$ for the transition probability that a population now carrying $k$ copies of the $A$ allele will carry $i$ copies in the next generation ($0 \leq i$,

$k \le 2N$). The single vertical bar is read "conditioned on," so $t(i|k)$ is the probability that a population will carry $i$ copies of the $A$ allele, assuming (or conditioned on the fact) that it is currently carrying $k$ copies. (We apologize for the potential for confusion between $t(\cdot|\cdot)$, a transition probability, and time $t$.)

Our model has $(2N+1)^2$ transition probabilities, corresponding to the $2N+1$ distinct values that each of $i$ and $k$ can assume. It proves convenient to organize these many transition probabilities in a single, two-dimensional, square matrix with $2N+1$ rows and $2N+1$ columns, written as

$$T = \begin{pmatrix} t(0|0) & t(0|1) & \cdots & t(0|2N) \\ t(1|0) & t(1|1) & \ldots & t(1|2N) \\ \vdots & \vdots & \ddots & \vdots \\ t(2N|0) & t(2N|1) & \cdots & t(2N|2N) \end{pmatrix}. \tag{2.8}$$

(Our convention will be that lowercase letters represent individual elements of a matrix and capital letters will stand for the full matrix.) Any particular $t(i|k)$ is found in the $i+1$st row and $k+1$st column: row and column numbers are offset by 1 from allele counts because the first row and column correspond to populations carrying zero copies of the $A$ allele. Each column in $T$ is a probability mass function, since for given number of $A$ alleles in the present generation, the corresponding column gives probabilities for all possible counts $i$ in the next. In the language of teachable moment 1.12, $\sum_{i=0}^{2N} t(i|k) = 1$ for all $k$. This makes $T$ a *stochastic matrix*. (Technically, $T$ is a left stochastic matrix; in right stochastic matrices, it is the row totals that sum to unity. The choice has no biological significance.)

$t(i|k)$ is the probability that there will be $i$ copies of the $A$ allele in the next generation conditioned on there currently being $k$ copies. Thus, all the biology must somehow be encoded in the entries in $T$, and we take that up in section 2.2.3. Indeed, the reader may appreciate glancing ahead to the first three paragraphs of section 2.2.3.1 to see a concrete example of the most commonly used $T$ matrix, called the Wright–Fisher model of random genetic drift.

---

**Power User Challenge 2.4**

What is the evolutionary significance of the second column and of the last row of the $T$ matrix? Hint: what values of the model's state variables do they correspond to?

---

Now given $T$, what are the corresponding probabilities two generations in the future? Introducing $p(i|j,2)$ for the probability of finding $i$ copies of the $A$ allele two generations after there were $j$ copies, we compute

$$p(i|j,2) = t(i|0)t(0|j) + t(i|1)t(1|j) + \cdots + t(i|2N)t(2N|j)$$

$$= \sum_{k=0}^{2N} t(i|k)t(k|j). \tag{2.9a}$$

Equation (2.9a) is a new recurrence equation. In English, it says that the probability of there being $i$ copies two generations after there were $j$ copies is the probability of transitioning from $j$ to $k$ copies in the first generation, multiplied by the probability of then transitioning from $k$ to $i$ copies in the second generation, summed over all possible counts $k$ in the intervening generation.

We similarly organize the $(2N+1)^2$ distinct $p(i|j,2)$ into a stochastic matrix, called $P(2)$. As we did for $T$, we arrange the $p(i|j,2)$ such that the element in the $i+1$st row and $j+1$st column is the probability that a population carrying $i$ copies of the $A$ allele carried $j$ copies two generations ago. This notation allows us to economically rewrite equation (2.9a) as

$$P(2) = TT = T^2. \tag{2.9b}$$

This tidy notation, which motivated our choice to write $T$ as a matrix in equation 2.8, is defined in teachable moment 2.10.

---

**Teachable Moment 2.10:** Matrix multiplication

Given matrices $X$, $Y$, and $Z$, write $x(i,j)$, $y(i,j)$, and $z(i,j)$ for their elements in the $i$th row and the $j$th column respectively. Matrix multiplication $Z = XY$ is defined as $z(i,j) = \sum_k x(i,k) y(k,j)$. Here, $k$ is an index that simultaneously runs through the $i$th column of $X$ and the $j$th row of $Y$. Thus, matrix multiplication is exactly what we need for the problem at hand. The reader is encouraged to write out $P(2) = T^2$ for $N = 2$ diploids to see that equation (2.9b) really does faithfully capture equation (2.9a).

Note that matrix multiplication is only defined when $X$ has the same number of columns as $Y$ has rows, although there are no constraints on the number of rows of $X$ and columns of $Y$. In our case, $T$ and $P$ are square matrices of the same size, trivially meeting this restriction. Finally, matrix multiplication is not commutative, which means that even if $XY$ and $YX$ are both defined, they need not be equal.

---

**Power User Challenge 2.5**

What is $P(0)$? Hint: the $2N+1$ entries of the $k$th column should be the probabilities of finding exactly $i$ copies of the $A$ allele in the population, given that there are exactly $k$ copies. The matrix $P(0)$ shares a deep similarity with the number 1. Explain.

---

Given $P(2)$, we can immediately write another recurrence equation for $P(3) = TP(2) = TT^2 = T^3$, and, in general, we have

$$P(t+1) = TP(t) = TT^t = T^{t+1}. \tag{2.10}$$

This is the Chapman–Kolmogorov equation, a big step forward. With each successive multiplication, $P$ captures the probability distribution of allele counts in generation $t+1$ in the future for given allele count in generation $t = 0$. These ideas are illustrated in the top panel of figure 2.5.

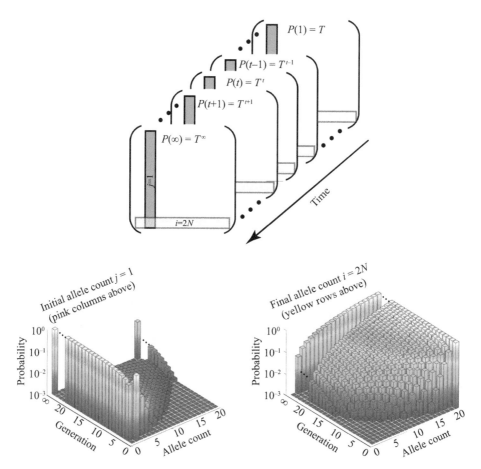

**Figure 2.5**
The Chapman–Kolmogorov equation. Top: the iterative multiplication of the $T$ matrix yields the $P(t)$ matrices (equation 2.10) whose entries $p(i|j, t)$ are the probabilities of finding $i$ copies of the $A$ allele in a population $t$ generations after there were $j$ copies. Rows correspond to values of $i$ and columns correspond to values of $j$ ($0 \leq i$, $j \leq 2N$). As in the answer to power user challenge 2.4, the second column ($j = 1$, shown in pink) of $P(t)$ captures the probability distribution among all conceivable allele counts ($i$) for a new allele that appeared $t$ generations ago. And its last row ($i = 2N$, yellow) captures probabilities of fixation, given each conceivable allele counts ($j$) $t$ generations ago. Bottom: $P(t) = T^t$ for the Wright–Fisher model ($T$ matrix given by equation 2.13) for $2N = 10$ and $0 \leq t \leq 20$ generations, together with $t = \infty$. Lower left: probability distributions for each conceivable current count $i$ given initial allele count $j = 1$ copies come from the second column of $P^t$ (pink in the upper panel). Notice that the distribution broadens over time, but its mean remains unchanged, reflecting the lack of any biased process in the Wright–Fisher model. In the language of teachable moment 2.7, the Wright–Fisher model causes diffusion but not translation. Also note that probability is accumulating at allele counts 0 and $2N$; this is the consequence of absorbing barriers (teachable moment 2.5) at these extremes. Lower right: probabilities of fixation ($i = 2N$) given all conceivable initial allele counts ($j$) come from the last row of $P$ (yellow in the upper panel). As noted in the text, these are not probability distributions; thus, they need not sum to 1. In both lower panels, values for finite $t$ generated by numerically computing $P(t) = T^t$, and values for $P(t = \infty)$ are from equation (2.21).

The columns of $P$ can be regarded as looking forward in time: each column ($j$) gives the probability of reaching each conceivable allele count $i$ in exactly $t$ generations. Simultaneously, its rows can be regarded as looking backward: each gives the probability of now being at each conceivable allele count $j$ given that the count was $i$ exactly $t$ generations ago. These two perspectives are illustrated in the lower two panels of figure 2.5 using the $T$ matrix for the widely used Wright–Fisher model of random genetic drift introduced in section 2.2.3.1. (Importantly, rows are not probability distributions. While columns contain probabilities for reaching all possible outcomes from a given starting configuration, rows contain probabilities for reaching one outcome from all possible starting configurations. To see the point, imagine a model in which, regardless of their current count, alleles are fixed in the next generation. The last row of that $P$ would contain a 1 in each column, giving a row sum of $2N+1$.)

---

**Teachable Moment 2.11a**

The Chapman–Kolmogorov equation yields probabilities for all conceivable allele counts in generation $t \geq 0$ as a function of the biology encoded in the $T$ matrix and allele count at time $t=0$. Columns of the $P$ matrix contain count probabilities in the future, taking $t=0$ as the present. Rows contain count probabilities in the present, taking $t=0$ sometime in the past. Care is warranted: rows do not contain count probabilities in the past as a function of known counts in the present.

---

### 2.2.2 The Diffusion Approximations of the Chapman–Kolmogorov Equation

The Chapman–Kolmogorov equation (equation 2.10) is a recurrence equation that describes the probability distribution over allele counts at time $t$ as a function of counts at $t=0$ and the biology encoded in the $T$ matrix. As such, it allows us to numerically calculate values such as those shown in figure 2.5. On the other hand, these calculations can be onerous, growing in magnitude with both population size and time. Indeed, the probability of fixation for a newly introduced allele is found in $p(i=2N \mid j=1, t=\infty)$, which obviously cannot be calculated by the direct application of the equation.

The intractability of the Chapman–Kolmogorov equation stems from the fact that it treats discrete generations and discrete allele counts. This renders it refractory to calculus, which demands continuous functions. Our objective now is to approximate the Chapman–Kolmogorov equation with expressions written in continuous time and continuous allele frequencies. On integration (when possible), these *diffusion approximations* will yield closed-form expressions for allele frequency probabilities at times $t$ for given transition matrix $T$. Conceptually, the discrete histograms in the two lower panels of figure 2.5 will be replaced by continuous surfaces, which may be amenable to calculus. (An analogous transition from discrete to continuous models was employed in getting from equation [1.1] to [1.2].)

We will accomplish this derivation in four steps. First, we rewrite the Chapman–Kolmogorov equation in a form that highlights how probability distributions change over time. More specifically, we will see how the $T$ matrix can be regarded as specifying the per generation rate at which probability mass moves among stochastic states (i.e., allele counts). Next, we make the simplifying approximation that in each time step, probability mass only moves between adjacent stochastic states. We then partition the per generation movement of probability mass encoded in the $T$ matrix into biased and unbiased components; that is, those responsible for translation and diffusion, respectively (teachable moment 2.7). Lastly, by simultaneously rescaling both time and allele counts and taking suitable limits, we arrive at the continuous diffusion approximations for the Chapman–Kolmogorov equation.

In fact, there are two approximations possible. The *forward diffusion approximation* focuses on the flow of probabilities within columns of the $T$ matrix. The *backward diffusion approximation* focuses on its flow within rows. We informally sketch the derivation of the former, closely following chapter 8 of Crow and Kimura (1970). (Interested readers will also find a more rigorous derivation there; see also teachable moment 2.13 for an entirely different path.) An analogous derivation of the backward diffusion approximation is possible. Both approximations teach us a great deal of biology.

**2.2.2.1 Probability mass moves through time**   The columns of the Chapman–Kolmogorov equation yield probability distributions over stochastic states (i.e., all conceivable allele counts in a population), as a function of three quantities: elapsed time $t$, allele counts at $t=0$, and whatever biology is encoded in the $T$ matrix. We now demonstrate that their entries also quantify how probability mass moves between generations within columns. These are closely related quantities but perhaps not trivially equal.

To begin, we explicitly write out the matrix multiplication in equation (2.10) for arbitrary $i$ and $j$

$$p(i\,|\,j,t+1) = \sum_{k=0}^{2N} t(i\,|\,k)\,p(k\,|\,j,t).$$

Subtracting $p(i\,|\,j, t)$ from both sides yields

$$\Delta p(i\,|\,j,t) = p(i\,|\,j,t+1) - p(i\,|\,j,t) = \sum_{k=0}^{2N} t(i\,|\,k)\,p(k\,|\,j,t) - p(i\,|\,j,t).$$

Next, because $T$ is a stochastic matrix, its columns sum to 1. Mathematically, $\sum_{k=0}^{2N} t(k\,|\,i) = 1$, and we now multiply the last term in the previous equation by this quantity

$$\Delta p(i\,|\,j,t) = \sum_{k=0}^{2N} t(i\,|\,k)\,p(k\,|\,j,t) - \sum_{k=0}^{2N} t(k\,|\,i)\,p(i\,|\,j,t).$$

Finally, since the two sums have the same limits, we can rewrite their difference as a single sum of differences

$$\Delta p(i\,|\,j,t) = \sum_{k=0}^{2N} \left[ t(i\,|\,k)\,p(k\,|\,j,t) - t(k\,|\,i)\,p(i\,|\,j,t) \right]. \tag{2.11a}$$

This perhaps obscure derivation actually yields an important biological advance: it decomposes the per generation change in probability that a population initially carrying $j$ copies of the $A$ allele now carries $i$ copies (i.e., is in state $i$, the expression on the left) into per generation gains from and losses to all possible states (on the right). More specifically, the first term in square brackets is the gain: the probability that the population will enter state $i$ from state $k$, multiplied by the probability that it was just in state $k$. The second term is the loss: the probability that the population will leave state $i$ and enter state $k$, multiplied by the probability that it was just in state $i$. This difference is then summed over all possible states $k$. This conception of the $T$ matrix is illustrated in the top panel of figure 2.6.

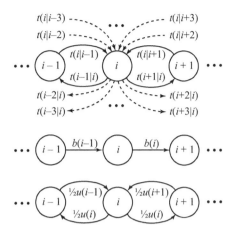

**Figure 2.6**
Representing the transition matrix $T$ as per generation movement of probability mass in matrix $P$. Circles represent elements in the state space (i.e., allele counts) and arrows capture the per generation movement of probability mass. Top: the per generation change in probability mass in state $i$ is due to gains from (above) and losses to (below) all $i-1$ other states (equation 2.11a). Focusing only on gains and losses from states $i \pm 1$ (solid lines) yields equation (2.11b). Middle: the biased component $b(i)$ of $T$ is, without loss of generality, assumed to be moving populations toward higher allele counts. Bottom: the unbiased component $u(i)$ is by definition equally split between adjacent states. The sum of biased and unbiased components is given by equation (2.11c). (Figure suggested by Julian Stamp.)

#### 2.2.2.2 Restricting movement of probability mass to adjacent stochastic states  As we shall see in section 2.2.3, most of the probability mass in population genetic $T$ matrices is concentrated in those elements $t(i|j)$ where $i$ is fairly close to $j$. In English, what we are saying is that in each generation, allele counts will generally only change by small amounts, or equivalently, that state transitions tend to be local. For example, a 1% selection coefficient will on expectation only change allele counts by 1%. This biological reasoning allows us to now approximate the Chapman–Kolmogorov equation by assuming that transitions only occur between adjacent states. This is equivalent to disregarding all dashed lines in figure 2.6, and algebraically reads

$$\Delta p(i|j, t) \approx$$

$$t(i|i-1)p(i-1|j, t) - t(i-1|i)p(i|j, t) + t(i|i+1)p(i+1|j, t) - t(i+1|i)p(i|j, t). \quad (2.11b)$$

The first two terms are the gain to and loss from the next-lower state $(i-1)$; the second two are the gain and loss to the next-higher state $(i+1)$.

#### 2.2.2.3 Partitioning movement of probability mass into biased and unbiased components  We now begin to address a question left open in section 2.2.1: how to represent biology in the $T$ matrix. Recall from teachable moment 2.7 that the temporal movement in a stochastic probability distribution can be described by translational (or biased) and diffusive (or unbiased) components. We now rewrite the elements of $T$ as the sum of these components. In section 2.2.3, we will see how to capture specific biological models in them.

We first introduce $b(i)$ to represent the per generation, biased probability loss from state $i$. Without loss of generality (see below), assume the bias is toward higher allele counts. (This $b$ is unrelated to that representing birth rate in sections 1.1.1 and 2.1.1.) The biased component of equation (2.11b) reads

$$b(i-1)p(i-1|j,t) - b(i)p(i|j,t).$$

The first term is the gain to state $i$ from the next-lower state: state $i-1$'s probability of loss multiplied by its probability of occupancy. The second term is the biased loss from state $i$ to the next-higher state; again, its probability of loss multiplied by its probability of occupancy. Our assumption that bias is upward renders the other two terms in equation (2.11b) zero. This reasoning is illustrated in the middle panel of figure 2.6.

Similarly, we introduce $u(i)$ to represent the per generation, unbiased probability of loss from state $i$. Since unbiased loss to adjacent states is by definition split equally between them, this component's contribution to equation (2.11b) is

$$\frac{1}{2}u(i-1)p(i-1|j,t) - \frac{1}{2}u(i)p(i|j,t) + \frac{1}{2}u(i+1)p(i+1|j,t) - \frac{1}{2}u(i)p(i|j,t).$$

Here, the first two terms represent the gain and loss from and to the next-lower state, and the second two terms represent gain and loss from and to the next-higher state. In each case, we again multiply the probability of loss $\left(\frac{1}{2}u\right)$ by the corresponding state's probability of occupancy (see bottom panel of figure 2.6).

Combining the above, the net per generation movement of probability mass into stochastic state $i$ in terms of functions $b$ and $u$ is

$$\Delta p(i|j,t) \approx b(i-1)p(i-1|j,t) - b(i)p(i|j,t) + \frac{1}{2}u(i-1)p(i-1|j,t)$$

$$-\frac{1}{2}u(i)p(i|j,t) + \frac{1}{2}u(i+1)p(i+1|j,t) - \frac{1}{2}u(i)p(i|j,t). \quad (2.11c)$$

The first two terms on the right represent the biased component, and the last four capture the unbiased component.

**2.2.2.4  Moving to continuous time and frequency**    Although equation (2.11c) substantially simplifies equation (2.11a), it still exhibits the fundamental defect of the Chapman–Kolmogorov equation: it is written in terms of discrete allele counts and discrete time steps. The final step is to simultaneously take limits that make the function continuous in both dimensions.

To connect allele counts to frequencies, note first that the Chapman–Kolmogorov equation's state space can be regarded equivalently as a space of counts $0 \leq i \leq 2N$ or of frequencies $0 \leq \frac{i}{2N} \leq 1$. We now introduce $\Delta x = \frac{1}{2N}$ to represent the granularity of this new but equivalent frequency space, and $x = i\Delta x$ and $x_0 = j\Delta x$ for the frequency of the $A$ allele in generations $t + 1$ and $0$, respectively. Correspondingly, we make small changes in notation, now writing $p(x|x_0,t)$ as shorthand for

$$p\left(\frac{x}{\Delta x} \Big| \frac{x_0}{\Delta x}, t\right), \, b(x) \text{ for } b\left(\frac{x}{\Delta x}\right), \text{ and } u(x) \text{ for } u\left(\frac{x}{\Delta x}\right).$$

We also introduce $M(x)$ and $V(x)$ for the mean and variance in change in allele frequency in a population in which the $A$ allele is currently at frequency $x$. Assuming the biased component pushes allele counts upward, equation (1.12) yields $M(x) = b(x)\Delta x + [1 - b(x)] \times 0 = b(x)\Delta x$, or equivalently, $b(x) = \dfrac{M(x)}{\Delta x}$. (Note now that if the biased probability loss represented by $b(x)$ is instead toward lower allele counts, we simply replace $\Delta x$ with $-\Delta x$ in the previous sentence, reversing the sign of $M(x)$.) Similarly, equation (1.13c) gives us $V(x) = \dfrac{1}{2}u(x)(\Delta x)^2 + \dfrac{1}{2}u(x)(-\Delta x)^2 - 0^2 = u(x)(\Delta x)^2$, or equivalently, $u(x) = \dfrac{V(x)}{(\Delta x)^2}$. (We subtract $0^2$ in computing $V(x)$ because the mean change due to unbiased components is zero by definition.)

Finally, we introduce $\Delta t$ to represent the granularity in time, replacing the per generation change $\Delta p(x|x_0, t)$ with the per $\Delta t$ change $\dfrac{\Delta p(x|x_0, t)}{\Delta t}$.

Writing equation (2.11c) in this new notation yields

$$\frac{\Delta p(x|x_0, t)}{\Delta t} \approx \frac{1}{2} \left[ \frac{\dfrac{V(x+\Delta x)p(x+\Delta x|x_0, t) - V(x)p(x|x_0, t)}{\Delta x} - \dfrac{V(x)p(x|x_0, t) - V(x-\Delta x)p(x-\Delta x|x_0, t)}{\Delta x}}{\Delta x} - \frac{M(x)p(x|x_0, t) - M(x-\Delta x)p(x-\Delta x|x_0, t)}{\Delta x} \right] \quad (2.11d)$$

after some modest rearrangement. Taking limits $\Delta x \to 0$ and $\Delta t \to 0$ brings us to the continuous analog

$$\lim_{\Delta t \to 0} \frac{\Delta p(x|x_0, t)}{\Delta t} = \frac{1}{2} \lim_{\Delta x \to 0} \left[ \frac{\dfrac{V(x+\Delta x)p(x+\Delta x|x_0, t) - V(x)p(x|x_0, t)}{\Delta x} - \dfrac{V(x)p(x|x_0, t) - V(x-\Delta x)p(x-\Delta x|x_0, t)}{\Delta x}}{\Delta x} \right] - \lim_{\Delta x \to 0} \frac{M(x)p(x|x_0, t) - M(x-\Delta x)p(x-\Delta x|x_0, t)}{\Delta x}.$$

As in section 1.1, we do not mean that either quantity literally is made equal to zero, but rather that they come sufficiently close to zero that we can discount the possibility of probability mass traversing more than one stochastic state in one-time step. Finally, recalling the definitions of the first and second derivatives (teachable moment 1.3) yields

$$\frac{\partial p(x|x_0, t)}{\partial t} = \frac{1}{2}\frac{\partial^2 [V(x)p(x|x_0, t)]}{\partial x^2} - \frac{\partial [M(x)p(x|x_0, t)]}{\partial x}. \tag{2.12a}$$

This is the forward diffusion approximation of the Chapman–Kolmogorov equation. Despite its apparent opacity, this result is of tremendous biological utility. After writing down the $T$ matrix for models of biological interest, we will compute its $M(x)$ and $V(x)$; that is, the matrix's mean and variance in per generation change in allele frequency. Substituting these expressions into equation (2.12a) (or [2.12b], introduced shortly) can yield access to at least some elements of $p(x|x_0, t)$.

But before proceeding, we make two comments. The first is technical: the discrete probability mass functions (teachable moment 1.12) in the columns of the Chapman–Kolmogorov equation have become continuous probability density functions (teachable moment 2.2) in the diffusion approximation. Thus, whereas $p(i|j, t)$ in the Chapman–Kolmogorov equation represents the probability of finding $i$ copies of the $A$ allele in the population $t$ generations after there were $j$ copies, $p(x|x_0, t)dxdt$ in the diffusion approximation is the probability of finding the $A$ allele in frequency interval $(x, x+dx)$ during time interval $(t, t+dt)$ given that it was at frequency $x_0$ at time $t=0$.

Second, an analogous path focusing on rows can be followed from the Chapman–Kolmogorov equation to find its backward diffusion approximation, which reads

$$\frac{\partial p(x|x_0, t)}{\partial t} = \frac{1}{2}V(x_0)\frac{\partial^2 p(x|x_0, t)}{\partial x_0^2} + M(x_0)\frac{\partial p(x|x_0, t)}{\partial x_0}. \tag{2.12b}$$

We will make good use of both approximations in sections 2.2.4 and 2.2.5. (The more abstract question of when to use which diffusion approximation is discussed in appendix 1 of Walsh and Lynch [2018].)

We emphasize that both diffusion approximations find allele frequency probability densities at time $t \geq 0$ as a function the biology and of allele frequency at time $t=0$. In the forward case, $t=0$ is the present moment. Consequently, $V$ and $M$ are functions of frequency $x$ in the future, and the derivates on the right are with respect to future frequency $x$. In contrast, the backward diffusion takes time $t=0$ at some point in the past, and $V$ and $M$ are functions of frequency $x_0$ in the past. (This reasoning also explains the reversal of sign on the biased component on the right in the backward form.) Similarly, the derivates on the right are with respect to past frequency $x_0$.

---

**Teachable Moment 2.11b**

Care is again warranted: the backward diffusion approximation does not provide allele frequency probabilities in the past as a function of current allele frequencies (Ewens 1979). Teachable moment 2.11a made the analogous point with respect to the Chapman–Kolmogorov equation.

---

**Teachable Moment 2.12**

In the Chapman–Kolmogorov equation, time $t$ was written in generations. In moving to the diffusion approximations, we set $\Delta x = \dfrac{1}{2N}$ and took the limit $\Delta x \rightarrow 0$. In order to preserve the scaling between time and counts given in the $T$ matrix (i.e., the aspect ratios in the lower two panels of figure 2.5), we (implicitly) set $\Delta t = \dfrac{1}{2N}$ in taking the limit $\Delta t \rightarrow 0$. Thus, time $t$ in the diffusion approximations is in units of $2N$ generations.

---

**Teachable Moment 2.13**

That the forward and backward diffusion approximations are a first derivative plus half a second derivative suggests they may be Taylor approximations. That intuition is correct. See chapter 4 in Ewens (1979) or appendix 1 in Walsh and Lynch (2018) to learn more.

---

### 2.2.3  Representing Biology in the Chapman–Kolmogorov and Diffusion Equations

The Chapman–Kolmogorov equation captures the influence of random genetic drift on a population's evolution. More specifically, the iterative multiplication of a $T$ matrix yields probability distributions over allele counts as a function of time and the biology encoded in $T$. But how do we capture the biology in $T$? Similarly, the diffusion approximations of the Chapman–Kolmogorov equation provide an analytically more tractable framework, where now the biology is captured in the functions $M(x)$ and $V(x)$. But again, how do we construct those functions? We now address these questions.

This section relies heavily on results from probability theory. None of the derivations are conceptually complex; indeed, most are nothing more than the judicious application of the distributive property of multiplication over addition (i.e., the fact that $ax + bx = (a+b)x$), occasionally combined with a Taylor approximation. However, they are often unwieldy, and in the interest of space, we suppress them all. Interested readers are encouraged to consult any introductory textbook in probability theory (or the excellent appendices in Gillespie [2004], Wakeley [2009] and Nielsen and Slatkin [2013]); despite sometimes arcane notation, the mathematics are actually quite accessible.

#### 2.2.3.1  The Wright–Fisher model of random genetic drift
The most widely used model of random genetic drift is called the *Wright–Fisher model* (both of these pioneers worked on the problem beginning in the 1920s). This model assumes a constant sized population of $N$ diploid organisms with nonoverlapping generations and the complete absence of any deterministic pressures on allele frequency. Each generation, $2N$ alleles are sampled at random from the $N$ parents, and then duplicated. This is done using *sampling with replacement*, meaning that the same allele can be chosen more than once. The parents die and these $2N$ alleles are then paired at random to form the $N$ offspring that comprise the next generation. The

Wright–Fisher model can equally be applied to a population of $N$ haploid organisms, in which case each generation is constructed by sampling $N$ alleles with replacement. These are each duplicated, after which all the parents die, leaving the $N$ haploid offspring.

(We note a subtle shift in our use of the word "allele." To this point, alleles have been assumed to be genetically distinct, but the alleles sampled in the Wright–Fisher model need only be physically distinct. Whether or not they are also genetically distinct will continue to be of utmost importance, and which sense of the word is intended will be clear from context.)

The transition matrix $T$ for the Wright–Fisher model of random genetic drift is found as follows. Assuming two genetically distinct alleles designated $A$ and $a$, we write $k$ for the number of $A$ alleles in the present generation. In the absence of any deterministic processes, the probability of sampling an $A$ allele is given by its frequency, namely $\frac{k}{2N}$. The number of $A$ alleles $i$ in a sample of $2N$ alleles is then *binomially distributed*, written mathematically as

$$t(i|k)=\binom{2N}{i}\binom{2N}{i}\left(\frac{k}{2N}\right)^{i}\left(1-\frac{k}{2N}\right)^{2N-i}. \tag{2.13}$$

Here, $\binom{2N}{i}=\frac{(2N)!}{(2N-i)!i!}$ and $x!$ is the factorial function introduced in teachable moment 2.8. In English, equation (2.13) gives the probability that there will be $i$ copies of the $A$ allele in the next generation, given that there are currently $k$ copies. The last two terms on the right are the probabilities respectively of drawing exactly $i$ copies of $A$ and $2N-i$ copies of $a$. And the first term is the number of (for us equivalent) temporal sequences in which that split of alleles can be sampled. The $T$ defined by equation (2.13) is a stochastic matrix: some number of $A$ alleles $0 \le i \le 2N$ must be observed in the sample of $2N$, so each of its columns will sum to 1. The lower two panels of figure 2.5 were computed by using the Chapman–Kolmogorov equation with this $T$ matrix.

To employ equation (2.13) in the diffusion approximations, we need its $M(x)$ and $V(x)$; in English, its mean and variance in per generation change in allele frequency. We apply equation (1.12) to find the expected number of $A$ alleles in the next generation (i.e., the mean averaged over the stochasticity, written E[·]; see teachable moment 2.1):

$$E[i]=\sum_{i=0}^{2N}\left[t(i|k)\times i\right]=\sum_{i=0}^{2N}\left[\binom{2N}{i}\left(\frac{k}{2N}\right)^{i}\left(1-\frac{k}{2N}\right)^{2N-i}\times i\right]=k.$$

(As noted, the interested reader is directed to appropriate textbooks to find all derivations from probability theory used here.) Next, writing frequencies $x=\frac{k}{2N}$ and $x'=\frac{i}{2N}$, we compute $M(x)$, the expected per generation change in allele frequency as

$$M(x)=E[x'-x]=E\left[\frac{i}{2N}-\frac{k}{2N}\right]=\frac{E[i]}{2N}-\frac{k}{2N}=0. \tag{2.14}$$

(Here, we take advantage of two facts from probability theory: for arbitrary constant $C$, $E[x-C]=E[x]-C$ and $E[xC]=E[x]C$.) This demonstrates that the Wright–Fisher model of random genetic drift exerts no biased, translational pressure on a population's allele frequency, as already seen in the lower left panel of figure 2.5.

Critically, however, there is now a nonzero probability that the number of $A$ alleles will change in the next generation. To not change would mean observing exactly $k$ $A$ alleles two generations in a row, not an impossibility but also not a certainty. We now apply equation (1.13a) to find the variance in the number of $A$ alleles in the next generation

$$\text{Var}(i)=\sum_{i=0}^{2N}\left[\binom{2N}{i}\left(\frac{k}{2N}\right)^{i}\left(1-\frac{k}{2N}\right)^{2N-i}(i-k)^2\right]-0=2N\left(\frac{k}{2N}\right)\left(1-\frac{k}{2N}\right).$$

This perhaps cryptic result immediate tells us that random genetic drift is absent in monomorphic populations (i.e., when $k=0$ or $2N$). (In the language of teachable moment 2.5, these are absorbing barriers.) To compute $V(x)$, the per generation variance in allele frequency change, we again write $x=\dfrac{k}{2N}$ and $x'=\dfrac{i}{2N}$ to find

$$V(x)=\text{Var}(x'-x)=\text{Var}\left(\frac{i}{2N}-\frac{k}{2N}\right)=\frac{V(i)}{(2N)^2}=\frac{x(1-x)}{2N}. \tag{2.15}$$

(Here, we take advantage of two more facts from probability theory: given constant $C$, $\text{Var}(x-C)=\text{Var}(x)$, and $\text{Var}[xC]=\text{Var}[x]C^2$.) In haploid populations, $V(x)$ is twice this value, since only $N$ alleles are sampled each generation. In either case, this result holds an important lesson: not only is genetic drift weaker in larger populations, it precisely scales inversely with population size, a finding that deeply informs much of stochastic population genetics.

---

**Teachable Moment 2.14**

Sampling with replacement gives rise to the stochasticity in allele frequency inherent in the Wright–Fisher model. More specifically, sampling with replacement means that some lineages present in the parental population can be sampled more than once, in which case other lineage will necessarily fail to survive. Conversely, if sampling were done without replacement, every lineage would be sampled exactly once each generation, with no net effect on allele frequencies. Note that this is quite a different source of stochasticity than that underlying the establishment of a beneficial allele.

### 2.2.3.2 Incorporating biased processes into the Wright–Fisher model

We can use natural selection to illustrate how to incorporate biased translational pressure on allele frequency into the Wright–Fisher model. Recall from chapter 1 that in diploids, setting dominance coefficient $h=\dfrac{1}{2}$ means that natural selection deterministically increases the proportion of $A$ alleles by the factor $\dfrac{1+s}{\bar{w}}$ each generation. (We arrived at the same result for haploids.) Thus, in the language of the previous section, selection increases the probability of drawing an $A$ allele by the factor $\dfrac{1+s}{\bar{w}}$.

The corresponding $T$ matrix reads

$$t\left(i|k\right)=\binom{2N}{i}\left(\frac{k}{2N}\frac{1+s}{\bar{w}}\right)^{i}\left(1-\frac{k}{2N}\frac{1+s}{\bar{w}}\right)^{N-i}. \tag{2.16}$$

Following the same path that led to equations (2.14) and (2.15), one can show that

$$M(x)=\frac{x(1-x)s}{\bar{w}}\approx x(1-x)s \tag{2.17}$$

and

$$V(x)=\frac{x(1-x)(1+s)^2}{2N\bar{w}^2}\approx\frac{x(1-x)}{2N} \tag{2.18}$$

for the Wright–Fisher model of selection and random genetic drift. (Both approximations assume $s$ is much less than 1.)

We now see that, like the establishment problem (section 2.1.1), the Wright–Fisher model with and without selection (equations 2.16 and 2.13) have the feature alluded to in teachable moment 2.7: the expected change in allele frequency is identical to that in the analogous, deterministic model. Or equivalently, the biased change in each is driven entirely by deterministic processes in these models. The claim is trivially true in the case without selection: the deterministic treatment predicts no change in allele frequency, in agreement with the stochastic expectation given by equation (2.14). In the case with selection, we see that equations (1.8) and (2.17) are equal.

---

**Power User Challenge 2.6**

Write down the $T$ matrix for the Wright–Fisher model of random genetic drift with recurrent two-way mutation. Again parameterize the per capita, per generation probabilities of $A\rightarrow a$ and $a\rightarrow A$ mutations as $\mu$ and $v$, respectively. Now derive $M(x)$ for this $T$ matrix. Hint: your answer to this second question should match the analogous, deterministic result in chapter 1. What does this tell us about this $T$ matrix?

---

**Power User Challenge 2.7**

Write down the $T$ matrix for the Wright–Fisher model of random genetic drift in an island population experiencing immigration (i.e., replacement) from a mainland population with per capita per generation probability $m$. You should assume that the frequency of the $A$ allele on the mainland remains $x_M$. Now derive its $M(x)$. What is the analogous, deterministic result in chapter 1?

---

**2.2.3.3 The Moran model of random genetic drift** The *Moran model* is another commonly used framework for studying random genetic drift. Recall that the Wright–Fisher model assumes nonoverlapping generations: each iteration consists of the production of $N$ offspring, followed by the death of all $N$ parents. In contrast, the Moran model is one of overlapping generations: each iteration consists of a single reproductive event and a single death.

More specifically, the Moran model assumes $N$ haploid organisms. In each iteration, two individuals are chosen with replacement. Of these, one is chosen at random to reproduce, and its offspring replaces the other. Assuming two genetically distinct alleles at a locus labeled $A$ and $a$, we again write $k$ for the number of $A$-bearing individuals in the population. In the absence of deterministic processes, the probability of sampling an $A$-bearing individual is again $\dfrac{k}{N}$.

---

**Power User Challenge 2.8**

What feature of the Moran model underlies random genetic drift? Hint: that individuals are chosen with replacement in each iteration of the model is of only very modest stochastic consequence.

---

Thus, in each iteration, the number of $A$ alleles will increase by one with probability $\dfrac{k}{N}\left(1-\dfrac{k}{N}\right)$. This is the product of the probabilities of picking an $A$-bearing individual to reproduce and an $a$-bearing individual to be replaced. Similar reasoning means that the probability that the number of $A$ alleles goes down by one is also $\dfrac{k}{N}\left(1-\dfrac{k}{N}\right)$: we now require sampling an $a$-bearing individual for reproduction and an $A$-bearing individual for death. The number of $A$ alleles remains unchanged if two $A$-bearing individuals or two $a$-bearing individuals are sampled, which happens with probability $\left(\dfrac{k}{N}\right)^2+\left(1-\dfrac{k}{N}\right)^2$. Finally, since each iteration of this model captures just one reproductive event, there is no possibility that the number of $A$ alleles can change by more than one. (That the sum of the above probabilities is 1 yields the same conclusion.) The $T$ matrix for the Moran model of random genetic drift thus reads

$$t(i|k) = \begin{cases} \dfrac{k}{N}\left(1-\dfrac{k}{N}\right) & \text{if } i=k+1 \\[2ex] \dfrac{k}{N}\left(1-\dfrac{k}{N}\right) & \text{if } i=k-1 \\[2ex] \left(\dfrac{k}{N}\right)^2 + \left(1-\dfrac{k}{N}\right)^2 & \text{if } i=k \\[2ex] 0 & \text{otherwise} \end{cases} \qquad (2.19)$$

---

**Power User Challenge 2.9**

Show that the per generation mean and variance in allele frequency change under the Moran model are $M(x)=0$ and $V(x)=\dfrac{2x(1-x)}{N}$, respectively. Hint: first compute the per time-step mean and variance. Then, because there are $N$ reproduction events per generation, multiply these quantities by $N$ to find per generation values.

---

**Teachable Moment 2.15**

Recalling that in the Wright–Fisher model $V(x)=\dfrac{x(1-x)}{N}$ for haploids (see comment below equation [2.15]), power user challenge 2.9 illustrates that random genetic drift is twice as strong in the Moran model as in the Wright–Fisher model. This is typical of stochastic models of overlapping generations, because organisms face an additional stochastic risk in each time step: in addition to the possibility of not reproducing, they are also susceptible to death. (See also power user challenge 2.8.) Consequently, while the mean number of offspring is 1 in both models, overlapping generations always increase the variance in reproductive success. See Felsenstein (2019, 274–275) to learn more.

---

**Power User Challenge 2.10**

Write down the $T$ matrix for the Moran model of random genetic drift with selection. First assume that selection enriches the probability of selecting an individual for reproduction. Then assume that it reduces the probability of selecting an individual for replacement. Do these two framings yield different answers?

---

While the Wright–Fisher model is perhaps more widely used, the Moran model has the virtue of admitting some closed-form results without appeal to the diffusion approximation. For example, consider $i$ copies of a mutant allele with Malthusian fitness $r$ in a population of $N$ haploids. It is thus competing with $N-i$ copies of a wild type; call their fitness 1. One can show (e.g., chapter 6 in Nowak 2006) that the mutant's probability of fixation is

$$P_{\text{fix}}^{\text{Moran}} = \frac{1-\left(\dfrac{1}{r}\right)^i}{1-\left(\dfrac{1}{r}\right)^N}. \tag{2.20}$$

This result is exact (i.e., not an approximation); but for our purposes, the more interesting point is that direct analytic progress is possible with the Moran model because it is a *one-step* or *birth–death process* allele counts cannot change by more than one in a single iteration (equation 2.19). Recall that the key breakthrough in the diffusion treatment of the Chapman–Kolmogorov equation was approximating the $T$ matrix as a one-step model (section 2.2.2.2). But unlike the Moran model, there is still some distance to travel before the diffusion treatment yields the Wright–Fisher analog of equation (2.20), as we see next.

### 2.2.4 Using the Backward Diffusion Approximation to Study Fixation Events

We motivated the Chapman–Kolmogorov equation and its diffusion approximations by noting our inability to compute probabilities of eventual fixation for selectively neutral and deleterious alleles [mathematically, $p(1|x_0, \infty)$]. Our problem can be restated as finding the probability with which a population now with allele frequency $x=1$ could have evolved from one in which its frequency was ever $x_0$. This is precisely the framing of the backward diffusion approximation.

#### 2.2.4.1 The probability of fixation for a selectively neutral allele    Under the Wright–Fisher model, $M(x_0)=0$ and $V(x_0)=\dfrac{x_0(1-x_0)}{2N}$ (section 2.2.3.1). Substituting these values into the backward diffusion (equation 2.12b) gives

$$\frac{\partial p(x|x_0, t)}{\partial t} = \frac{1}{2}\frac{x_0(1-x_0)}{2N}\frac{\partial^2 p(x|x_0, t)}{\partial x_0^2} + 0.$$

Next, because fixation is an absorbing barrier, it is an equilibrium state (teachable moment 1.17). Mathematically, this lets us represent fixation of a neutral allele as

$$0 = \frac{1}{2}\frac{x_0\left(1-x_0\right)}{2N}\frac{\partial^2 p(x=1|x_0, t=\infty)}{\partial x_0^2} = \frac{\partial^2 p(x=1|x_0, t=\infty)}{\partial x_0^2}.$$

Note that in the second step, we assume $x_0$ is neither 0 nor 1. Were it either value, the solution would trivially be that fixation is impossible or assured, respectively. We now integrate twice, first finding $\dfrac{\partial p(x=1|x_0, t=\infty)}{\partial x_0}=C$, and then

$$p(x=1\,|\,x_0, t=\infty)=Cx_0+D.$$

Here $C$ and $D$ are constants of integration (see teachable moment 1.4), which again acknowledge the fact that a family of equations satisfy $\dfrac{\partial^2 p(x=1\,|\,x_0, t=\infty)}{\partial^2 x_0}=0$. The specific values of $C$ and $D$ come from boundary conditions (analogous to what we called initial conditions in chapter 1). In the present case, the boundary conditions correspond the absorbing barriers at $x_0=0$ and 1, mathematically represented $p(x=1\,|\,x_0=0, t=\infty)=0$ and $p(x=1\,|\,x_0=1, t=\infty)=1$, respectively. Applying the first boundary condition (the absorbing barrier at $x_0=0$) yields $D=0$, and the second gives us $C=1$. Introducing $P_{\text{fix}}(x_0)$ as a shorthand for $p(x=1\,|\,x_0, t=\infty)$, the probability of eventual fixation for a selectively neutral allele under the Wright–Fisher model of random genetic drift thus reads

$$P_{\text{fix}}(x_0)=x_0. \tag{2.21}$$

And indeed, equation (2.21) follows for any $T$ matrix in which $M(x)=0$, since $V(x)$ disappeared in the first step of the derivation. We should not be surprised: in the absence of any stochastic bias, the expected allele frequency must remain unchanged, as per the deterministic treatment. Coupled with the fact that 0 and 1 are absorbing barriers, it follows that if that allele frequency begins at $x_0$, $x_0$ is the only possible solution for the probability of eventual fixation. (Writing things mathematically, we require $E[x]=x_0=[1-P_{\text{fix}}(x_0)]\times 0+P_{\text{fix}}(x_0)\times 1$, which gives equation [2.21].)

That selectively neutral alleles can ever reach fixation represents a key biological novelty: this possibility is outside the realm of deterministic models. But in fact, few population genetic data can be explained without assuming that many (often, the majority) of segregating and fixed alleles are selectively neutral; one such example is illustrated in empirical aside 2.1. This conclusion motivated Motoo Kimura's revolutionary and profoundly influential neutral theory of molecular evolution described in empirical aside 2.2 and fully developed in Kimura (1983). (Or at least, most segregating and fixed alleles must be nearly neutral: as we show shortly, alleles whose selection coefficients are closer to zero than the reciprocal of population size are also essentially invisible to natural selection. See also empirical aside 2.3.)

**Empirical Aside 2.1:** The molecular clock

Perhaps the most important biological implication of equation (2.21) is that the probability of fixation for a new, selectively neutral allele in a population of $N$ diploids is $P_{\text{fix}}\left(x_0=\dfrac{1}{2N}\right)=\dfrac{1}{2N}$.

Writing $\mu$ for the per copy, per generation probability of selectively neutral mutations, the expected, population-wide number of neutral alleles created each generation in a population of $N$ diploids is $2N\mu$. Introducing $\rho$ for the per generation rate of fixation of neutral alleles, we have

$$E[\rho]=\frac{2N\mu}{2N}=\mu. \tag{2.22}$$

Rather remarkably then, the rate of fixation of neutral mutations is expected to be independent of population size, implying that fixed differences between copies of the same gene in two species should simply accumulate linearly with time since their divergence. This is the theoretical foundation for the so-called the *molecular clock hypothesis* (Zuckerkandl and Pauling [1962] and Margoliash [1963]; reviewed in Kumar [2005]), and while this clock only keeps time stochastically (neither mutation nor fixation are deterministic), there is good empirical support for it. Figure 2.7 (after Dickerson [1971]) is an early, compelling illustration. Gene-specific differences in slope are interpreted as reflecting differences in the neutral mutation rate $\mu$, in turn suggesting differences in functional constraints. (Looking ahead, Dickerson employed a technique analogous to that described in teachable moment 4.6 to correct for the possibility of multiple fixation events at the same site.)

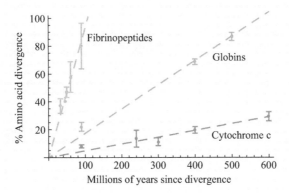

**Figure 2.7**
Proportion of amino acids that have been mutated at least once between homologs for three different proteins (*x*-axis) as a function of time since species divergence (*y*-axis). Dashed lines are linear models fit through the origin. (After Dickerson [1971].)

Gene-specific rates of the molecular clock can be calibrated in years by dividing the number of fixations between two species in a gene by twice an estimate of divergence time, such as a fossil. (The factor of two reflects the fact that mutations are fixing in both lineages.) One can also go in the opposite direction: given a known gene-specific fixation rate calibrated in time, one can estimate divergence times between pairs of species for which we only have sequence data.

**Empirical Aside 2.2:** The neutral theory of molecular evolution

The neutral theory of molecular evolution (Kimura 1983) hypothesizes that most mutations segregating in nature are selectively neutral (or nearly so; see next section). It doesn't claim that deleterious and beneficial mutations don't exist. Rather, it holds that natural selection will keep the former at low frequency before they are purged, and that the latter are both rare and only segregate briefly before fixation. This theory can explain three empirical observations in sequence data that are incompatible with the hypothesis that most segregating mutations are beneficial.

First, the molecular clock (empirical aside 2.1) is most easily explained if most mutations are selectively neutral. Second, models of substitutional load (teachable moment 1.19) suggest that the required number of selective deaths required to drive empirically observed fixation rates would far exceed most species' reproductive capacity.

But the earliest evidence for the neutral theory of molecular evolution was the unexpected high amount of genetic variation found in fruit flies (Lewontin and Hubby 1966), a phenomenon sustained in all subsequent surveys of natural populations (see also empirical aside 2.5).

While these could formally be understood by models of overdominance (section 1.3.2), the resulting segregation load (teachable moment 1.19) appears untenable. Moreover, similar levels of genetic variation were subsequently demonstrated in haploids, in which overdominance is impossible. No such difficulty exists if most segregating variation is selectively neutral.

**2.2.4.2  The probability of fixation for a selected allele**   Following the same path that brought us to equation (2.21), we now substitute $M(x)$ and $V(x)$ given by equations (2.17) and (2.18) into the backward diffusion approximation (equation 2.12b). This lets us represent the eventual fixation of a selected allele in the Wright–Fisher model as

$$0 = \frac{1}{2} \frac{x_0(1-x_0)}{2N} \frac{\partial^2 p(x=1|x_0, t=\infty)}{\partial x_0^2} + sx_0(1-x_0) \frac{\partial p(x=1|x_0 t=\infty)}{\partial x_0}.$$

(Recall that our $M(x)$ assumes the absence of dominance; that is, that $h = \frac{1}{2}$. This assumption can be relaxed, and the interested reader is directed to Crow and Kimura [1970] to learn more.) Next, again assuming $x_0$ is neither zero nor one lets us write

$$0 = \frac{\partial^2 p(x=1|x_0, t=\infty)}{\partial x_0^2} + 4Ns \frac{\partial p(x=1|x_0, t=\infty)}{\partial x_0},$$

which is integrated to find

$$p(x=1|x_0, t=\infty) = \frac{C}{-4Ns} e^{-4Nsx_0} + D$$

(teachable moment 2.16). Appealing to the same boundary conditions as before, we now find $C = \dfrac{4Ns}{1 - e^{-4Ns}}$ and $D = \dfrac{1}{1 - e^{-4Ns}}$, which, after a bit of simplification yield

$$P_{\text{fix}}(x_0, N, s) = p(x=1|x_0, t=\infty) = \frac{1 - e^{-4Nsx_0}}{1 - e^{-4Ns}}. \tag{2.23}$$

This equation gives the probability of eventual fixation of an allele responding to natural selection and genetic drift, making it among the most important and interesting results in stochastic one-locus population genetics.

**Teachable Moment 2.16**

Solving the backward diffusion equation for the Wright–Fisher model of random genetic drift with selection.

The mathematics are made easier by first introducing a new function $Q(x_0) = \dfrac{\partial p(x=1|x_0, t=\infty)}{\partial x_0}$. Now substituting, our equation becomes $0 = \dfrac{\partial Q(x_0)}{\partial x_0} + 4NsQ(x_0)$.

Rearranging it as $\dfrac{\partial Q(x_0)}{Q(x_0)} = -4Ns\,\partial x_0$ reveals the same differential form we encountered in our deterministic model of population growth (there written $\dfrac{\partial N(t)}{N(t)} = rdt$), and the solution is again logarithmic: $\ln[Q(x_0)] = -4Nsx_0 + C$, yielding $Q(x_0) = Ce^{-4Nsx_0}$. Finally, substituting back for $Q$ yields $\dfrac{\partial p(x=1|x_0, t=\infty)}{\partial x_0} = Ce^{-4Nsx_0}$, which integrates to $\dfrac{C}{-4Ns}e^{-4Nsx_0} + D$, as required.

We note first that equation (2.23) depends entirely on the compound parameter $Ns$. Figure 2.8 illustrates the behavior of this equation as a function of $x_0$ for a range of values of $Ns$. Naturally, beneficial alleles are far more likely to fix than are neutral alleles, while the fixation of deleterious alleles is far less likely. (Although the contrast with deterministic results—where beneficial alleles always fix and selectively neutral and deleterious alleles never do—is of course striking.)

**Power User Challenge 2.11**

Sketch the left panel of figure 2.8 under deterministic assumptions. Note how this is the limiting case of the stochastic results shown in the figure.

The right panel in figure 2.8 highlights that for given $Ns > 0$, there is a rather sharp transition in initial allele frequency $x_0$, above which fixation is almost assured. This

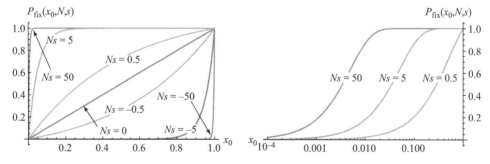

**Figure 2.8**
Probability of fixation for alleles as a function of starting frequency for different values of $Ns$ under the diffusion approximation of the Wright–Fisher model of genetic drift with natural selection (equation 2.23). The $x$-axis on the right has been log transformed to better illustrate behavior for beneficial alleles when rare.

is establishment: once there are enough copies of a beneficial allele in the population, the law of large numbers renders stochastic fluctuations in individual offspring counts almost irrelevant. As promised in teachable moment 2.9, equation (2.23) allows us to roughly quantify this threshold. Note first that both numerator and denominator are of the form $1 - e^{-Z}$, which converges to one as $Z$ grows to be much larger than unity. So, for $Ns = 5$ and 50, the denominator is essentially one. Similarly, once $4Nsx_0$ is much larger than 1, so too is the numerator. Thus, a beneficial allele is established once its frequency well exceeds $\dfrac{1}{4Ns}$ or, equivalently, once there are many more than roughly $2N\dfrac{1}{4Ns} = \dfrac{1}{2s}$ copies of the allele. This underscores the population-size independence of stochasticity for beneficial alleles observed in our branching process treatment. Of course, this transition isn't absolute, but the probability of being lost quickly collapses for frequencies or counts above these values.

Returning attention to the panel on the left, results for $Ns = 0.5$ illustrate that beneficial alleles can fix before being established, as already suggested in teachable moment 2.9. Since frequencies cannot exceed 1, establishment approximately requires $Ns$ to well exceed 0.25. Most practitioners take $Ns \approx 1$ as the threshold.

---

**Teachable Moment 2.17:** Scaling relationships

Examination of figure 2.8 and equation (2.23) illustrates that a beneficial allele becomes established once its frequency is roughly larger than $\dfrac{1}{4Ns}$. But that the transition isn't absolute: technically, beneficial alleles are at nonzero risk of loss until the moment of their fixation. The real lesson is that the threshold scales inversely with the compound parameter $Ns$. We use a tilde to signal scaling relationships, and say that beneficial alleles become established at a critical frequency $x_0 \sim \dfrac{1}{Ns}$, read as "a critical frequency $x_0$ that scales with $\dfrac{1}{Ns}$." And similarly, the critical count scales with $\dfrac{1}{s}$, now thinking of stochastic states as counts rather than frequencies. These expressions capture the model's dependence on parameter values; the constant coefficients are less important. We will encounter many such scaling relationships in this book, which demarcate boundaries between regimes of stochastic model behavior.

---

As before, we are perhaps most often interested in the fate of a new allele at the moment of its appearance, meaning $x_0 = \dfrac{1}{2N}$, in which case, equation (2.23) now reads

$$P_{\text{fix}}\left(x_0 = \frac{1}{2N}, N, s\right) = p\left(x = 1 \,\middle|\, \frac{1}{2N}, t = \infty\right) = \frac{1 - e^{-2s}}{1 - e^{-4Ns}}. \qquad (2.24)$$

Figure 2.9 illustrates the behavior of $P_{\text{fix}}\left(x_0 = \dfrac{1}{2N}, N, s\right)$ as a function of $N$ and $s$. Several regimes are evident. First, if $Ns \gg 1$, the denominator of equation (2.24) is

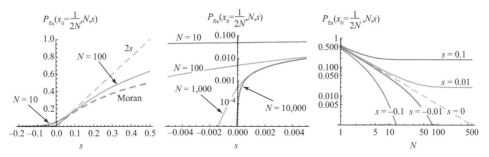

**Figure 2.9**
Probability of fixation of a single copy of an allele under the Wright–Fisher model of natural selection and genetic drift (equation 2.24, except as noted). Left and center: as a function of $s$ for several values of $N$. Left: $N = 10$ (blue), $N = 100$ (mustard; results in yet-larger populations are indistinguishable). The branching-process approximation (green dashed) comes from equation (2.7). The Moran solution (red) comes from equation (2.20), computing $r$ in terms of $s$ as described in power user challenge 1.10, and doubling $s$ to correct for that model's two-fold stronger drift (teachable moment 2.15). Center: $x$- and $y$-axes expanded to more clearly reveal dependence on population size for $|Ns| \ll 1$. Each line crosses the $y$-axis at $\frac{1}{2N}$, as expected. Right: as a function of $N$ for several values of $s$.

1, and if $0 < s \ll 1$, then its numerator is $\approx 2s$ via the Taylor approximation. This recovers our branching process result (equation 2.7), which is represented by the dashed green line in the left panel of figure 2.9. In this regime, fixation probability is independent of $N$ but the process is still stochastic, as already seen.

As $s$ grows beyond $s \ll 1$ for beneficial alleles, the Taylor approximation that yielded

$$P_{\text{fix}}\left(x_0 = \frac{1}{2N},\, N,\, s\right) \approx 2s \text{ breaks down. Fixation probability is still independent of } N,$$

but beneficial alleles establish with probability less than $2s$. In answer to power user challenge 2.2, this is because $A$-carrying individuals now increase quickly enough that they can interfere with one another's success. This is an example of the Hill–Robertson effect, a topic to which we return in section 3.2.4. Moreover, once $s$ exceeds roughly 0.1, we undermine the diffusion approximation's restriction that probability movement be restricted to adjacent states (section 2.2.2.2), thereby over-estimating the probability of fixation. This is seen by comparison with the Moran (exact) solution, shown by the red dashed line in the left panel of the figure.

Next, if $|Ns| \ll 1$ (in which case also $|s| \ll 1$), separately computing Taylor approxi-

mations of the numerator and the denominator yields $P_{\text{fix}}\left(x_0 = \frac{1}{2N},\, N,\, s\right) \approx \frac{s}{2Ns} = \frac{1}{2N}$,

as already seen (equation 2.21). Here, drift dominates selection, regardless of the sign of $s$ (figure 2.9, center panel). This effect is also seen for $N = 10$ in the left panel, and also in the left panel of figure 2.8 for $Ns = \pm 0.5$.

The right panel of figure 2.9 illustrates behavior as a function of $N$ for representative values of $s$. Several features are noteworthy. First focusing on beneficial alleles, probabilities of fixation asymptotically approach $2s$, as expected. But perhaps unexpected, they do so from above. As noted in teachable moment 2.9, beneficial alleles can fix by drift or selection. The figure illustrates that at sufficiently small $N$, the

probability of former $\left(\frac{1}{2N}\right)$ is larger than of the latter ($2s$). The transition between regimes can be found by equating terms, yielding $N \sim \frac{1}{s}$; that is, when $N$ is large enough to sustain establishment of the beneficial allele. (This is another path to the scaling relationship discussed in teachable moment 2.17.)

The right panel also Illustrates that the probability of fixation for selectively neutral and deleterious alleles never stops declining with population size. This is easily understood. Such fixations require a succession of unlikely events: carriers of these alleles must have more offspring than their competitors by chance alone. A minute's reflection makes clear that the number of such unlikely events required for the allele to reach fixation grows with population size.

Finally, we see that the probability of fixation for deleterious alleles soon plummets as $N$ grows; one can show that the threshold population size scales with $\frac{1}{|s|}$ (power user challenge 2.12). In populations smaller than this, deleterious alleles are sometimes called *nearly neutral*, because their fate is largely governed by random genetic drift. Above this threshold, behavior quickly converges to our deterministic result. This population-size dependence motivated Tomoko Ohta's nearly neutral theory of molecular evolution (Ohta 1973), described in empirical aside 2.3.

---

**Empirical Aside 2.3:** The nearly neutral model of molecular evolution and the drift-barrier hypothesis

The molecular clock (empirical aside 2.1) exhibits a rate constancy in units of time. In contrast, the neutral theory of molecular evolution (empirical aside 2.2) predicts a constant per generation rate of fixation. Thus, in units of time, it predicts that the molecular clock should run faster in short-lived organisms, which experience more generations per unit time—and thus more opportunities for mutation—than do long-lived organisms.

To address this difficulty, Tomoko Ohta proposed the nearly neutral theory of molecular evolution (Ohta 1973). She began from the empirical observation that population size tends to be inversely correlated with generation time. Together with the prediction that purifying selection is less effective in small populations (figure 2.9, right panel), she suggested that the net per generation fixation rate would consequently be higher in long-lived organisms. This hypothesis may explain the approximate rate constancy of the molecular clock calibrated in time.

More recently, this same theory has been used to explain the inverse relationship between mutation rate and population size observed in nature (Lynch 2010). On the premise that most mutations are deleterious, natural selection is expected to eliminate lineages that have heritably higher mutation rate. But as just noted, such lineages will be more effectively eliminated in larger populations, potentially explaining their lower mutation rates. This is called the *drift-barrier hypothesis* (Lynch et al. 2016). We return to the theory of mutation rate evolution in section 3.3.2.

---

**Power User Challenge 2.12**

Use equation (2.24) to show that deleterious mutations are nearly neutral in populations smaller than $\frac{1}{|s|}$.

**2.2.4.3  Time to fixation for a new allele**  Diffusion theory yields access to many other interesting and important population genetic quantities, including times to fixation. This work has a technical complexity that is beyond our scope, and we instead present two key results without derivation. Some intuition into methods for approaching these questions is developed in teachable moment 2.18.

---

**Teachable Moment 2.18:** Computing times to fixation and loss in diffusion theory

$P(x|x_0, t)dxdt$ is the probability of finding an allele in frequency interval $(x, x+dx)$ during time interval $(t, t+dt)$ given that it was at frequency $x_0$ at time $t=0$. Thus, the expected total time an allele spends in frequency interval $(a, b)$, written as $t_{(a, b)}$, is this probability integrated over all frequencies between $0 \leq a < b \leq 1$ and times $0 \leq t \leq \infty$. Mathematically, we have

$$E[t_{(a,b)}] = \int_0^\infty \int_a^b p(x|x_0, t)\,dxdt.$$

Of particular interest, setting $a$ and $b$ to 0 and 1, respectively, gives the expected time that a new allele spends segregating in the population before fixation or loss.

This calculation can also be conditioned on eventual fixation of the allele by computing

$$E[t_{\text{fix}}] = \int_0^\infty \int_0^1 p(x|x_0, t) \frac{p(1|x, t)}{p(1|\frac{1}{2N}, t)}\,dxdt.$$

The new factor is the probability of fixation from each frequency $x$, conditioned on fixation from frequency $\frac{1}{2N}$. The expected time to loss is similarly written, now weighting $p(x|x_0, t)$ inside the integral by the conditioned probability of loss.

Of course, we don't have direct access to $p(x|x_0, t)$, which makes solutions to these more sophisticated questions quite technical. The interested reader is directed to chapters 8 and 9 in Crow and Kimura (1970), chapters 3 and 4 in Ewens (1979), and appendix A in Walsh and Lynch (2018) for more rigorous discussions of methods and for derivations of many results, including the next two.

---

First, the approach outlined in teachable moment 2.18 can be used to show that the expected time to fixation for a new, selectively neutral allele in a population of $N$ diploids is $4N$ generations. (We derive a closely related result in section 2.4 using coalescent theory.) On the other hand, the expected time to loss is much less: approximately $\ln(2N)$.

And second, the expected time to fixation for a new, selected allele depends only on population size and the absolute value of its selection coefficient but not its sign (Maruyama and Kimura 1974). Beneficial alleles fix much more quickly than do neutral alleles. Since the fixation of a deleterious allele relies on a sequence of unlikely events, we perhaps ought not be surprised that that process also occurs quickly. But that the two quantities should be equal is intriguing. Moreover, the same is true for time to loss: it also depends only on the absolute value of the compound parameter $Ns$.

We conclude this section with a heuristic treatment for the time to fixation for a new beneficial allele. Recall from section 2.1.1 that established populations tend to be larger than predicted by deterministic theory (figure 2.2). This is because, averaged across stochasticity, the mean population size should match the deterministic prediction (teachable moment 2.6), and since many realizations go extinct, those that establish must be anomalously large. On the other hand, once established, their growth rate is given by deterministic theory, as per the law of large numbers.

Something similar occurs for beneficial alleles. Since most are stochastically lost, those that become establishment generally enjoy an initial, anomalously large jump in frequency. This is true since, unconditioned on fixation or loss, the expected frequency should again match the deterministic prediction. And once established, a beneficial allele's subsequent frequency trajectory closely matches the deterministic prediction, again as per the law of large numbers.

Recalling that beneficial alleles are established at frequency $\sim \dfrac{1}{Ns}$, the reasoning in the previous paragraph suggests that the expected time to fixation for a beneficial allele can be approximated by following the path in teachable moment 1.9, only now assuming a starting frequency of $\dfrac{1}{Ns}$ rather than $\dfrac{1}{N}$. This yields

$$t_{\text{fix}}^{\text{stochastic}} \approx \frac{2\ln(Ns)}{s}. \tag{2.25}$$

This treatment disregards the time for the initial jump to frequency $\sim \dfrac{1}{Ns}$, which, however, we have argued will be very short.

### 2.2.5 Using the Forward Diffusion Approximation to Study Internal Equilibrium Allele Frequencies

We used the backward diffusion approximation to study populations encountering their absorbing barriers: eventual allele fixation and loss. By definition (teachable moment 2.5), the probability density at those equilibria is zero at all frequencies except $x = 0$ or 1. In contrast, the forward diffusion approximation can be used to solve models with internal equilibrium probability density; that is, equilibria with nonzero densities between 0 and 1. Since drift inexorably drives populations toward absorbing barriers, internal equilibria require a deterministic process that restores variation. Recurrent two-way mutation is one such process. In chapter 1, we wrote $\mu$ and $v$ for the instantaneous rates of $A \rightarrow a$ and $a \rightarrow A$ mutations, finding that an evolving population will equilibrate to $\tilde{p}_A = \dfrac{v}{\mu + v}$ (equation 1.18). We now ask how random genetic drift influences this result.

In answer to power user challenge 2.6, recasting $\mu$ and $v$ as the per copy, per generation probabilities of $A \rightarrow a$ and $a \rightarrow A$ mutations, respectively, we find $M(x) = -\mu x + v(1-x)$. Still using $V(x) = \dfrac{x(1-x)}{2N}$ (warranted if $\mu$ and $v$ are much less than 1), the corresponding forward diffusion approximation reads

$$\frac{\partial p(x|x_0,t)}{\partial t}=\frac{1}{2}\frac{\partial^2[\frac{x(1-x)}{2N}p(x|x_0,t)]}{\partial^2 x}-\frac{\partial[(-\mu x+v(1-x)p(x|x_0,t))]}{\partial x}.$$

Setting this expression to zero describes the equilibrium allele frequency probability density in a population at *mutation/drift equilibrium*.

The solution is found by integrating twice. The first integration yields

$$\frac{\partial[x(1-x)p(x|x_0,t=\infty)]}{\partial x}=4Nv(1-x)p(x|x_0,t=\infty)-4N\mu xp(x|x_0,t=\infty).$$

The next integration requires a few more steps (see teachable moment 2.19) but we soon reach

$$\tilde{p}(x)=p(x|x_0,t=\infty)=Cx^{4Nv-1}(1-x)^{4N\mu-1}. \tag{2.26}$$

We introduce $\tilde{p}(x)$ for the allele frequency probability density function in a population at mutation/drift equilibrium. $C$ represents another constant of integration, which in this case is found by appeal to the fact that, because $\tilde{p}(x)$ is a probability density function (teachable moment 2.2), it must integrate to unity. Mathematically, $\int_0^1 \tilde{p}(x)dx=1$, giving $C=\dfrac{1}{\displaystyle\int_0^1 x^{4Nv-1}(1-x)^{4N\mu-1}dx}$. Finally, note that population size and mutation rates only appear as compound parameters in equation (2.26). Figure 2.10 illustrates the behavior of equation (2.26).

---

**Teachable Moment 2.19:** Finding the equilibrium solution for the forward diffusion equation with mutation

This trick is only slightly more subtle than that employed in teachable moment 2.16. We first introduce a new $Q(x|x_0,t)=x(1-x)p(x|x_0,t)$. Thus, $(1-x)p(x|x_0,t)=\dfrac{Q(x|x_0,t)}{x}$ and $xp(x|x_0,t)=\dfrac{Q(x|x_0,t)}{1-x}$. Substituting these three equalities into the first, second, and third terms of the equation at hand yields

$$\frac{\partial Q(x|x_0,t)}{\partial x}=\frac{4NvQ(x|x_0,t)}{x}-\frac{4N\mu Q(x|x_0,t)}{1-x}.$$

In a maneuver again reminiscent of that used to solve the first model in chapter 1, we divide through by $Q(x|x_0,t)$ and multiply through by $\partial x$ to arrive at

$$\frac{\partial Q(x|x_0,t)}{Q(x|x_0,t)}=\frac{4Nv\partial x}{x}-\frac{4N\mu\partial x}{1-x},$$

which readily integrates to

$$\ln[Q(x|x_0,t)]=4Nv\ln(x)+4N\mu\ln(1-x)+C.$$

Now we take advantage of a property of logarithms used in the paragraph immediately below power user challenge 1.9 to write

$$\ln[Q(x|x_0, t)] = \ln(x^{4Nv}) + \ln[(1-x)^{4N\mu}] + C.$$

Next, we exponentiate and simplify

$$Q(x|x_0, t) = [x^{4Nv}(1-x)^{4N\mu}]e^C = Cx^{4Nv}(1-x)^{4N\mu}.$$

(Since $C$ is an as-yet-unspecified constant, we can replace it with $e^C$ without loss of generality.) Finally, substituting back for $Q(x|x_0, t) = x(1-x)\ \tilde{p}(x)$ yields $\tilde{p}(x) = Cx^{4Nv-1}(1-x)^{4N\mu-1}$ as required.

Inspection of equation (2.26) reveals an interesting threshold: when $4N\mu$ and $4Nv$ are both less than 1, the exponents in $\tilde{p}(x)$ are both negative, meaning that the function is maximal at extreme allele frequencies and declines toward intermediate values. In English, random genetic drift dominates mutation, in which case the $A$ allele is very likely to be nearly fixed or nearly lost. Conversely, when $4N\mu$ and $4Nv$ are both greater than 1, the exponents in $\tilde{p}(x)$ are positive, meaning that the function is minimal at extreme allele frequencies and increases toward intermediate values. In this case, mutation dominates random genetic drift and the $A$ allele is very likely to be present at some intermediate frequency.

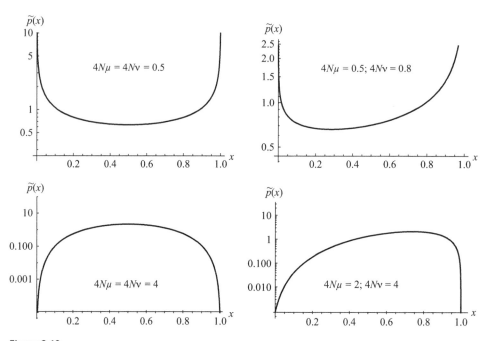

**Figure 2.10**
Allele frequency probability distributions at mutation/drift equilibrium under the Wright–Fisher model with mutation (equation 2.26). In all panels, the $x$-axis is allele frequency and the $y$-axis is the probability density for given frequencies of the $A$ allele. In the upper panels, probability density is concentrated near allele frequency 0 and 1 because $4N\mu$ and $4Nv$ are both less than 1. In the lower panels, probability density is concentrated at intermediate allele frequencies because $4N\mu$ and $4Nv$ are both greater than 1 (see text for more). The panels on the left assume symmetric mutation rates; hence those distributions are symmetric. In the panels on the right, the $a \rightarrow A$ mutation rate is greater than the $A \rightarrow a$ rate (mathematically, $v > \mu$), accounting for the rightward skew of both distributions.

**Power User Challenge 2.13:** Allele frequencies at *migration/drift equilibrium*

In answer to power user challenge 2.7, the mean per generation change in allele frequency is $M(x) = m(x_M - x)$ on an island with allele frequency $x$ subject to a per copy, per generation probability of replacement $m$ from a mainland population in which allele frequency is a constant $x_M$. Assuming $V(x) = \dfrac{x(1-x)}{2N}$ for this model (reasonable if $m \ll 1$), follow the same path used to reach equation (2.26) to find

$$\tilde{p}(x) = p(x|x_0, t) = Cx^{4Nmx_M - 1}(1-x)^{4Nm(1-x_M)-1} \tag{2.27}$$

for the allele frequency probability density in a population at migration/drift equilibrium. (The coefficient of integration $C$ is again given by the requirement that $\int \tilde{p}(x) = 1$.)

The strong formal parallels between mutation and migration noted in chapter 1 are again evident here. In some sense, $mx_M$ and $m(1-x_M)$ are analogous to $v$ and $\mu$, respectively. As before, assuming symmetry (now meaning $x_M = \frac{1}{2}$) reveals that when $4Nm$ is below a threshold (again 1), random genetic drift will dominate and the island population will most likely be fixed for one allele or the other. Conversely, when $4Nm$ is above the threshold, migration will dominate and allele frequency will be some intermediate value rather close to $x_M$. The threshold of $4Nm \approx 1$ corresponds to roughly one immigrant per generation, meaning that only a few migrants per generation are enough genetically homogenize a subdivided population.

Finally substitute $v = mx_M$ and $\mu = m(1-x_M)$ into equation (2.26) to derive equation (2.27). Make the same substitution into equation (1.17) to find the deterministic equilibrium allele frequency on the island. Are you surprised by your answer? Why or why not?

## 2.3   Random Genetic Drift and Heterozygosity

Thus far, we have considered the influence of stochasticity on allele frequencies. We now explore the stochasticity of heterozygosity given by equation (1.23d). One could imagine bringing the full might of the Chapman–Kolmogorov and diffusion approaches to bear on the problem. For example, we might seek the probability distribution of all conceivable genotype frequencies in the future as a function of their frequencies in the present. This is possible, and the interested reader is directed to appendix A in Walsh and Lynch (2018) for an introduction to this approach, and to chapters 8 and 9 in Crow and Kimura (1970) for many results.

Our goal is more modest: we shall content ourselves with the expected value of this distribution. And importantly, the remainder of this chapter assumes that all alleles are selectively neutral. We relax this limitation when we take up stochastic multilocus theory in chapter 3.

### 2.3.1   The Rate of Decay in Heterozygosity under the Wright–Fisher Model
Recall from chapter 1 that deviations from Hardy–Weinberg frequencies can be quantified by comparing observed heterozygosities with those expected under

some model. Mathematically, writing $H_{obs}$ and $H_{exp}$ for observed and expected heterozygosities, respectively, such discrepancies are captured by fixation indices $F = 1 - \dfrac{H_{obs}}{H_{exp}}$, which differ only in their conceptions of what benchmark frequencies to place in the denominator (teachable moment 1.27a). We now introduce a new fixation index, $F_t = 1 - \dfrac{H_t}{H_0}$, where $H_0$ and $H_t$, respectively, are population heterozygosities in generation 0 and after $t$ generations of drift. (The reader should note the distinction between $H_t$, heterozygosity after $t$ generations of drift, and $H_T$, total heterozygosity in a metapopulation, employed in our treatment of the Wahlund effect in section 1.4.4.3.)

We can compute $F_t$ under the Wright–Fisher model as follows. Recall that in this model, $N$ diploid genotypes are formed each generation by the random paring of alleles, which themselves have been randomly sampled from alleles in the previous, parental generation. What is the per capita probability of pairing two genetically distinct alleles? Naively, we might imagine it is equal to the heterozygosity in the parental generation, which after all is exactly the frequency with which genetically distinct alleles were paired in the parents. But this reasoning overlooks the fact that allelic sampling, which precedes pairing, is performed with replacement (thereby driving stochasticity via random genetic drift; teachable moment 2.14).

Thus, we decompose the probability of pairing two genetically distinct alleles in the present generation as follows. First, as there are $2N$ physically distinct alleles, with probability $\dfrac{1}{2N}$, the paired alleles will be copies of the same *parental allele* (i.e., copies of the same allele in the parental population). In this case they are genetically identical. Otherwise, they will be copies of physically distinct parental alleles, which will be genetically different with probability given by the heterozygosity in the previous generation. Mathematically, the expected heterozygosity in the next generation is thus given by the recurrence equation

$$E[H_{t+1}] = \frac{1}{2N} \times 0 + \left(1 - \frac{1}{2N}\right) H_t = \left(1 - \frac{1}{2N}\right) H_t,$$

which immediately gives

$$E[H_t] = H_0 \left(1 - \frac{1}{2N}\right)^t \approx H_0 e^{-\frac{t}{2N}}. \tag{2.28}$$

(Our approximation uses the first-order Taylor approximation $e^{-\frac{1}{2N}} = 1 - \frac{1}{2N} + O\left(\dfrac{1}{(2N)^2}\right)$ developed in teachable moment 2.8, which is extremely accurate if population size is at all large. In the language of probability theory, the exact expression is a *geometric distribution* with parameter $\dfrac{1}{2N}$. What we now see is that a geometric distribution is well-approximated by an *exponential distribution* with the same parameter, if said parameter is much less than one.)

Thus, our new $F$ statistic reads

$$F_t = 1 - \frac{E[H_t]}{H_0} = 1 - \left(1 - \frac{1}{2N}\right)^t \approx 1 - e^{-\frac{t}{2N}}.$$

In English, the stochasticity of random mating preferentially favors the formation of homozygotes over heterozygotes. In spite of its name, random mating now induces a form of inbreeding (an excess of pairings of like alleles). Sharpening this point, the deterministic analog to equation (2.28) is the Hardy–Weinberg model, which predicts that heterozygosity equilibrates to $2p_A(1-p_A)$. That random mating induces inbreeding in the Wright–Fisher model illustrates a point made in teachable moment 2.6: stochastic expectations do not always match deterministic predictions. On the contrary, in the language of teachable moment 2.7, stochasticity induces a biased effect on heterozygosity.

---

**Power User Challenge 2.14**

What is the expected frequency of $AA$ genotypes in a population of $N$ diploids after $t$ generations of random genetic drift? You should assume that all other Hardy–Weinberg assumptions are met and that the frequency of $A$ alleles at time $t=0$ was $p_A$.

---

**Teachable Moment 2.20:** The loss of heterozygosity induced by random mating within populations is exactly offset by the Wahlund effect

To see this, imagine a vast number of replicate diploid populations of size $N$ that each honor all Hardy–Weinberg assumptions. Suppose allele frequencies are all initially $p_A$; heterozygosity at the outset will thus uniformly be $H_0 = 2p_A(1-p_A)$. By definition, drift has no influence on expected allele frequency, but over time it causes them to diverge among replicates. As this happens, the Wahlund effect (section 1.4.4.3) means that the expected heterozygosity must become less than $H_0$. More specifically, consider the situation after $t$ generations. Equation (1.33) can be rewritten as

$$H_0 - E[H_t] = 2E[\text{Var}(p_t)],$$

where the $p_t$ are now the realized allele frequencies among replicate population in generation $t$. Substituting equation (2.28) and simplifying yields

$$E[\text{Var}(p_t)] = \frac{H_0\left(1 - e^{-\frac{t}{2N}}\right)}{2}. \tag{2.29}$$

This result quantifies the rate of divergence in allele frequencies among replicate populations due to drift. But it does more. Since the variance in allelic identity within any population is exactly half its heterozygosity (as for example in equation 1.34c), we also have

$$\mathrm{E}[\mathrm{Var}(I_t)] = \frac{E[H_t]}{2} = \frac{H_0 \mathrm{e}^{-\frac{t}{2N}}}{2}, \qquad (2.30)$$

where the $I_t$ are now the realized indicator function values (teachable moment 1.15) among individuals in any replicate population in generation $t$. Summing equations (2.29) and (2.30) yields

$$\mathrm{E}[\mathrm{Var}(I_t)] + \mathrm{E}[\mathrm{Var}(p_t)] = \frac{H_0\left(1 - \mathrm{e}^{-\frac{t}{2N}}\right)}{2} + \frac{H_0 \mathrm{e}^{-\frac{t}{2N}}}{2} = \frac{H_0}{2} = Var(I_0).$$

In English, random genetic drift converts within-population variance [Var($I_t$)] into between replicate variance [Var($p_t$)], but the total variance is conserved. This conservation of variance offers a new perspective on the apparent asymmetry in heterozygosity induced by drift, and is illustrated in figure 2.11. (It is also reminiscent of the fact that the probability of fixation for a neutral allele must equal its starting frequency; see the paragraph immediately below equation [2.21].)

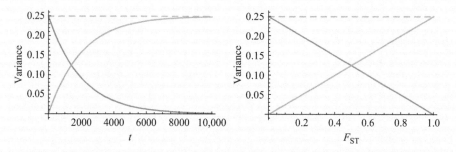

**Figure 2.11**
Population subdivision converts within-population genetic diversity into between-population genetic diversity. Left: expected allelic variance within a population (blue, equation [2.30]) and expected variance in allele frequencies among replicates (mustard, equation [2.29]) under the Wright–Fisher model of random genetic drift are shown as a function of time. Total variance (green dashed, sum of equations [2.29] and [2.30]) is constant. Here, $N = 1,000$ and $p_0 = 0.5$. Right: the exact same relationship applies in the deterministic treatment of population subdivision developed in section 1.4.4.3, except that there, $F_{ST}$ rather than time is the independent variable. But again, as subdivision increases, within-deme allelic variance (blue) is converted into between-deme variance in allele frequency (mustard) with no change in total variance.

### 2.3.2 The Infinite Alleles Model and Heterozygosity at Mutation/Drift Equilibrium under the Wright–Fisher Model

Despite the foregoing analysis, heterozygosity is decidedly nonzero in all natural populations. Why? Mutation is an obvious answer. Even as random genetic drift eliminates lineages and thus genetic diversity, mutation creates it. Where does the balance between these two forces lie?

Our first step is to define a new model of mutation. To this point, we have imagined loci carrying one of just two alleles, but whole genes admit hundreds or thousands of variants. This perspective motivates the *infinite alleles model* (Kimura and Crow 1964), which postulates that each mutation creates a new, genetically distinct allele. While obviously not strictly correct, it is not a bad approximation for loci of any appreciable length. For example, a gene encoding a 100 amino acid protein has

$3 \times 3 \times 1{,}000 = 900$ conceivable single-mutant variants. (Three nucleotides encode each amino acid, and each can mutate to any of three alternate nucleotides. That only a subset of mutations changes the encoded amino acid [empirical aside 4.4] only modestly affects the matter.) And many genes are much longer than this.

With this model in hand, we begin as above and ask, given heterozygosity $H_t$ in generation $t$, what is $E[H_{t+1}]$, the expected heterozygosity in generation $t+1$? The answer can again be decomposed according to the parentage of two alleles under the Wright–Fisher model. As before, any two alleles will share a common parental allele with probability $\frac{1}{2N}$. Now writing $\mu$ for the per capita, per generation probability of mutation, these two alleles will differ in the offspring if either is mutated when copied, which happens with probability $2\mu$. (Strictly speaking, the probability that at least one copy is mutated is $2\mu + \mu^2$; $\mu^2$ is the probability that both copies were mutated. But as $\mu$ is generally much less than one—see empirical aside 1.4—this simplification introduces only negligible error.)

On the other hand, the two alleles may be the offspring of different parents (with probability $1 - \frac{1}{2N}$). In this case again, the two alleles will be different if the two parental alleles were already different (with probability $H_t$). But even if the parental alleles are the same (with probability $1 - H_t$), the copies can now differ if either copy was mutated when copied, again with approximate probability $2\mu$. Putting these ideas mathematically, we have the recurrence equation

$$E[H_{t+1}] = \frac{1}{2N} 2\mu + \left(1 - \frac{1}{2N}\right)[H_t + (1 - H_t)2\mu] \approx H_t\left[1 - \frac{1}{2N} - 2\mu\right] + 2\mu, \quad (2.31)$$

where the approximation follows if the compound parameter $\frac{\mu}{N} \ll 1$; that is, unless population size $N$ is very small.

The equilibrium between mutation and random genetic drift is found by setting $E[H_{t+1}] = H_t = \tilde{H}$ in equation (2.31), yielding

$$\tilde{H} = \frac{4N\mu}{1 + 4N\mu} = \frac{\theta}{1 + \theta}. \quad (2.32)$$

We introduce $\theta = 4N\mu$, which, as shown next, captures the balance between mutation and random genetic drift. For haploid populations, $\tilde{H} = \frac{2N\mu}{1 + 2N\mu}$, since $N$ haploids only carry $N$ alleles.

Expected heterozygosity in a population at mutation/drift equilibrium under the infinite alleles model (equation 2.32) is shown in figure 2.12, which illustrates a somewhat sharp transition from 0 to 1 as $\theta = 4N\mu$ grows: intermediate values are only observed when $\theta$ is roughly between 0.1 and 10. In English, what the figure tells us is that when the per locus mutation rate is much lower than the reciprocal of population size, random genetic drift dominates the process and genetically, the population is almost monomorphic. Conversely, when mutation rate is much higher

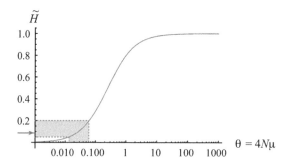

**Figure 2.12**
Expected heterozygosity at mutation/drift equilibrium under the Wright–Fisher and infinite alleles models as a
function of $\theta=4N\mu$. Solid line given by equation (2.32). Horizontal dashed lines indicate approximate range of
heterozygosity observed in 13 eukaryotes from nature (Lewontin 1974). Vertical dashed lines indicate range of
values of $\theta=4N\mu$ consistent with those data. Red arrow indicates maximum heterozygosity observed in a much
larger survey of eukaryotes (Leffler et al. 2012). See empirical aside 2.4 for more. Figure after Lewontin (1974).

than this threshold, almost no two alleles are identical. We observed a similar threshold behavior for a two-allele model in our diffusion treatment of mutation/drift equilibrium in section 2.2.5. Assuming symmetric mutation rates $\mu=\nu$, we found that
when $\theta$ is less than 1, probability mass in figure 2.10 is concentrated around $x=0$
and $x=1$, while when it is larger than 1, it is much closer to $x=0.5$.

As first noted by Lewontin (1974), equation (2.32) poses a serious challenge to the
hypothesis that random genetic drift is the dominant process driving allelic evolution. As shown in figure 2.12, heterozygosity is rarely greater than 20% empirically,
despite the fact that population sizes in nature surely range over several log orders.
See empirical aside 2.4 for more.

---

**Empirical Aside 2.4:** Lewontin's paradox

Lewontin's survey of published heterozygosity data (Lewontin 1974) showed that empirically, values of $H$ lie between roughly 5% and 20%. A much larger survey of 167 eukaryotes
lowered the upper bound to just 8% (Leffler et al. 2012; strictly speaking, these are nucleotide heterozygosities, but the difference is immaterial here, as explained immediately below
equation [2.42]). These observations represent a serious empirical challenge to the theory
leading to equation (2.32): while eukaryotic mutation rates are fairly constant (empirical
aside 1.4) population sizes $N$ in nature surely range over several log orders. Why then do
empirical heterozygosities fall in such a narrow range?

The resolution likely reflects several facts. For example, selection coefficients on deleterious mutations must have some probability distribution (empirical aside 1.3). If this distribution
is constant across species, the proportion with selection coefficients $0>s>-\dfrac{1}{N}$ will become
smaller as population size $N$ increases. Consequently, the effectively neutral mutation rate will
inherently be lower in larger population sizes, since fewer mutations will escape selective
elimination. This is the nearly neutral model, restated (empirical aside 2.3). Periodic *population bottlenecks* (i.e., contractions in population size) can also dramatically reduce heterozygosity (section 2.5). And in section 3.2, we will see that selection acting at one locus introduces
a population size-independent source of stochasticity at genetically linked, selectively neutral
loci. This, in turn, also predicts an upper bound on heterozygosity less than 1.

### 2.3.3  Heterozygosity at Migration/Drift Equilibrium under the Wright–Fisher Model

Another plausible mechanism for persistent, nonzero heterozygosity in natural populations is recurrent migration from elsewhere. Adopting an island model (section 1.4.1) and assuming infinite alleles, any immigrant allele will be different from all those already in the focal deme. Suppose demes each carry $N$ diploids with a per capita, per generation probability of replacement written $m$. Given heterozygosity $H_t$ in generation $t$ in a focal deme, we again decompose the parentage of two alleles to compute $\mathrm{E}[H_{t+1}]$. As before, two alleles will share a common parental allele with probability $\dfrac{1}{2N}$, in which case they are identical. Otherwise (with probability $1-\dfrac{1}{2N}$), they may be genetically distinct for one of two reasons. Either they were both already present in the previous generation and genetically distinct (with probability $H_t$) or at least one of them just immigrated (with probability $1-(1-m)^2 \approx 2m$ if $m \ll 1$), in which case it will be genetically distinct by our infinite alleles assumption. Putting these ideas mathematically yields another recurrence equation

$$\mathrm{E}[H_{t+1}] = \frac{1}{2N} \times 0 + \left(1 - \frac{1}{2N}\right)[H_t + (1 - H_t)2m],$$

and setting $\mathrm{E}[H_{t+1}] = H_t = \tilde{H}$ now yields

$$\tilde{H} = \frac{4Nm}{1 + 4Nm} \tag{2.33}$$

under the Wright–Fisher model, here assuming compound parameter $\dfrac{m}{N} \ll 1$ We note again the strong formal parallels between mutation and migration.

### 2.4  Coalescent Theory

We now turn to a somewhat different question: how does random genetic drift influence the structure of genetic lineages in a population? This is the domain of *coalescent theory*. (Interestingly, these ideas were originally developed to model the extinction of surnames in patronymic societies in late nineteenth century England under the name of the Galton–Watson process.) As we shall see in chapter 4, coalescent theory's ability to model the structure of mutations in genetic sequence data makes this framework a workhorse of modern population genetics.

Coalescent theory is concerned with the sampling of haploid alleles rather than of diploid genotypes. Two alleles are said to *coalesce* in the previous generation if they are copies of the same parental allele. (This is also sometimes called a *coalescent event*.) Thereafter (looking backward in time), their two lineages will be one. (In the language of section 1.4.4.2, their coalescence renders two alleles identical by descent.) By extension, the *genealogy* defined by some sample of $n$ alleles is a succession of $n-1$ coalescent events among lineages leading from the present time

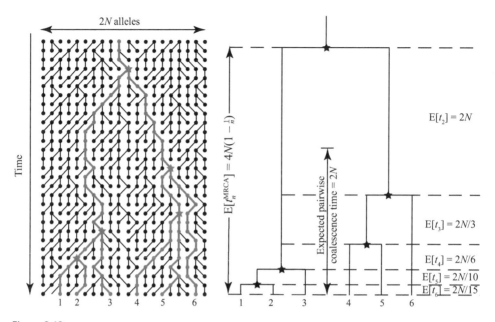

**Figure 2.13**
Sample genealogy defined by $n = 6$ physically distinct alleles in a population of $N$ diploids. Both panels assume nonoverlapping generations, although this is not an intrinsic feature of coalescence theory. Left: the sample's genealogy embedded in the population. Each circle is an allele, arranged in rows according to generation, from the past (top) to the present (bottom). Lines link alleles to their parental alleles. A sample of $n = 6$ alleles numbered at the bottom of the figure defines a genealogy (heavy red lines) that consists of a succession of $n - 1$ coalescent events (red stars), the last of which (working backward in time) defines the sample's MRCA. Right: only coalescent events are shown (stars), linked by *branches*, which represent the succession of parent/offspring pairs between coalescences. The expected time back to each successive coalescent event under the Wright–Fisher model (equation 2.38) is shown at right, as is the expected total time to the MRCA (equation 2.39) at left. The right panel is drawn to scale vertically; the left is not.

back to the sample's *most recent common ancestor* (or *MRCA*). These ideas are illustrated in the left panel of figure 2.13.

Following Hudson (1990), we partition coalescent theory into two subdomains: modeling a sample's genealogy (which we take up now) and modeling the resulting pattern of mutations found in that sample (section 2.4.2). Importantly, this dichotomy assumes mutations are selectively neutral, in which case we are entitled to model genealogies using the Wright–Fisher model. It breaks down otherwise, since selected mutations will themselves influence genealogies. We return to this point in chapter 3.

### 2.4.1   Modeling Genealogies under the Wright–Fisher Model

To begin, we ask a simple question: assuming the Wright–Fisher model of random genetic drift, how many generations ago did the lineages defined by two particular alleles in the present generation share a parent, or coalesce, in a population of $N$ diploids? It being a stochastic model, we can't say for sure, but we can find the expectation. We first have

$$\text{Probability}\left(\text{pairwise coalescence in the previous generation}\right) = \frac{1}{2N}. \quad (2.34)$$

Again, given the first allele's parental allele, this is the probability that the second has the same parental allele. Thus, the probability that a pair of lineages coalesced exactly $t+1$ generations ago is the probability of not coalescing for the first $t$ generations, multiplied by the probability that they do coalesce in the $t+1$st generation. Putting this mathematically, we have

Probability (pairwise coalescence exactly $t+1$ generations ago)

$$= \left(1 - \frac{1}{2N}\right)^t \frac{1}{2N} \approx \frac{1}{2N} e^{-\frac{t}{2N}}, \tag{2.35}$$

again approximating a geometric probability distribution with an exponential. The expected value of equation (2.35) is $2N$ generations. (We continue to suppress derivations from probability theory.) We now introduce $t_2$ for the time to coalesce between two lineages in a population. Thus, we have

$$\mathrm{E}[t_2] = 2N \tag{2.36}$$

under the Wright–Fisher model. Importantly, one can show that the variance in this estimate is $4N^2$ generations, a point to which we return in teachable moment 3.13.

---

**Power User Challenge 2.15**

What is $\mathrm{E}[t_2]$ in generations under the Moran model in a population of $N$ haploids? Your answer should conform to the point made in teachable moment 2.15. Hint: one of two things must be true for two random alleles to coalesce in the previous timestep. What are they, and what is the probability that either occurs?

---

**Teachable Moment 2.21**

The coalescent framework provides another path to the heterozygosity in a Wright–Fisher population at mutation/drift equilibrium (equation 2.32). But the infinite alleles assumption requires a slightly circuitous route: we shall find $\tilde{H}$ as one minus the probability that two alleles are identical. This captures the possibility that they differ by at least one mutation (see also teachable moment 2.4).

   The probability that two alleles are identical is equal to the probability that there were no mutations in their shared genealogy before they coalesced. Working backward in time, the per generation probability of mutation in the genealogy is $2\mu$ (since there are two lineages) and the per generation probability of coalescence is $\frac{1}{2N}$. Thus, the probability that two alleles are identical is the probability of coalescence before mutation, written as

$$\frac{\frac{1}{2N}}{\frac{1}{2N} + 2\mu} = \frac{1}{1 + 4N\mu}$$ (The denominator on the left is the probability of either event; the ratio

is the proportion of that total probability associated with coalescence.) The probability that they are different is thus again $1 - \frac{1}{1+4N\mu} = \frac{4N\mu}{1+4N\mu} = \frac{\theta}{1+\theta}$.

---

**Power User Challenge 2.16**

Following the same path used in teachable moment 2.21, derive expected heterozygosity in a population at migration/drift equilibrium (equation 2.33) in the coalescent framework assuming the infinite alleles model.

---

We can extend this reasoning to find the expected time to the first (i.e., most recent) coalescence among lineages defined by a sample of $n$ alleles. The probability that none of the $n$ coalesce in the previous generation reads $\left(1-\frac{1}{2N}\right)\left(1-\frac{2}{2N}\right)\cdots\left(1-\frac{n-1}{2N}\right)=\prod_{i=1}^{n-1}\left(1-\frac{i}{2N}\right)$. The first factor is again the probability that the first two alleles in the sample don't share a parental allele. The second is the probability that the third doesn't share a parental allele with either of the first two, and so on. (We make the simplifying assumption that $2N$ is large enough that at most one coalescent event happens in any one generation.) We approximate this as

Probability (no coalescence among $n$ lineages in the previous generation)

$$=\prod_{i=1}^{n-1}\left(1-\frac{i}{2N}\right)=1-\frac{1+2+\cdots+n-1}{2N}+O\left(\frac{n}{4N^2}\right)\approx 1-\frac{\binom{n}{2}}{2N}.$$

(See equation [2.13] to be reminded what $\binom{n}{2}$ means. The interested student may enjoy writing out and expanding the product for, say $n=4$ to see how it unfolds, and recall from teachable moment 2.8 that $O\left(\frac{n}{4N^2}\right)$ means that the error is of order no larger than $\frac{n}{4N^2}$. Since $n\leq 2N\ll 4N^2$, the approximation is extremely good.) Following the reasoning that yielded equation (2.35), we now find

Probability(first coalescence among $n$ lineages is

$$\text{exactly } t+1 \text{ generations ago})=\left(1-\frac{\binom{n}{2}}{2N}\right)^t\frac{\binom{n}{2}}{2N}\approx\frac{\binom{n}{2}}{2N}e^{-t\frac{\binom{n}{2}}{2N}}. \quad (2.37)$$

Thus, the expected time to the first coalescence among a sample of $n$ lineages is $E[t_n]=\frac{2N}{\binom{n}{2}}$ The only novelty compared to equation (2.35) is that we have expanded attention to coalescence among the $\binom{n}{2}$ pairs of lineages defined by our sample of $n$ alleles, shortening the expected time to the next coalescence by the factor $\binom{n}{2}$.

After the first coalescence, our sample will be represented by $n-1$ lineages (see figure 2.13), and by the same reasoning, the expected time to the second coalescence will be $\dfrac{2N}{\binom{n-1}{2}}$, and so on after each successive coalescence. Thus, we now write

$$E[t_i]=\frac{2N}{\binom{i}{2}} \tag{2.38}$$

for the expected number of generations during which the genealogy defined by $n$ alleles in the current generation carries $1 < i \le n$ distinct lineages. Not surprisingly, equation (2.38) recovers equation (2.36) when $i=2$. These expectations are shown in the right panel of figure 2.13.

---

**Power User Challenge 2.17**

The Wright–Fisher assumption of selective neutrality means that we can model the expected times to coalescent events, but not the identities of the next pair of lineages to coalesce. How many (equally likely) topologically distinct genealogies are possible for a sample of $n$ alleles? Hint: how many distinct pairs can coalesce first? Second?

---

Two other important statistics about coalescent times are now easily derived. First, the expected number of generations backward in time to the MRCA of a sample of $n$ alleles is

$$E\left[t_n^{\mathrm{MRCA}}\right]=\sum_{i=2}^{n}E[t_i]=2N\sum_{i=2}^{n}\frac{1}{\binom{i}{2}}=2N\sum_{i=2}^{n}\frac{2}{i(i-1)}$$

$$=4N\sum_{i=2}^{n}\left(\frac{1}{i-1}-\frac{1}{i}\right)=4N\left(1-\frac{1}{n}\right). \tag{2.39}$$

(The second-to-last step is a trick: work backward to see $\dfrac{1}{i(i-1)}=\dfrac{1}{i-1}-\dfrac{1}{i}$. This then then yields a "telescoping" sum that readily collapses, as can be seen by writing it out for some small $n$.)

Incidentally, setting $n=2N$ gives us $4N\left(1-\dfrac{1}{2N}\right)\approx 4N$ for the expected time to the most recent common ancestor of the entire population (Wakeley 2009), consistent with the expected time to fixation for a new neutral mutation, a result we presented without derivation in section 2.2.4.3. (Interestingly, these are subtly different quantities: the mutation that fixes must have occurred in the lineage leading to the MRCA sometime before $t_{2N}^{\mathrm{MRCA}}$).

Finally, the expected total number of generations on the entire genealogy is the sum of the $E[t_i]$, each now multiplied by $i$, the number of lineages present during that time interval. Mathematically, we have

$$E\left[t_n^{\mathrm{total}}\right]=\sum_{i=2}^{n}i\times E[t_i]=2N\sum_{i=2}^{n}\frac{i}{\binom{i}{2}}=2N\sum_{i=2}^{n}\frac{2i}{i(i-1)}$$

$$=4N\sum_{i=2}^{n}\frac{1}{i-1}=4N\sum_{i=1}^{n-1}\frac{1}{i}.$$

(2.40)

### 2.4.2   The Infinite Sites Model and the Site Frequency Spectrum

With expected coalescent times in hand, we now incorporate neutral mutations into our theory. We introduce the *infinite sites model* of mutation (Kimura 1969) to represent gene and genome sequence data. It assumes each mutation targets a never-before-mutated nucleotide site in a gene, a refinement of the infinite alleles model of section 2.3.2, which only focused on whether two alleles were different (recall the modest gymnastics of teachable moment 2.21). Like its predecessor, the infinite sites model is also biologically tenable. Critically, its condition need only hold since the most recent population-wide common ancestor. Thus, in expectation, the total number of generations during which mutation would have to target the same site twice is

$$E\left[t_N^{\mathrm{total}}\right]=4N\sum_{i=1}^{2N-1}\frac{1}{i}\approx 4N\times\ln(2N) \text{ generations under Wright–Fisher assumptions}$$

(equation 2.40; see teachable moment 2.22 for the approximation).

---

**Teachable Moment 2.22**

The approximation $\sum_{i=1}^{2N-1}\dfrac{1}{i}\approx\ln(2N)$ reflects the fact that $\displaystyle\int_{x=1}^{Z}\frac{dx}{x}=\ln(Z)$ as first noted in section 1.1.1, coupled with the very close connection between summation and integration. More specifically, integration is the sum of infinitely many infinitely small terms, meaning that this approximation becomes exact as $N\to\infty$. Lastly, we disregard the difference between $\ln(2N-1)$ and $\ln(2N)$ as, for example, was also done in our derivation of $t_{\mathrm{fix}}$ in chapter 1.

---

In cellular organisms, per nucleotide mutation rates are in the neighborhood of $10^{-8}-10^{-10}$ per generation (empirical aside 1.4). So even if population size $N=10^5$

(likely a large value as we shall see beginning in section 2.5) and $\mu = 10^{-8}$ (also likely large), the probability of two mutations at the same site ($\approx [4N \times \ln(N) \times \mu]^2$) is less than 0.25%. On the other hand, the infinite sites assumption is also sometimes violated, as for example in the case of HIV, both because of its mutation rate ($10^{-3}$ per site per generation) and because of its immense population size (perhaps $10^6$/ml peripheral blood). We return to this point in section 4.4.4.

We begin by finding the balance between mutation under the infinite sites model and random genetic drift, now quantified as per site equilibrium heterozygosity. Thus, in place of $H$ (now recognized as the per locus heterozygosity), we introduce $\pi$, the per site heterozygosity. Writing $\mu$ for the per generation, per site probability of mutation, we immediately have

$$\tilde{\pi} = 2\mu E[t_2]. \tag{2.41}$$

In English, the equilibrium pairwise per site heterozygosity is given by twice the per generation mutation rate (since mutation can occur on either lineage), multiplied by the expected number of generations since the two sites' coalescence. Now substituting $E[t_2]$ under the Wright–Fisher model (equation 2.36) yields

$$\tilde{\pi} = 4N\mu = \theta. \tag{2.42}$$

It is satisfying that the allelic equilibrium heterozygosity $\tilde{H} = \dfrac{\theta}{1+\theta}$ converges to the single-site analog as $\mu$ (and thus $\theta$) become smaller than 1. This simply reflects the fact that we have replaced the per allele mutation rate with its (much smaller) per site analog.

---

**Empirical Aside 2.5:** What are values of $\pi$ in nature?

Approximately 0.1% of homologous sites in a random pair of humans differ. This is roughly 10 times less than that in *Drosophila melanogaster*, and 25–400 times less than in many single-celled organisms (Lynch and Conery 2003). Interestingly, considerable population-level variation also exists. For example, in humans, $\pi$ is highest in individuals of African ancestry and lowest in those of Native American ancestry (Rosenberg et al. 2002). It thus seems reasonable to expect similar variability among populations in other species. As per equation (2.42), we might commonly regard differences in $\pi$ as reflecting differences in population size. But see also empirical aside 2.4.

---

Another important result that follows immediately from the infinite sites assumption is the number of variable sites expected in a sample of $n$ alleles, written $S_n$. This is simply the total time on the sample's genealogy (equation 2.40) multiplied by the per site, per generation mutation rate. Mathematically, this reads

$$E[S_n] = \mu E\left[t_n^{\text{total}}\right] = 4N\mu \sum_{i=1}^{n-1} \frac{1}{i} = \theta \sum_{i=1}^{n-1} \frac{1}{i}. \tag{2.43}$$

Going further, imagine a sample of $n$ alleles that are polymorphic at some site. That site partitions the sample into two fractions: $1 \leq j \leq n-1$ of them will carry the mutant or *derived nucleotide* while the remaining $n-j$ carry the wild type or *ancestral nucleotide*. We now define the *site frequency spectrum* (or *SFS*) as the proportion of polymorphic sites that partitions a sample of $n$ alleles into fractions of size $j$ and $n-j$ for each $1 \leq j \leq n-1$.

To find the expected site frequency spectrum under Wright–Fisher and infinite sites assumptions, we introduce $\text{SFS}_n^j$ for the probability that the derived nucleotide at a site appears $j$ times in a sample of $n$ alleles. (As for any probability distribution, we have $\sum_{j=1}^{n-1} \text{SFS}_n^j = 1$.) Next, note that every branch in the genealogy of $n$ alleles can be characterized by an integer $1 \leq j \leq n-1$, corresponding to the number of alleles in the sample that are derived from that branch (see, for example, the numbers adjacent to branches in figure 2.14, left). Having assumed the infinite sites model, a mutation occurring on a branch leading to exactly $j$ alleles will be present in exactly those alleles and in no others (see orange arrow and asterisks in figure 2.14, left). Thus, the probability of finding a derived mutation in exactly $j$ of $n$ alleles in a sample is equal to the expected proportion of the total branch lengths in its genealogy that carries the number $j$. Now writing $\text{E}[t_n^j]$ for the expected total length in generations of all branches characterized by integer $j$ in a genealogy of size $n$, we will shortly prove inductively that under the Wright–Fisher model

$$\text{E}[t_n^j] = \frac{4N}{j} \tag{2.44}$$

in a population of $2N$ haploids. Finally, recalling that the expected total branch length in the genealogy for a sample of $n$ alleles is $\text{E}\left[t_n^{\text{total}}\right] = 4N \sum_{i=1}^{n-1} \frac{1}{i}$ (equation 2.40), we have

$$\text{SFS}_n^j = \frac{\text{E}\left[t_n^j\right]}{\text{E}\left[t_n^{\text{total}}\right]} = \frac{\dfrac{4N}{j}}{4N \sum_{i=1}^{n-1} \frac{1}{i}} = \frac{\dfrac{1}{j}}{\sum_{i=1}^{n-1} \frac{1}{i}}. \tag{2.45}$$

---

**Teachable Moment 2.23:** Proofs by induction

Inductive proofs always have two steps. The first is to prove the statement for some base case (here, when sample size $n=2$). We then assume that the statement is true for some arbitrary case $n$, and prove that on that premise, it is true for case $n+1$. Together with the base case, this inductive step yields a telescoping proof for all $n$. We immediately know it is true for the first case larger than the base case, which, in turn, proves that it is true for the second case larger than the base case, and so on.

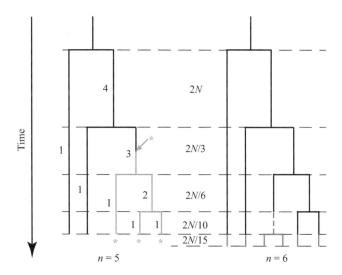

**Figure 2.14**
Deriving the SFS under Wright–Fisher assumptions. Left: a possible genealogy defined by a sample of $n = 5$ alleles. Numbers to left of each branch give the number $j$ of alleles in the sample descended from that branch. Under the infinite sites model, a mutation that occurs on a branch labeled $j$ will be present in exactly $j$ alleles in the sample and no other. For example, the orange arrow represents a mutation on a branch labeled $j = 3$ that is faithfully transmitted to exactly three alleles in the sample: those labeled with orange asterisks. Thus, the probability that a site partitions the sample into $j$ derived and $n - j$ ancestral alleles is proportional to the length in generations of branches labeled $j$. Right: illustration of inductive proof (teachable moment 2.23) that $E[t_{n+1}^{j=1}] = 4N$ for the case $n = 5$, drawn after figure 1 in Hudson (2015). The interested reader is directed to that paper for a proof for arbitrary $j$. The base case in this proof is for a sample size $n = 2$ alleles. In this case, the mutation is necessarily a singleton (mathematically, recall that $1 \leq j \leq n - 1$), so $\mathrm{SFS}_2^1$ must be 1. Setting equation (2.45) equal to 1 and solving for $E[t_n^{j=1}]$ (here, the sum in the denominator is 1) yields $4N$, as required. We now assume that $E[t_n^{j=1}] = 4N$, and seek to show that $E[t_{n+1}^{j=1}]$ has the same value. First, in adding an $n + 1$st allele to the sample, we add $(n+1)E[t_{n+1}] = (n+1)\dfrac{2N}{\binom{n+1}{2}} = \dfrac{4N}{n}$ to the expected total length of singleton branches (solid red lines). Why? Because there are now $n + 1$ singleton branches, and $E[t_{n+1}]$ (given in equation 2.38) is the time back to the new, now-first coalescence. But at the same time, a portion of what had been one of the singleton branches before we added the new sample no longer contributes to $E[t_{n+1}^{j=1}]$. Instead, it now leads to the new coalescence and thus to two alleles in the sample (dashed red line). While we don't know which singleton branch is affected when we add the $n + 1$st allele to the sample (see power user challenge 2.17), we know that the expected length to be subtracted is $\dfrac{4N}{n}$. Why? By our assumption that $E\left[t_n^{j=1}\right] = 4N$, the total length of singleton branches for a sample of $n$ alleles (i.e., before adding the $n + 1$st) is $4N$, but only one of the $n$ such branches is involved. Putting things all together, we have $E\left[t_{n+1}^{j=1}\right] = E\left[t_n^{j=1}\right] + \dfrac{4N}{n} - \dfrac{4N}{n} = 4N$.

Q.E.D. (See teachable moment 1.14.)

The SFS for a sample of $n = 10$ alleles under the Wright–Fisher and infinite sites assumptions (equation 2.45) is shown in figure 2.15. By comparing a site frequency spectrum observed in nature with that expected by our theory, we can test the hypothesis that a population honors our assumptions. We return to and expand on this idea in chapter 4.

Before leaving the theory of site frequency spectra, we note one final refinement that is sometimes required. Consider a polymorphic site that partitions $n$ alleles into

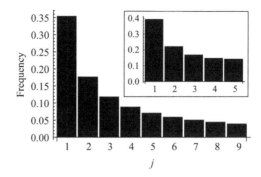

**Figure 2.15**
The expected SFS for a sample of 10 alleles under Wright–Fisher assumptions. The $x$-axis is $j$, the number of alleles in which the derived mutation is found, and the $y$-axis is the probability that the derived allele will be seen in that number of alleles (equation 2.45). Inset: folded SFS for the same sample size (see text for details).

subsets of size $j$ and $n-j$. Does that site then contribute to the $j$ class or the $n-j$ class in the SFS? Putting things slightly differently, which nucleotide at the site is derived, and which is ancestral? Extrinsic information about the allele in the MRCA of our sample often allows us to *polarize* mutations; that is, determine which is which. (Such information can be found in sequence data from some closely related but distinct population or species.) However, absent the ability to label the ancestral and derived nucleotides, we can only compute the so-called *folded SFS* from the data. Here, each site's frequency is tabulated as the minimum of $j$ and $n-j$. In this case, data will then be compared with the expected, neutral, folded SFS, which is shown for a sample of $n=10$ alleles in the inset of figure 2.15.

### 2.4.3   Population Subdivision and Incomplete Lineage Sorting
Genealogies describe the history of lineages corresponding to some sample of alleles. To this point, we have assumed that these alleles come from a single, well-mixed population. Correspondingly, the expectations we have developed for their genealogies have only incorporated the influence of random genetic drift. But population subdivision represents another important process with direct implications for the structure of genealogies. Coalescence between lineages in different demes is impossible in the absence of gene flow, and even in its presence, probabilities of coalescence must somehow be less than in the models developed thus far.

Power user challenge 2.13 explores the equilibrium balance between migration and drift in the diffusion framework, and we will consider the same question with the coalescent theory at the end of section 2.5. But for the moment, we investigate the transient process of isolation between demes in a newly subdivided population. To be more concrete, imagine a population that fully honors Wright–Fisher assumptions. Now suppose that at some moment in time, the population is split into two demes with no migration between them. At that moment, all alleles in both demes share a single MRCA, and allelic genealogies will be as described above. Conversely, after a "very" long time, all alleles in each deme will trace their ancestry to two, deme-specific MRCAs. This state is called *reciprocal monophyly*, and at this point, genealogies within demes will again

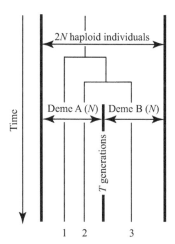

**Figure 2.16**
Incomplete lineage sorting. A population of $2N$ alleles is subdivided into two demes of $N$ alleles each $T$ generations in the past. Heavy vertical lines represent population boundaries. ILS reflects the lag between population
subdivision and partitioning of the genealogy of a sample. In this case, individuals 1 and 2 are sampled from
Deme A and individual 3 from Deme B, yet 2 and 3 coalesce before 1 and 2.

be as described above. But between these times, alleles in one deme may share a MRCA
with alleles in the other that is more recent than their own deme's MRCA. This situation, called *incomplete lineage sorting* (or *ILS*), is illustrated in figure 2.16. The name
reflects the lag in time between populations subdivision and the corresponding partitioning of their constituent allelic genealogies.

We can use coalescent theory to compute the probability of ILS. Imagine a population of $2N$ alleles that was partitioned into demes A and B, each of size $N$, $T$ generations in the past. Now suppose we sample two alleles from deme A and one from
deme B. What is the probability that one of the two alleles from deme A coalesces
with the allele from deme B more recently than it does with the other allele from its
own deme? ILS first requires that the two lineages in deme A fail to coalesce during
those $T$ generations. Since the per generation probability of pairwise coalescence in a
deme of $N$ haploid individuals is $\left(1-\dfrac{1}{N}\right)$, two lineages will fail to coalesce for $T$
generations with probability $\left(1-\dfrac{1}{N}\right)^{T}\approx e^{-\frac{T}{N}}$. In this event, all three lineages will
have been present the original population (before subdivision): two from deme A and
one from deme B. ILS further requires that the first coalescence (working backward
in time as always) be between one of the two lineages from deme A and the one deme
B lineage. This happens with probability $\dfrac{2}{3}$, since it corresponds to two of the three
possible first coalescence (see also power user challenge 2.17). Putting these two
ideas together, we have

$$\Pr(\text{ILS})\approx\frac{2}{3}e^{-\frac{T}{N}}. \tag{2.46}$$

In English, the probability of ILS depends on the compound parameter $\frac{T}{N}$, the number of generations since population subdivision, expressed in units of demic population size. This is easily understood: the expected time to coalescence is proportional to effective population size, so as that quantity goes up, the per generation probability of within-deme coalescence does too.

## 2.5   The Effective Size of a Population

Under both the Wright–Fisher and Moran models of random genetic drift, the strength of stochastic effects scales inversely with population size $N$. To this point, $N$ has been the number of individuals in the population, or its *census size*. A population's *demography* is its size and structure, and both models assume a well-mixed, constant-sized population. What happens to the strength of random genetic drift under other demographic assumptions? As we shall see, it usually becomes stronger. To be more quantitative, we define any demographic model's *effective population size* ($N_e$) as the census size of a Wright–Fisher population in which the strength of drift is equal to that in the given model. What we are saying, mathematically, is that usually, $N_e < N$.

To illustrate, imagine a population whose census size in each generation is $N_1$ with probability $p$ and $N_2$ with probability $1-p$, but which otherwise honors all Wright–Fisher assumptions. First, recall that in the context of the diffusion approximation, random genetic drift is quantified by $V(x)$, the per generation variance in allele frequency change. Under the Wright–Fisher model, this is $\frac{x(1-x)}{2N_e}$ (equation [2.15]; by definition, $N_e = N$ in a Wright–Fisher population). In our population the per generation variance in allele frequency change will with probability $p$ be $\frac{x(1-x)}{2N_1}$, and with probability $1-p$ it will be $\frac{x(1-x)}{2N_2}$. Thus, we have

$$V(x) = \left( \frac{x(1-x)}{2} \right) \left( \frac{p}{N_1} + \frac{1-p}{N_2} \right)$$ for this population. Equating this to the Wright–

Fisher analog and solving, we find its $N_e = \dfrac{1}{\dfrac{p}{N_1} + \dfrac{(1-p)}{N_2}}$. This result generalizes to

any probability distribution of population census sizes $p_i$:

$$N_e = \frac{1}{\sum_i \dfrac{p_i}{N_i}}. \tag{2.47}$$

Equation (2.47) tells us that the effective population size is the *harmonic mean* of population sizes averaged over time. Importantly, even very occasional population bottlenecks have a much stronger influence on the harmonic mean than they do on the more familiar arithmetic mean (given by equation 1.12). For example, suppose that $N_1 = 100$, $N_2 = 1,000,000$, and $p = 0.01$. While the arithmetic mean is $0.01 \times 100 + 0.99 \times 1,000,000 = 990,001$, the harmonic mean is

$$\frac{1}{\dfrac{0.01}{100} + \dfrac{0.99}{1,000,000}} \approx 9,902.$$ This much sharper dependence on small values in

equation (2.47) might not surprise us: the strength of drift goes as the reciprocal of population size, so its influence grows much more quickly than linearly as it gets small.

As noted in empirical aside 2.4, equation (2.47) suggests a resolution to Lewontin's paradox. Since effective population size is disproportionately sensitive to even rare population bottlenecks, such events could mask the impact of larger differences in census sizes on heterozygosity.

We next compute the rate at which allelic heterozygosity decays under this same demographic model. Its analog to the Wright–Fisher expression (equation 2.28) is

$$E[H_t] = \left[ \left(1 - \frac{1}{2N_1}\right)^{pt} \left(1 - \frac{1}{2N_2}\right)^{(1-p)t} \right] H_0.$$

Setting this equal to the Wright–Fisher result (equation 2.28) and solving yields

$$N_e = \frac{1}{\dfrac{p}{N_1} + \dfrac{(1-p)}{N_2}} + O\left(\frac{1}{N_i^2}\right).$$ Thus, the effective population size for this model is

again (approximately) the harmonic mean of population sizes averaged over time. (Perhaps not surprisingly, allelic heterozygosity at mutation/drift equilibrium is

now $\tilde{H} = \dfrac{4N_e\mu}{1 + 4N_e\mu}$.) Finally, the expected time to coalescence between two alleles

in this model is also easily found. In any generation, the probability of coalescence

now reads $\dfrac{p}{2N_1} + \dfrac{1-p}{2N_1}$, which is equal to the Wright–Fisher result ($\dfrac{1}{2N_e}$, equation

2.34) when $N_e = \dfrac{1}{\dfrac{p}{N_1} + \dfrac{1-p}{N_2}}$.

---

**Power User Challenge 2.18**

Prove that under this demographic model, $N_e < N$ unless $N$ is invariant using a geometric approach analogous to that in figure 1.11.

The foregoing illustrates the approach to finding the effective population size for any demographic model. First, write one of the equations that captures the strength of random genetic drift under the Wright–Fisher model in terms of $N_e$. (This is the definition of $N_e$.) Then, setting this expression equal to the analog computed under the model of interest, solve for $N_e$. It also illustrates a modest complexity: we have developed three expressions for the strength of random genetic drift under the Wright–Fisher model (equations 2.15, 2.28, and 2.35), with which we can define three effective population sizes. These are called the *variance, inbreeding,* and *coalescent effective population sizes*, respectively. Fortunately, these rarely differ greatly. (We direct the interested reader to Crow and Kimura [1970], Caballero [1994], and Wang [2005] to learn more. Indeed, a fourth, the *eigenvalue effective population size*, is also found in the literature.) Because the coalescent framework together with the infinite sites model lends itself extremely well to genetic sequence data, we follow much of the literature and focus exclusively on the coalescent effective population size for the remainder of the book.

We conclude by computing the coalescent effective population size for two other demographic models. First, imagine a diploid, dioecious population, meaning each offspring is the product of a mating between two parents of different sexes (see empirical aside 1.10). Call these male and female, and suppose further that there are $N_m$ males and $N_f$ females (where $N_m + N_f = N$). We first seek the probability that two alleles chosen at random coalesce in the previous generation. To coalesce, the two alleles must have come from parents of the same sex. With $\frac{1}{4}$ probability, their parental alleles were both in a female, in which case, with probability $\frac{1}{2N_f}$ they will be copies of the same parental allele. Similarly, with $\frac{1}{4}$ probability, their parental alleles were both in a male, and will with probability $\frac{1}{2N_m}$ be copies of the same parental allele. Thus, in this model, the (coalescent) effective population size is the solution to

$$\frac{1}{2N_e} = \frac{1}{4}\left( \frac{1}{2N_f} + \frac{1}{2N_m} \right),$$

which reads

$$N_e = \frac{4N_f N_m}{N_f + N_m}.$$

When there are an equal number of males and females, dioecy has no influence on effective population size. (Mathematically, if $N_f = N_m$, then $N_e = \frac{4N_f^2}{2N_f} = 2N_f = N_f + N_m = N$.) But in general, unequal sex ratio reduces effective population size. (This is also immediately obvious in the solution to power user challenge 2.18.)

This result illustrates a general principle: anything that increases the variance in reproductive success among organisms reduces the effective population size. We can amplify this point with reference to variance effective population size. Recall that for haploids, $V(x) = \dfrac{x(1-x)}{N_e}$ under the Wright–Fisher model. Now writing $\sigma^2$ for the variance in offspring number, one can show that $V(x) = \dfrac{x(1-x)\sigma^2}{N}$ (see §2.8 in Gillespie [2004]), or equivalently, that $N_e = \dfrac{N}{\sigma^2}$.

Finally, we consider the influence of population subdivision on coalescent effective population size. Perhaps not surprisingly, subdivision can cause effective population size to exceed the census size, since coalescence of two lineages currently in different demes is impossible. More specifically, imagine a population of $N$ diploid individuals in two, equal-sized demes, each of which honors all Wright–Fisher assumptions, and between which there is a symmetric, per capita, per generation probability of migration written as $m$. Following the same path developed in our treatment of coalescent times (section 2.4.1), we quickly find

Probability (given lineage migrated exactly $t$ generations ago)
$$m(1-m)^t \approx m e^{-mt}$$

and thus that

$$\mathrm{E}\big[\text{ time since last migration on a given lineage }\big] = \frac{1}{m}\,\text{generations.}\qquad(2.48)$$

Following Nielsen and Slatkin (2013), we examine two cases to find the expected time to coalescence between two alleles sampled at random in this subdivided population: the two alleles can have been sampled from the same deme or from different demes. We thus introduce $E\big[t_2^{\mathrm{D}}\big]$ and $E\big[t_2^{\mathrm{S}}\big]$ to represent the expected time to coalescence between two alleles sampled in different (D) or the same (S) demes, respectively. Two alleles sampled from different demes cannot coalesce until one of their lineages migrates, after which their coalescent time will be captured by our expression for two alleles in the same deme. Mathematically, this yields

$$\mathrm{E}\big[t_2^{\mathrm{D}}\big] = \frac{1}{2m} + \mathrm{E}\big[t_2^{\mathrm{S}}\big].$$

The time until migration is half the value given by equation (2.48), since either lineage may migrate into the other deme.

The situation for alleles sampled from the same deme is a bit more complicated. Looking backward in time, the two lineages will persist in the deme until one of two things happens: coalescence or migration. The expected times to these are $\mathrm{E}[t_2] = N$ (because each deme is of size $\dfrac{N}{2}$) and $\dfrac{1}{2m}$ generations, respectively. Thus, the probabilities of coalescence and migration are given by the relative rates of the two processes: $\dfrac{\dfrac{1}{N}}{\dfrac{1}{N}+2m}=\dfrac{1}{1+2Nm}$ and $\dfrac{2Nm}{1+2Nm}$, respectively. (See teachable moment 2.21.) Putting these ideas together, we have

$$E\left[t_2^S\right]=\frac{1}{1+2Nm}N+\frac{2Nm}{1+2Nm}E\left[t_2^D\right].$$

Substituting our previous expression for $E\left[t_2^D\right]$ into this last expression and simplifying, we find

$$E\left[t_2^S\right]=2N$$

and thus that

$$E\left[t_2^D\right]=\frac{1}{2m}+2N.$$

Finally, since the two demes are of equal size, a random pair of alleles are equally likely to be sampled from the same deme as from different demes, yielding

$$E\left[t_2^{\text{subdivided}}\right]=\frac{1}{2}E\left[t_2^s\right]+\frac{1}{2}E\left[t_2^D\right]=2N+\frac{1}{4m}.$$

We now proceed as above: set $E\left[t_2^{\text{subdivided}}\right]=E[t_2]$ for a Wright–Fisher population whose size is that of the total population (equation 2.36), which yields

$$N_e=N+\frac{1}{8m}=N\left[1+\frac{1}{8Nm}\right]$$

for the coalescent effective population size in this model. As expected, subdivision causes effective population size to exceed census size, but the threshold for this effect (roughly $Nm=1$) is consistent with what was seen via the diffusion approach in power user challenge 2.13.

## 2.6  Chapter Summary

- Stochastic population genetic models capture random fluctuations in the number of offspring born to genetically identical parents. There are two forms of stochasticity: beneficial alleles with fitness advantage $s$ are at risk of loss until their establishment, and all alleles are subject to a second, population-size dependent form of stochasticity called random genetic drift.

- The Chapman–Kolmogorov equation captures the full stochastic complexity of random genetic drift, but analytic and computational progress is difficult. Diffusion approximations can yield further progress.

- The Wright–Fisher model of random genetic drift assumes nonoverlapping generations and a well-mixed population. Together with the infinite alleles model of mutation, we can compute the equilibrium between the loss of selectively neutral alleles via random genetic drift and their introduction via new mutations.

- The coalescent framework allows us to model the genealogical history of a sample of alleles. Together with the infinite sites model of mutation, we can compute the equilibrium between the loss of selectively neutral mutations via random genetic drift and their introduction via new mutations.

- Results for selectively neutral alleles and mutations developed under the Wright–Fisher framework can be applied to arbitrary demographic models after rescaling population size so that the strength of random genetic drift is matched. This rescaled population size is called the model's effective population size.

# 3

## Multilocus Population Genetics

In the first two chapters of this book, we explored how the genetic composition of a population evolves on the premise that organisms carry just one locus. We now relax that assumption, and imagine organisms that cotransmit ensembles of alleles at two or more loci to their offspring. The ensemble of alleles found in a multilocus organism is called its *multilocus genotype*. Critically, because survival and reproduction occur at the level of whole organisms, the evolutionary fate of an allele at one locus will often be closely correlated to that of other loci.

Before beginning, we make two comments. First, from an applied perspective, the correlation in evolutionary fates among alleles in the genome make this content among the most important in the field. Phenotypic effects of an allele at one locus can give rise to population genetic signals at other loci. In chapter 4, we will invert this reasoning to see how data can be used to physically localize the positions of loci carrying alleles of biological interest in the genome. Second, the attentive reader may find that this chapter is less thoroughly developed than were the first two. This is correct: a literally exponential explosion in model complexity emerges in multilocus theory, requiring a corresponding reduction in the completeness of our models. As a result, we will now be forced to rely more on inductive generalizations from unrealistically simplified models, something already anticipated in teachable moment 1.1.

Multilocus genotypes generalize the one-locus case employed in our treatment of This and the next request sharpen meaning for the reader sexual reproduction in chapter 1, and the two complications that emerged there (segregation and dominance) are also central to this chapter. First, even in the absence of mutation, multilocus offspring can again differ genetically from their parents, now via a genetic process called recombination. Second, multilocus organismal fitness can be influenced by interactions among loci; this is called epistasis.

We frame the first of these complications in terms of sexually reproducing organisms (although see empirical aside 3.1), whose life cycle consists of diploid and haploid phases (figure 1.8). To review, the diploid phase begins at syngamy: the fusion of two haploid gametes to form a diploid zygote. The haploid phase begins at meiosis: the division of a diploid cell to form two haploid gametes. These go on to form diploid

offspring in the next generation. In chapter 1, we considered the evolutionary consequences of meiotic segregation at one locus: the transmission of just one of the two
alleles from the diploid into the haploid. In multilocus theory, it is whole chromosomes that segregate, but we now must also account for a second meiotic process:
*genetic recombination* between loci. Each haploid gamete receives an allele at each
locus from the diploid, and genetic recombination means that the particular allelic
ensemble found in successive generations of gametes can change. For example, imagine two loci with alleles $A$ and $a$, and $B$ and $b$, respectively. Gametic fusion of $Ab$ and
$aB$ haploid gametes will produce an $\dfrac{Ab}{aB}$ diploid zygote. At meiosis, this diploid will
cotransmit the alleles as inherited to its gamete unless recombination separates
them. Gametes in which alleles are coupled as they were inherited by the diploid
(here, as $Ab$ and $aB$ gametes) are called *nonrecombinants*; when the inherited cou-
pling is disrupted (here, pairing $AB$ and $ab$ alleles), the resulting gametes are called
*recombinants*. We now require a population genetic model that captures this
possibility.

---

**Empirical Aside 3.1:** Genetic recombination is not restricted to sexual diploids

All sexually reproducing organisms are eukaryotes. Some of these (including most animals) are *obligately sexual*: sexual reproduction (and usually also sexual recombination)
occurs in each generation. Others are *facultatively sexual*, meaning that, while capable of
sexual reproduction, they can also reproduce clonally. In addition, genetic recombination is
possible in many noneukaryotic organisms, but is almost always decoupled from reproduction. For example, it appears that most microbes are haploids capable of recombining portions of their genomes in a reproduction-independent manner (e.g., via horizontal genetic
transfer in bacteria). Models of reproduction-independent recombination differ from others
only in how recombination rates are quantified. Thus, because qualitative conclusions are
unchanged, this book will equate sexual and genetic recombination and focus only on the
former.

---

The second population genetic complexity that emerges in multilocus models is
that organismal fitness can now reflect interactions among alleles at different loci.
Recall from chapter 1 that allelic fitness interaction at a single diploid locus is
called dominance: the fitness effect of the $A \rightarrow a$ mutation can differ in homozygotes and heterozygotes. Mathematically, dominance means that $\dfrac{w_{AA}}{w_{Aa}}$ need not
equal $\dfrac{w_{Aa}}{w_{aa}}$. The analog in multilocus models is allelic interaction among loci, called
*epistasis*. Epistasis means that the fitness effect the $A \rightarrow a$ mutation can differ
depending on the allelic identity at the $B/b$ locus. Mathematically, epistasis means
that $\dfrac{w_{AB}}{w_{aB}}$ need not equal $\dfrac{w_{Ab}}{w_{ab}}$, again demanding new theory.

> **Teachable Moment 3.1:** Haploid or diploid models?
>
> We began chapter 1 with selection in haploids because it is simpler mathematically. We then saw that diploid and haploid models are equivalent assuming Mendelian segregation, random mating, and no dominance ($h=\frac{1}{2}$). Theory developed in chapter 2 maintained those three assumptions; as noted, in the Wright–Fisher model, stochastic consequences in a diploid population of $N$ individuals are identical to those in a haploid population of size $2N$. We continue to maintain those three assumptions in this chapter (though see power user challenge 3.1), again allowing us to only track haploid genotype (or *haplotype*) frequencies. As a consequence, our theory will be equally applicable to all multilocus organisms, be they sexual diploids or not. (Readers interested in the joint effects of dominance and epistasis are directed to Crow and Kimura [1970].)

As we did for our one-locus models, we begin by developing deterministic multilocus theory. The influence of stochastic sampling in multilocus models will be explored in section 3.2, leading us to the Hill–Robertson effect. Conceptually, this is perhaps the most important lesson in the chapter: as we shall see, recombination almost always makes selection more efficient in the presence of random genetic drift. This may explain the fact that genetic recombination is so widespread in nature, a point we take up (and then generalize) in the last section of this chapter.

### 3.1 Deterministic Multilocus Theory

As we did in section 1.3, we simplify by assuming nonoverlapping generations, allowing us to model syngamy and meiosis as occurring synchronously in the population. As noted in teachable moment 3.1, by assuming random mating, Mendelian segregation, and that all dominance coefficients are $h=\frac{1}{2}$, we can focus only on haplotype frequencies.

#### 3.1.1 Pairwise Linkage Disequilibrium and Genetic Recombination

We begin by developing the mathematical connection between haplotype frequencies and allele frequencies in recombining populations. Consider a population of two-locus, haploid organisms with alleles $A$ and $a$ at the first locus, and $B$ and $b$ at the second. These define four possible haplotypes: $AB$, $Ab$, $aB$, and $ab$. Call their frequencies $p_{AB}$, $p_{Ab}$, $p_{aB}$, and $p_{ab}$. (These of course sum to unity.) We can immediately compute allele frequencies from haplotype frequencies as

$$p_A = p_{AB} + p_{Ab}$$
$$p_B = p_{AB} + p_{aB}. \tag{3.1}$$

In general, however, the reverse calculation is underspecified: we lack information about the statistical associations between alleles across loci in the population. For example, allele frequencies $p_A = p_B = \frac{1}{2}$ are consistent with haplotype frequencies

$p_{AB} = p_{ab} = \dfrac{1}{2}$ and $p_{Ab} = p_{aB} = 0$, in which case the $A$ and $B$ alleles are perfectly associated. But those same allele frequencies are also consistent with $p_{AB} = p_{Ab} = p_{aB} = \dfrac{1}{4}$, in which case the $A$ and $B$ alleles are randomly associated. There are uncountably many other haplotype frequencies consistent with these same allele frequencies.

We call the statistical association between alleles at two loci in a population their *linkage disequilibrium*. The word "linkage" refers to statistically unexpected allelic cooccurrence in the population. Genetic recombination slowly randomizes (or equilibrates) allelic associations in a population, and the word "disequilibrium" refers to the possibility that linkage can be present in spite of this effect. Importantly, because each recombination event yields two recombinant and two nonrecombinant products, linkage disequilibrium can also be observed between loci with no genetic linkage, as for example, between loci on different chromosomes.

The coefficient of linkage disequilibrium $D$ quantitatively captures the discrepancy between observed haplotype frequencies and those predicted in the absence of statistical association. For example, focusing, on the $AB$ haplotype, we can define it as

$$D = p_{AB} - p_A p_B.$$

Substituting equation (3.1) and simplifying (recall that $p_{AB} + p_{Ab} + p_{aB} = 1 - p_{ab}$), yields

$$D = p_{AB} p_{ab} - p_{Ab} p_{aB}. \tag{3.2}$$

This, in turn, yields

$$
\begin{aligned}
p_{AB} &= p_A p_B + D, \\
p_{Ab} &= p_A p_b - D, \\
p_{aB} &= p_a p_B - D, \text{ and} \\
p_{ab} &= p_a p_b + D.
\end{aligned}
\tag{3.3}
$$

Thus, the coefficient of linkage disequilibrium $D$ allows us to uniquely specify haplotype frequencies from allele frequencies. For example, in the two examples offered in the paragraph before last, $D = \dfrac{1}{4}$ and 0, respectively. (The attentive reader may wonder about our seemingly arbitrary choice to define $D$ in terms of the $AB$ haplotype. Starting from the $ab$ haplotype also brings us to equation (3.2), while $D$ derived in terms of the $Ab$ or $aB$ haplotype is of equal magnitude but of opposite sign.)

---

**Teachable Moment 3.2:** Counting degrees of freedom provides another way of understanding why allele frequencies alone underspecify haplotype frequencies

The number of *degrees of freedom* in a system is equal to the number of quantities in the system that can vary independently. In this case, there are four haplotypes, but since their frequencies must sum to unity, there are three degrees of freedom. Converting haplotype frequencies to allele frequencies capture two of the degrees of freedom, and linkage disequilibrium captures the third.

We encountered the same situation in section 1.3.1: three one-locus, diploid genotype frequencies again sum to 1, leaving two degrees of freedom, which can equivalently be described by one allele frequency and a fixation index.

**Teachable Moment 3.3:** Other measures of linkage disequilibrium

Although $D$ captures the last degree of freedom required to compute two-locus haplotype frequencies, its dynamic range varies with allele frequencies. Specifically

$$-\mathrm{Min}[p_A p_B, (1-p_A)(1-p_B)] \le D \le \mathrm{Min}[p_A(1-p_B), (1-p_A)p_B].$$

These bounds define $p_A \times D$-space, as shown in figure 3.1.

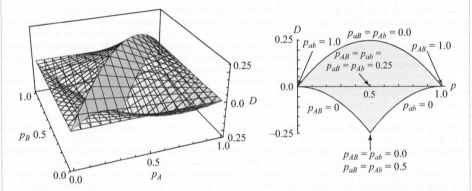

**Figure 3.1**
Two locus $p_A \times D$-space represents a population's genetic configuration as a function of two allele frequencies (bounded by 0 and 1) and linkage disequilibria (bounded by the above expression). Left: orange plane is a transect through the space assuming $p_A = p_B$. Right: this transect, together with various landmark genotypic frequencies. (Figure from Weinreich, Sindi, et al. 2013. © SISSA Medialab. Reproduced from the *Journal of Statistical Mechanics* 2013: P01001 with permission. All rights reserved.)

Two approaches are commonly employed to resolve this issue. Some authors use Lewontin's $D' = \dfrac{|D|}{D_{\max}}$, where $D_{\max}$ is given by the bounds in the previous equation, and others use Pearson's correlation coefficient between alleles, namely $r_P = \dfrac{D}{\sqrt{p_A p_B(1-p_A)(1-p_B)}}$. (We employ the subscript "P" for Pearson, to distinguish this quantity from the recombination rate $r$ introduced below. We return to this measure in section 3.2.1.2 and again in chapter 4.)

Finally, note that higher-order linkage disequilibria emerge in models with more than two loci. See teachable moment 3.11 to learn more.

We now model recombination between two loci, writing $r$ for the recombination rate (or *genetic map distance*; see teachable moment 3.4) between two loci; that is, the per meiosis probability of recombination between cotransmitted alleles. (The reader is cautioned to distinguish this $r$ from the correlation coefficient $r_P$ introduced in teachable moment 3.3. It is also unrelated to the Malthusian parameter $r$ used in overlapping generation models.) Recombination rates between two loci can be no larger than $\frac{1}{2}$, because at most half of the gametes produced during meiosis are recombinants. *Free recombination* between loci means that their $r=\frac{1}{2}$. Almost all eukaryotes have more than one pair of chromosomes, and two loci on different pairs experience free recombination. Moreover, as there is often at least one recombination event per chromosome per meiosis (empirical aside 3.2), the same is usually true for pairs of loci at opposite ends of the same chromosome. But $r<\frac{1}{2}$ for many pairs of loci on the same chromosome, and approaches zero for some.

---

**Teachable Moment 3.4:** The discovery of genetic recombination, genetic and physical mapping

All the traits that Mendel examined were transmitted independently (i.e., their genetic map distances $r$ were all $\frac{1}{2}$). That all pairs of traits are independently transmitted has been canonized as *Mendel's second law of independent assortment* (see empirical aside 1.11 to be reminded of his first law). In the first decade of the twentieth century, Thomas Hunt Morgan and his research group discovered that many pairs of traits in the fruit fly *Drosophila melanogaster* violate Mendel's second law. They then found that recombination rates between pairs of trait-determining loci are roughly additive. That is, given three loci $A$, $B$, and $C$, $r_{AB}+r_{BC} \approx r_{AC}$. (See any elementary genetics textbook to understand the basis of this approximation.) This suggested a linear arrangement of loci, which Morgan's group then located genetically (or *mapped*) relative to one another, in a linear space whose distance metric is defined on the premise that recombination events are uniformly distributed.

*Linkage groups* are groups of loci that each have genetic map distances $r$ less than $\frac{1}{2}$ to others in the group. One of the first steps in describing an organism's genome using genetic data is counting its linkage groups, each of which likely corresponds to a different *chromosome*, subcellular objects visible under a microscope. Visible differences within and between individual chromosomes that cosegregate with traits provided the first opportunity to physically map loci. The earliest example was Morgan's; he demonstrated that sex is genetically determined in the fruit fly by showing that it is cotransmitted with a cytologically distinct chromosome.

---

Genetic map distances are often expressed in centimorgans (cM), where 1 cM corresponds to a recombination rate of 1%. An organism's *genetic map length* is the sum of genetic map distances between all pairs of adjacent loci in its genome. Thus, a genetic map length of 100 cM corresponds to an average of 1 recombination event per generation.

---

**Empirical Aside 3.2:** What are genetic map lengths in nature?

In a survey of mammals (Dumont and Payseur 2008), mean genetic map length was approximately 2,400 cM. This figure is approximately 3,600 cM in humans, 1,500 cM in rats and mice, and less than 750 cM in two marsupial species. Not surprisingly, much of this variability is correlated with the number of chromosomes, each of which (after the first) contributes 50 cM of free recombination map length. The mean genetic map length per chromosome in mammals is very nearly 100 cM, and a comparable figure was observed in a broader survey of eukaryotes (Stapley et al. 2017).

With this bookkeeping out of the way, we explore how recombination affects haplotype frequencies and linkage disequilibrium. Table 3.1 computes the proportion of each haplotype in the population after one generation of reproduction under random mating. Substituting haplotype frequencies from the last line of table 3.1 into equation (3.2) allows us to compute the linkage disequilibrium after one generation:

$$D' = (p_{AB} - rD)(p_{ab} - rD) - (p_{Ab} + rD)(p_{aB} + rD)$$

$$= (p_{AB}p_{ab} - p_{Ab}p_{aB}) - (p_{AB} + p_{ab} + p_{Ab} + p_{aB})rD \qquad (3.4)$$

$$+ (rD)^2 - (rD)^2 = D - rD = (1 - r)D.$$

In English, the linkage disequilibrium between two loci is reduced by the proportion $1 - r$ each generation, yielding

$$D_t = (1 - r)^t D_0 \approx D_0 e^{-rt} \qquad (3.5)$$

for the linkage disequilibrium after $t$ generations of recombination. This confirms and quantifies the claim made in the paragraph below equation (3.1).

---

**Power User Challenge 3.1**

Derive an expression for $D_t$ in terms of $r$, $t$, and $D_0$ in a two-locus population undergoing perfect assortative mating.

---

**Power User Challenge 3.2**

What does mutation do to linkage disequilibrium in a deterministic two-locus model? Does it increase it, decrease it, or leave it unchanged?

---

### 3.1.2   The Two-Locus Wahlund Effect

Recall from chapter 1 that in a single-locus model, metapopulation-wide heterozygosity will be depressed if allele frequencies differ among demes, even if demes all undergo random mating. This is the Wahlund effect. An analogous effect occurs with linkage disequilibrium in a two-locus model. Even assuming no within-deme linkage disequilibrium (as for example with random mating within demes and free recombination between loci), metapopulation-wide linkage disequilibrium can

**Table 3.1**
Frequencies of two-locus haplotype pairings and of offspring produced

| Mating pair | Probability[a] | Proportion of offspring of given haplotype[b] | | | |
|---|---|---|---|---|---|
| | | $AB$ | $Ab$ | $aB$ | $ab$ |
| $AB \times AB$ | $p_{AB}^2$ | $p_{AB}^2$ | 0 | 0 | 0 |
| $AB \times Ab$ | $2p_{AB}p_{Ab}$ | $p_{AB}p_{Ab}$ | $p_{AB}p_{Ab}$ | 0 | 0 |
| $AB \times aB$ | $2p_{AB}p_{aB}$ | $p_{AB}p_{aB}$ | 0 | $p_{AB}p_{aB}$ | 0 |
| $AB \times ab$[c] | $2p_{AB}p_{ab}$ | $(1-r)$ $p_{AB}p_{ab}$ | $r$ $p_{AB}p_{ab}$ | $r$ $p_{AB}p_{ab}$ | $(1-r)$ $p_{AB}p_{ab}$ |
| $Ab \times Ab$ | $p_{Ab}^2$ | 0 | $p_{Ab}^2$ | 0 | 0 |
| $Ab \times aB$[c] | $2p_{Ab}p_{aB}$ | $r$ $p_{Ab}p_{aB}$ | $(1-r)$ $p_{Ab}p_{aB}$ | $(1-r)$ $p_{Ab}p_{aB}$ | $r$ $p_{Ab}p_{aB}$ |
| $Ab \times ab$ | $2p_{Ab}p_{ab}$ | 0 | $p_{Ab}p_{ab}$ | 0 | $p_{Ab}p_{ab}$ |
| $aB \times aB$ | $p_{aB}^2$ | 0 | 0 | $p_{aB}^2$ | 0 |
| $aB \times ab$ | $2p_{aB}p_{ab}$ | 0 | 0 | $p_{aB}p_{ab}$ | $p_{aB}p_{ab}$ |
| $ab \times ab$ | $p_{ab}^2$ | 0 | 0 | 0 | $p_{ab}^2$ |
| Totals[d] | 1 | $p_{AB}-rD$ | $p_{Ab}+rD$ | $p_{aB}+rD$ | $p_{ab}-rD$ |

[a] Assuming random mating. The leading 2 on the second, third, fourth, sixth, seventh, and ninth lines reflects the fact that two pairings are represented: one in which the maternally contributed gamete is on the left and one in which it is on the right.
[b] Assuming that the recombination rate between loci is $0 \leq r \leq \frac{1}{2}$, where the upper bound follows from the fact that at most half the gametes produced during meiosis are recombinant.
[c] Note that recombination only generates haplotypes different from the parents in double-heterozygote matings; that is, on the fourth and sixth lines of the table.
[d] Column totals found by factoring and substituting equation (3.2).

persist. This is called the *multilocus Wahlund effect*, which emerges if allele frequencies covary among demes. For example, imagine two equal-sized demes, one fixed for the *Ab* haplotype and the other for the *aB*. In this case, allelic frequencies covary negatively: the *A* allele is fixed in the first deme and absent in the second, while the *B* allele is absent in the first and fixed in the second. And while $D=0$ in each deme, metapopulation haplotype frequencies are $p_{AB} = p_{ab} = 0$ and $p_{Ab} = p_{aB} = \frac{1}{2}$, meaning that the metapopulation $D = 0 \times 0 - \frac{1}{2} \times \frac{1}{2} = -\frac{1}{4}$.

More generally, writing $N_i$, $D_i$, $p_{A,i}$, and $p_{B,i}$, respectively, for the size, linkage disequilibrium and frequencies of the *A* and *B* alleles in the *i*th deme and $N = \sum_i N_i$ for the metapopulation size, the demic average linkage disequilibrium is $\bar{D} = \sum_i \frac{N_i}{N} D_i$. Now defining covariance between variables $x$ and $y$ as

$$\text{Cov}(x, y) = \sum_i p_i x_i y_i - \bar{x}\,\bar{y},$$

we write the covariance in allele frequencies as

$$\text{Cov}(p_A, p_B) = \sum_i \frac{N_i}{N} p_{A,i}\, p_{B,i} - \bar{p}_A \bar{p}_B.$$

Putting these ideas together, one can show

$$D_{metapopulation} = \sum_i \frac{N_i}{\sum N_i} D_i + \mathrm{Cov}(p_A, p_B) = \bar{D} + \mathrm{Cov}(p_A, p_B). \qquad (3.6)$$

In English, the two-locus Wahlund effect tells us that a positive correlation in allele frequencies across demes inflates the metapopulation linkage disequilibrium compared to the demic mean, and a negative correlation deflates it. Power user challenge 3.3 provides some quantitative intuition into the effect, and the interested reader will find a full derivation of equation (3.6) in Nei and Li (1973).

---

**Teachable Moment 3.5**

Note the strong formal parallels between the definition of covariances and variances (equation 1.13b). In particular, the covariance of a variable with itself is its variance.

---

**Power User Challenge 3.3**

Imagine two equal-sized demes in which $p_{A,1} = p_{B,1} = p_1$, $p_{A,2} = p_{B,2} = p_2$ and $D_1 = D_2 = 0$. (This is the symmetry assumption of the right-hand panel of figure 3.1, in which these two populations are represented by points at $(p_1, 0)$ and $(p_2, 0)$, respectively.) Use equation (3.3) to find within-deme haplotype frequencies and then equation (3.2) to show that the metapopulation-wide linkage disequilibrium will be $\frac{1}{4}(p_1 - p_2)^2$. Now show that this answer is consistent with equation (3.6). Finally, recalling that $\mathrm{Cov}(p_A, p_B) = \mathrm{Var}(p)$ under these assumptions (teachable moment 3.5) and that $H_T - H_S = 2\mathrm{Var}(p)$ (equation 1.33) reveals an interesting perspective on the parallels between these two forms of the Wahlund effect. Explain.

---

### 3.1.3 Pairwise Epistasis, Natural Selection, and Recombination

In section 3.1.1, we confronted the fact that haplotype frequencies are not uniquely specified by allele frequencies. Something formally analogous is at work in thinking about natural selection. For example, two-locus organisms define four haplotypes and, thus, four fitness values. (As noted in teachable moment 3.1, we assume that natural selection acts only in the haploid phase of an organism's life cycle.) In keeping with our assumption of nonoverlapping generations, these are parameterized as relative Wrightian fitnesses $w_{AB}$, $w_{Ab}$, $w_{aB}$, and

$$w_{ab} = 1. \qquad (3.7a)$$

Following our haploid treatment (section 1.1.5), we next define Wrightian selection coefficients, $s_A = w_{Ab} - w_{ab} = w_{Ab} - 1$ and $s_B = w_{aB} - w_{ab} = w_{aB} - 1$, which yield

$$w_{Ab} = 1 + s_A, \text{ and} \qquad (3.7b)$$

$$w_{aB} = 1 + s_B. \qquad (3.7c)$$

But what is $w_{AB}$ in terms of $s_A$ and $s_B$? We know that the proportional fitness effect of the $a \to A$ mutation on the $ab$ haplotype is $1 + s_A$. So by analogy, we might guess that it has the same effect on the $aB$ haplotype. Algebraically, this would give us $w_{AB} = w_{aB}(1+s_A) = (1+s_B)(1+s_A) = w_{aB}w_{Ab}$, which could be true, but needn't be: recall that $w_{AB}$ can assume any nonnegative value. (Also note the symmetry in the situation: we might as well have written our guess as $w_{AB} = w_{Ab}(1+s_B) = (1+s_A)(1+s_B) = w_{aB}w_{Ab}$.)

We are again confronted by an underspecification of the problem. In the language of teachable moment 3.2, we have three independent relative fitness values, meaning three degrees of freedom. Two are captured by our two selection coefficients, but to capture the third, we now introduce the pairwise *Wrightian epistasis coefficient* between the $A$ and $B$ mutations, written $\varepsilon$. This allows us to compute the fitness of the $AB$ haplotype in terms of selection coefficients and epistasis as

$$w_{AB} = w_{Ab}w_{aB} + \varepsilon = (1+s_A)(1+s_B) + \varepsilon. \tag{3.7d}$$

In English, epistasis quantitatively captures the discrepancy between our naïve guess based on the fitness effects of each mutation in isolation and the true fitness of the double mutant. Inverting (3.7d) yields

$$\varepsilon = w_{AB} - w_{Ab}w_{aB}. \tag{3.8}$$

We note in passing that this framework also allows the quantification of allelic interactions for any other phenotype, although in the absence of a rationale for normalizing values, this more commonly reads $\varepsilon = z_{AB}z_{ab} - z_{Ab}z_{aB}$, where $z$ represents the trait of interest.

---

**Power User Challenge 3.4**

What is the analog to equation (3.8) in a model of overlapping generations; that is, in terms of Malthusian fitness values $r_{ab}$, $r_{Ab}$, $r_{aB}$, and $r_{AB}$? Take a guess, and then follow the path laid out in power user challenge 1.10 to see if your guess is right. And relatedly, in chapter 1 we defined the Wrightian dominance coefficient $h = \dfrac{w_{Aa}-1}{2s}$. What is its analog in terms of Malthusian fitness values?

---

With this bookkeeping out of the way, we next ask how natural selection affects linkage disequilibrium. Since selection acts on haplotypes, we generalize our deterministic model of selection in nonoverlapping generations from section 1.1.5 to accommodate four genetically distinct types. Combining this with equation (3.7), we have

$$
\begin{aligned}
p'_{AB} &= \frac{p_{AB}w_{AB}}{\bar{w}} = \frac{p_{AB}[(1+s_A)(1+s_B)+\varepsilon]}{\bar{w}} \\[2mm]
p'_{Ab} &= \frac{p_{Ab}w_{Ab}}{\bar{w}} = \frac{p_{Ab}(1+s_A)}{\bar{w}}, \\[2mm]
p'_{aB} &= \frac{p_{aB}w_{aB}}{\bar{w}} = \frac{p_{aB}(1+s_B)}{\bar{w}},
\end{aligned}
\tag{3.9}
$$

and

$$p'_{ab} = \frac{p_{ab}}{\bar{w}},$$

where $\bar{w} = p_{AB}[(1+s_A)(1+s_B)+\varepsilon] + p_{Ab}(1+s_A) + p_{aB}(1+s_B) + p_{ab}$. Substituting these values into the definition of linkage disequilibrium (equation 3.2) yields

$$
\begin{aligned}
D' &= p'_{AB} p'_{ab} - p'_{Ab} p'_{aB} \\
&= \frac{p_{AB}[(1+s_A)(1+s_B)+\varepsilon]}{\bar{w}} \frac{p_{ab}}{\bar{w}} - \frac{p_{Ab}(1+s_A)}{\bar{w}} \frac{p_{aB}(1+s_B)}{\bar{w}} \\
&= \frac{p_{AB} p_{ab}[(1+s_A)(1+s_B)+\varepsilon]}{\bar{w}^2} - \frac{p_{Ab} p_{aB}(1+s_A)(1+s_B)}{\bar{w}^2} \\
&= \frac{D(1+s_A)(1+s_B) + p_{AB} p_{ab}\varepsilon}{\bar{w}^2}.
\end{aligned}
\tag{3.10}
$$

This equation reveals several important biological conclusions. First, we see that if linkage disequilibrium is absent, it can only be created by selection if epistasis is nonzero, and the sign of the resulting linkage disequilibrium will be the same as the sign of epistasis. (Mathematically, if $D=0$ then $D' = \frac{p_{AB} p_{ab}\varepsilon}{\bar{w}^2}$.) Conversely, if linkage disequilibrium is present but epistasis is not, then the sign of linkage disequilibrium cannot change. These points are illustrated in the left panel of figure 3.2, which assumes symmetry between loci; that is, assuming that $p_A = p_B$ and $s_A = s_B$. (Relaxing this assumption introduces no qualitative novelties.)

The left panel of figure 3.2 focuses our attention on the influence that natural selection has on linkage disequilibrium; as predicted, its sign is given by the sign of epistasis. But the right panel illustrates a perhaps more important result: the influence that epistasis has on the speed with which the $AB$ haplotype goes to fixation. Specifically, negative epistasis (red) slows the pace of its fixation relative to the no-epistasis case (black), while positive epistasis (blue) has the opposite effect.

As we saw in chapter 1, analytic progress in models of selection in nonoverlapping generations is difficult, but some qualitative understanding is possible. By definition,

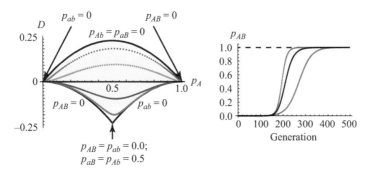

**Figure 3.2**
Two-locus evolutionary dynamics driving fixation of the $AB$ haplotype with epistasis and no recombination. In both panels, trajectories iteratively computed over 500 generations with equations (3.1), (3.9), and (3.10) starting from $p_A = 0.001$ and $D=0$ under symmetry assumptions $p_A = p_B$ and $w_{Ab} = w_{aB}$. We hold $w_{ab} = 1$ and $w_{AB} = 1.1$ so that the total selective pressure driving fixation of the $AB$ haplotype is constant. Solving equation (3.7d), we find $w_{Ab} = w_{aB} = (1.1-\varepsilon)^{\frac{1}{2}}$. Left: trajectories through $p_A \times D$-space (see teachable moment 3.3) with $\varepsilon = +0.25$ (blue), $\varepsilon = +0.1$ (orange), $\varepsilon = 0.0$ (black), $\varepsilon = -0.1$ (purple), and $\varepsilon = -0.25$ (red). Right, $p_{AB}$ over 500 generations for $\varepsilon = +0.25$ (blue), $\varepsilon = 0$ (black), and $\varepsilon = -0.25$ (red).

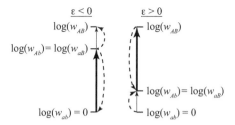

**Figure 3.3**
Graphical illustration of the influence of epistasis and recombination on the sign of linkage disequilibrium. As in figure 3.2, we maintain symmetry with the assumption $w_{Ab} = w_{aB}$, and hold $w_{AB}$ fixed and $w_{ab} = 1$, so that the total selective pressure driving fixation of the $AB$ haplotype is constant. As in figure 3.2, solving equation (3.7d) gives $w_{Ab} = w_{aB} = (w_{AB} - \varepsilon)^{\frac{1}{2}}$ or $\log(w_{Ab}) = \log(w_{aB}) = \dfrac{\log(w_{AB} - \varepsilon)}{2}$. In English, the log fitness of the single-mutant haplotypes is half that of double-mutant in the absence of epistasis. Negative epistasis increases this value and positive epistasis reduces it. Left: when epistasis is negative, the rate of selective enrichment of single mutants ($Ab$ and $aB$) at the expense of wild type ($ab$) haplotypes is faster than the rate of enrichment of double mutants ($AB$) at the expense of single mutants. (Weights of vertical arrows corresponds to the strength of selection.) This gives rise to negative linkage disequilibrium and reduces fitness variance in the population. Right: this reasoning is exactly reversed when epistasis is positive. In both cases, recombination (dashed lines) reduces linkage disequilibrium. Thus, when $\varepsilon < 0$, it recreates the comparatively underrepresented wild type and double mutants. Conversely, when $\varepsilon > 0$, it recreates the comparatively underrepresented single mutant haplotypes.

when epistasis is negative, the fitness gain due to the first beneficial allele is greater than that due to the second beneficial allele. This is illustrated graphically in the left panel of figure 3.3 (note the differing weights of the vertical arrows). In this case, haplotypes with one beneficial allele are being selectively enriched at the expense of those with none faster than they are being replaced by those with two beneficial alleles. Compared to the case with no epistasis, the net effect is the transient accumulation of single-mutant haplotypes in the population, explaining why linkage disequilibrium is negative. This same effect also lowers fitness variance in the population, since the population is concentrated in single-mutant individuals of intermediate fitness. By Fisher's fundamental theorem, this slows the pace of adaptation, as seen in figure 3.2. This reasoning is reversed when epistasis is positive, as illustrated in the right panel of figure 3.3.

Having examined the influence first of recombination and then of selection on linkage disequilibrium, we now put the two processes together. This is easily done by combining equations (3.4) with (3.10) to find

$$D' = \frac{D(1+s_A)(1+s_B) + p_{AB}p_{ab}\varepsilon}{\overline{w}^2}(1-r). \tag{3.11}$$

In English, whatever effect natural selection has on linkage disequilibrium, recombination will push the value closer to zero. These results are seen in figure 3.4. Not surprisingly, in the left panel, recombination pushes trajectories toward the x-axis regardless of the sign of epistasis (and thus, of linkage disequilibrium).

Since recombination reduces linkage disequilibrium, we might also not be surprised to see that it also affects the speed with which the $AB$ haplotype goes to fixation. Namely, it speeds it when negative (figure 3.4, middle panel; note arrow) and

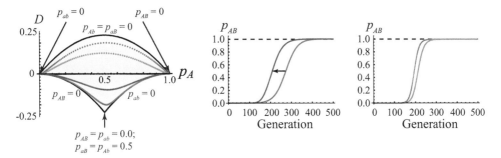

**Figure 3.4**
Two-locus evolutionary dynamics driving fixation of the $AB$ haplotype with epistasis and recombination. In all panels, trajectories iteratively computed as in figure 3.2. Left: trajectories through $p_A \times D$-space $\varepsilon = +0.25$ and $r = 0$ (blue) or $r = 0.1$ (orange); $\varepsilon = 0.0$ and $r = 0$ (black); and $\varepsilon = -0.25$ and $r = 0.1$ (purple) or $\varepsilon = -0.25$ and $r = 0$ (red). Center: frequency of double mutant ($AB$) haplotypes for $\varepsilon = -0.25$ with $r = 0.1$ (purple) and $r = 0.0$ (red). Horizontal arrow represents the impact of recombination. Right: frequency of double mutant ($AB$) haplotypes for $\varepsilon = +0.25$ and $r = 0$ (blue) and $r = 0.1$ (orange). Arrow omitted for clarity but would point to the right. In all three panels, blue, black, and red traces are identical to those in figure 3.2.

slows it when positive (right panel). Recall that when epistasis is negative, so too is linkage disequilibrium. In this case, recombination is on balance moving beneficial alleles from single-mutant haplotypes to wild type and double-mutants (see dashed lines in left panel of figure 3.3). This speeds fixation of the double-mutant because it increases fitness variance in the population. Conversely, recombination slows the process when epistasis is positive. Now, linkage disequilibrium will be positive, so recombination is on balance reducing fitness variance (see dashed lines in right panel of figure 3.3).

An analogous situation exists with respect to the ability of natural selection to purge recurrent deleterious mutations. Selection is again made more efficient by recombination when deleterious mutations exhibit negative epistasis because it again increases variance for fitness. That recombination facilitates more efficient natural selection has been proposed as a possible evolutionary explanation for its widespread observation of sexual reproduction in nature. We take up that question in section 3.3.1.

---

**Power User Challenge 3.5**

Sexual reproduction speeds or slows the pace of adaptation in a deterministic two-locus, haploid model as a function of the sign of epistasis. Otto and Lenormand (2002) point out that it has the same effect in a one-locus, diploid model, now as a function of dominance. Explain. Hint: think about how the dominance coefficient $h$ influences the inbreeding coefficient $F$. For example, if $h > \frac{1}{2}$, is $F$ positive or negative after selection acts? Next, imagine a diploid population that reproduces clonally. How does the pace of adaptation depend on $F$, and thus, on $h$? Finally, what features of our model of sexual reproduction in one-locus diploids influence $F$ at syngamy? How do these features affect the pace of adaptation as a function of dominance? Power user challenge 3.24 gives the stochastic analog to this question.

### 3.1.4   Multilocus Mutation/Selection Equilibrium

We now move from two-locus models, turning first to the equilibrium between natural selection and recurrent deleterious mutation in multilocus models. Recall from chapter 1 that in our overlapping generations treatment, a one-locus population will equilibrate at allele frequency $\tilde{p}_A = 1 - \dfrac{\mu}{s}$ if $\mu < s$, where $\mu$ and $s$ are the deleterious mutation rate and selection coefficient, respectively (equation 1.20). Under these conditions, the population's equilibrium mean fitness will be $\tilde{r}_{eq} = 1 - \mu$ (equation 1.21). That this last result is independent of selection coefficient is a form of the Haldane–Muller principle.

---

**Power User Challenge 3.6**

Our treatment of the one-locus Haldane–Muller principle assumed overlapping generations. Recover equations (1.20) and (1.21) in a one-locus model with nonoverlapping generations.

---

**Teachable Moment 3.6**

Felsenstein (2019) gives a nice intuition for the mutation/selection equilibrium in a one-locus model of nonoverlapping generations. By definition, a fraction $s$ of the deleterious $a$ alleles are selectively eliminated each generation. That means that each deleterious allele will persist in the population for an average $\dfrac{1}{s}$ generations (just as equation [2.36] followed from [2.34]). Thus, the fraction of $a$ alleles in the population at equilibrium is the proportion of $A$ individuals mutated in the last $\dfrac{1}{s}$ generations. Since a fraction $\mu$ of $A$ individuals are mutated each generation, $\dfrac{\mu}{s}$ is the equilibrium frequency of the deleterious $a$ allele.

---

We now seek analogous results for the multilocus case. Imagine an $L$-locus, haploid genome with two alleles at each locus. To simplify, we assume nonoverlapping generations, no dominance, and no recombination (we relax this last assumption in power user challenge 3.9). Writing $\mu$ for the per locus per generation deleterious mutation rate, we introduce $U = \mu L$ for the genome-wide mutation rate in an $L$-locus haplotype. As we did in our one-locus treatment, we disregard back mutations, again reasoning that selection will keep low-fitness haplotypes at low frequency. We model the number of new mutations in an offspring by a Poisson distribution. (Recall from section 2.1.2 that this distribution gives the probability of each conceivable number of events (now, mutations) in an interval (now, the genome), on the assumption that events occur independently at some rate that is constant over the interval.) Writing $j$ for the number of new deleterious alleles inherited by each offspring, this distribution reads

$$\text{Probability}(j \text{ new mutations}) = \frac{U^j e^{-U}}{j!}.$$

Remarkably, we can already find the population's mean fitness at mutation/selection equilibrium, even without specifying the fitness effects of deleterious alleles. The Poisson model means that the probability of exactly $j = 0$ new deleterious alleles in an offspring is $e^{-U}$. Writing $W_0$ for the absolute fitness of haplotypes with zero deleterious alleles, we have

$$p_0' = \frac{p_0 W_0}{\bar{W}} e^{-U},$$

where $p_0$ and $p_0'$ are the frequencies of haplotypes carrying zero deleterious alleles in the current and next generations, respectively. Note that while absolute mean fitness $\bar{W}$ is implicitly a function of the frequencies and fitness values of all possible haplotypes, for the moment, our treatment remains independent of those details.

Again without loss of generality, we rescale all fitness values, this time by $W_0$ to find

$$p_0' = \frac{p_0}{\bar{w}} e^{-U}.$$

Finally, at equilibrium $\tilde{p}_0 = p_0' = p_0$, which gives us $1 = \dfrac{e^{-U}}{\bar{w}_{eq}}$ or that mean relative fitness will equilibrate to

$$\bar{w}_{eq} = e^{-U} \qquad\qquad (3.12)$$

assuming $\tilde{p}_0 \neq 0$. (We made this same assumption in the one-locus derivation in chapter 1. As there, its relaxation yields the error catastrophe, a point we take up shortly.)

Equation (3.12) extends our one-locus observation of the Haldane–Muller principle (equation 1.21): in this parameter regime (i.e., assuming $\tilde{p}_0 \neq 0$), mean fitness is again dependent only on the deleterious mutation rate. In particular, it is independent of haplotypic fitness values. As in the one-locus case, the larger reduction in mean fitness due to haplotypes of particularly low fitness is exactly offset by their lower frequency in the population. Or equivalently, mutation load again depends on mutation rate but is independent of fitness values.

---

**Power User Challenge 3.7**

Use the Taylor approximation of equation (3.12) around $U=0$ to rederive equilibrium mean fitness under recurrent deleterious mutation in a one-locus model (equation [1.21]).

---

In chapter 1, we derived equilibrium mean fitness via the equilibrium frequency of the deleterious allele. By contrast, we just found the multilocus analog (equation 3.12) without saying anything about the equilibrium distribution of deleterious haplotypes (and indeed, without even specifying their fitness values). We now seek that distribution, which will clearly depend on those details.

---

**Power User Challenge 3.8**

Follow the path that yielded equation (3.12) in the one-locus model to recover equation (1.21), that is, without specifying the fitness effect of the deleterious allele.

---

Following Haigh (1978), we continue to assume nonoverlapping generations and no recombination or epistasis, and further, assume that the *fitness function* (or mapping from haplotype to fitness) is

$$w_k = (1-s)^k \qquad (3.13)$$

for all haplotypes carrying exactly $k$ deleterious alleles. We subtract $s > 0$ from 1 because these are deleterious mutations, and because our fitness function is multiplicative, it has no epistasis. (Setting $s_A = s_B$ and $\varepsilon = 0$ in equation [3.7d] yields the two-locus analog.) Anticipating a graphical representation of fitness functions in which each locus corresponds to a different dimension [section 3.1.6], we designate fitness functions in which alleles have the same selection coefficient regardless of locus as *rotationally symmetric*. Equivalently, loci are interchangeable in rotationally symmetric fitness functions.

Now extending our earlier notation, we introduce $p_k$ for the frequency of haplotypes carrying exactly $k$ deleterious alleles. And extending the reasoning that earlier yielded $p_0' = \frac{p_0}{\bar{w}} e^{-U}$ for the frequency of mutation-free haplotypes after one generation of selection and mutation, we write

$$p_1' = \frac{p_0}{\bar{w}} \frac{e^{-U} U^1}{1!} + \frac{p_1(1-s)}{\bar{w}} e^{-U}$$

for the frequency of the single-mutant haplotypes in the next generation. Here, the first term is the contribution from the mutation-free class, namely its frequency after selection multiplied by the probability that its offspring have exactly one deleterious allele. The second term is what remains of the single-mutant class after accounting for selection and mutation. (Recall again that we disregard back mutations, reasoning that selection will keep lower-fitness haplotypes at comparatively much low frequencies.) Similarly, the frequency of double mutants in the next generation is

$$p_2' = \frac{p_0}{\bar{w}} \frac{e^{-U} U^2}{2!} + \frac{p_1(1-s)}{\bar{w}} \frac{e^{-U} U^1}{1!} + \frac{p_2(1-s)^2}{\bar{w}} e^{-U}.$$

Here, the first two terms are contributions from the mutation-free and single-mutant classes, and the third term is what remains of the double-mutants after selection and mutation. In general, we have

$$p_k' = \sum_{j=0}^{k} \frac{p_j(1-s)^j}{\bar{w}} \frac{e^{-U} U^{(k-j)}}{(k-j)!}, \qquad (3.14)$$

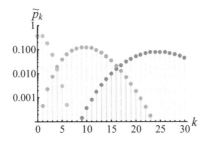

**Figure 3.5**
Probability distribution of haplotypes carrying $k$ deleterious alleles in a nonrecombining population at mutation/selection equilibrium. Values given by equation (3.15) for $\frac{U}{s}=1$ (green), 10 (mustard), and 25 (blue). Note that the $y$-axis is log transformed.

where $\bar{w}=\sum_{j=0}^{\infty}p_j(1-s)^j$. Here, the frequency of each fitness class (i.e., for each value of $k$) in the next generation is a function of the mutational input from the $k-1$ higher-fitness classes, plus the fraction of nonmutated members of the $k$-mutant class that survive selection. One mutation/selection equilibrium solution to equation (3.14) is

$$\tilde{p}_k=\frac{e^{-U/s}\left(\dfrac{U}{s}\right)^k}{k!}. \tag{3.15}$$

In English, equation (3.15) has the property that $p_k{}'=p_k=\tilde{p}_k$ for all $k\geq 0$ in equation (3.14). That result is proven in teachable moment 3.7, and illustrated in figure 3.5 for several values of $\frac{U}{s}$.

---

**Teachable Moment 3.7:** Proof that equation (3.15) is an equilibrium solution to equation (3.14)

To see this, we need to show that substituting equation (3.15) into equation (3.14) recovers equation (3.15). Recalling that $\bar{w}_{eq}=e^{-U}$ for all fitness functions (equation 3.12), this reads

$$p_k{}'=\sum_{j=0}^{k}\frac{\dfrac{e^{-U/s}\left(\dfrac{U}{s}\right)^j}{j!}(1-s)^j}{e^{-U}}\frac{e^{-U}U^{(k-j)}}{(k-j)!}$$

$$=\sum_{j=0}^{k}e^{-U/s}\frac{\left(\dfrac{U}{s}\right)^j}{j!}(1-s)^j\frac{U^{(k-j)}}{(k-j)!}=e^{-U/s}U^k\sum_{j=0}^{k}\left(\frac{1-s}{s}\right)^j\frac{1}{j!(k-j)!}.$$

Next, we write $\left(\dfrac{1}{s}\right)^k=\left(\dfrac{1}{s}-1+1\right)^k=\left(\dfrac{1-s}{s}+1\right)^k$. The binomial theorem gives us $(a+b)^k=$

$$\sum_{j=0}^{k}\binom{k}{j}a^jb^{k-j}. \text{ Setting } a=\left(\dfrac{1-s}{s}\right) \text{ and } b=1 \text{ lets us write}$$

$$\left(\frac{1}{s}\right)^k = \left(\frac{1-s}{s}+1\right)^k = \sum_{j=0}^k \binom{k}{j}\left(\frac{1-s}{s}\right)^j 1^{k-j} = \sum_{j=0}^k \frac{k!}{(k-j)!\,j!}\left(\frac{1-s}{s}\right)^j.$$

Thus $\sum_{j=0}^k \left(\frac{1-s}{s}\right)^j \frac{1}{j!(k-j)!} = \frac{\left(\frac{1}{s}\right)^k}{k!}$, which on substitution into our previous expression yields

$$p_k{}' = \frac{e^{-U/s}\left(\dfrac{U}{s}\right)^k}{k!} = p_k = \tilde{p}_k$$

Q.E.D.

---

**Power User Challenge 3.9**

Our two-locus treatment demonstrated that recombination has no evolutionary consequence in the absence of epistasis. (E.g., compare black traces in figures 3.2 and 3.4 or equivalently, note that equation 3.10 is zero if $D = \varepsilon = 0$. Though as we see in section 3.2.4, stochasticity complicates matters). Now show that in the absence of epistasis, mean fitness at mutation/selection equilibrium (equation 3.12), and the probability distribution of haplotypes carrying $k$ deleterious alleles (equation 3.15) are also unaffected by genetic recombination. Equivalently, assume each locus evolves independently. Hint: writing $k$ for the number of loci carrying deleterious alleles, first show that $E[k] = \dfrac{U}{s}$ at equilibrium. (Hint: $U = \mu L$.) In the absence of epistasis and assuming loci evolve independently, this result yields equation (3.12) after substituting it into equation (3.13) and recalling that $1 - x \approx e^{-x}$ for small $x$. Next, again assuming loci evolve independently (and no epistasis), use the binomial theorem to find an expression analogous to equation (3.14). Finally, another result from probability theory called the Poisson limit theorem states that $\lim_{n\to\infty}\binom{n}{k}p^k(1-p)^{n-k} = \dfrac{e^{-\lambda}\lambda^k}{k!}$, where $\lambda = np$.

Use this to show that your result converges to equation (3.15) as the number of loci in the genome grows large, still assuming the per locus mutation rate is constant.

---

**Teachable Moment 3.8:** Alleles at loci or mutations at nucleotide sites?

To this point, chapter 3 has considered haplotypes composed of loci carrying alleles. But gene and genome sequence data motivate attention to nucleotide sites carrying mutations. Importantly, much of population genetics is agnostic toward this distinction. We hope the reader will recognize these equivalences; our usage in particular settings reflect tradition or pedagogical convenience.

### 3.1.5   The Multilocus Error Catastrophe

We now return to the assumption that $\tilde{p}_0 \neq 0$, which was required to derive mean fitness under recurrent deleterious mutations (equation 3.12). As noted, failure of that condition gives rise to the multilocus error catastrophe, which we now explore. We disregard recombination (but see Boerlijst et al. [1996]) and largely follow Bull et al. (2005).

We first recover our one-locus results using fitness function

$$w_i = \begin{cases} 1 & \text{if } i = 0 \\ 1-s & \text{otherwise} \end{cases}, \tag{3.16}$$

where $0 < s < 1$, together with the assumption that the zero-mutation haplotypes mutate to the single-mutant class with genome-wide rate $U$. We disregard mutations among haplotypes carrying one or more mutations, since we assume their fitness values are identical. This is the discrete generation analog to the one-locus treatment in chapter 1. We also continue to disregard back mutations; power user challenge 3.10 relaxes that assumption.

We now locate the error threshold by finding the deleterious mutation rate that causes the per capita replacement rate of the more-fit haplotype to equal that of the less-fit haplotype, fitness advantage notwithstanding. Mathematically, we have

$$N_0' = N_0 e^{-U}$$

and

$$N_{i>0}' = N_{i>0}(1-s)$$

for the number of more- and less-fit haplotypes in the next generation, respectively. $e^{-U}$ is again the proportion of more-fit haplotypes that do not mutate, and by equation (3.16), their relative fitness is unity. The relative fitness of all other haplotypes is $1-s$. (We discount mutational input from the more-fit class because we seek the threshold mutation rate where that class disappears.) Now setting per capita replacement rates equal (mathematically, $\dfrac{N_0'}{N_0} = \dfrac{N_{i>0}'}{N_{i>0}}$) and solving for $U$ yields an error threshold when mutation rate $U^* = -\ln(1-s)$. (The reader shouldn't be concerned by the suggestion of a negative error threshold: since $0 < s < 1$, $\ln(1-s)$ is itself negative, making $U^*$ positive.) Once again, the fittest haplotype disappears at mutation rates above this threshold: purifying selection is too weak in the face of such a high mutation rate. This threshold is illustrated in the left panel of figure 3.6. (Reassuringly, the Taylor approximation of $U^*$ around $s = 0$ recovers the one locus result. See also power user challenge 3.7.)

This line of thinking suggests that more than one error threshold may be possible. For example, suppose

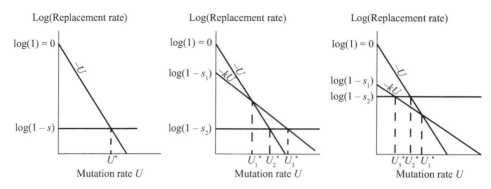

**Figure 3.6**
Multilocus error thresholds. Left: results for the two-class fitness function given by equation (3.16). Center and right: results for the three-class fitness function given by equation (3.17), together with the assumption that the deleterious mutation rate in the second-fittest class is $kU$, where $0 < k < 1$. Center: because the error threshold corresponding to the loss of the fittest class to the second-fittest class ($U_1^*$) is less than that corresponding the loss to the third-fittest class ($U_2^*$), two transitions will be observed. Right: in the opposite case, only one transition is observed: the fittest and second-fittest classes are both lost at the first error threshold ($U_2^*$). Figure drawn after Bull et al. (2005).

$$w_i = \begin{cases} 1 & \text{if } i = 0 \\ 1 - s_1 & \text{if } i = 1 \\ 1 - s_2 & \text{otherwise} \end{cases}, \tag{3.17}$$

where $0 < s_1 < s_2 < 1$. We further assume that, while zero-mutation haplotypes mutate to the single-mutant class with rate $U$, members of the single-mutant class mutate to the two-mutant class with rate $kU$, where $0 < k < 1$. That is, we assume that the deleterious mutation rate declines with fitness. (Biologically, this means that higher-fitness organisms are more susceptible to mutational disruption than are lower-fitness organisms. See power user challenge 3.10 and empirical aside 3.3 to learn more.) Following the approach used in the two-fitness-class case, we have

$$N_0' = N_0 e^{-U},$$
$$N_1' = N_1(1 - s_1)e^{-kU}$$

and

$$N_{i>1}' = N_{i>1}(1 - s_2),$$

yielding per capita replacement rates $e^{-U}$, $(1 - s_1)e^{-kU}$ and $(1 - s_2)$, respectively, for the fittest, second-fittest, and least-fit classes. Now, the fittest class disappears when its replacement rate is equal to the greater of the replacement rates for the other two, or mathematically, when $e^{-U} = \max[(1 - s_1)e^{-kU}, (1 - s_2)]$.

There are thus two scenarios. Setting per capita replacement rates for the fittest- and second-fittest classes equal yields an error threshold when $U$ is greater than $U_1^* = -[\ln(1 - s_1)]/(1 - k)$. Setting the first- and third-fittest classes' per capita replacement rates equal yields an error threshold when $U$ is greater than $U_2^* = -\ln(1 - s_2)$. If

$U_1^* < U_2^*$, then the fittest class will be lost to the second-fittest class when $U > U_1^*$. In

this case, a second error threshold will occur when $U$ is greater than $U_3^* = -\dfrac{\ln\left[\dfrac{1-s_2}{1-s_1}\right]}{k}$,

found by setting the second- and third-fittest class's replacement rates equal. This situation is illustrated in the middle panel of figure 3.6. But if $U_2^* < U_1^*$, then the fittest and second fittest classes will both be lost to the third-fittest class when $U$ is greater than $U_2^*$, as shown in the right panel of figure 3.6.

Can we extend these ideas to define fitness functions that can sustain more than two error thresholds? Yes, but interestingly, not every fitness function is susceptible. Critically, the error catastrophe requires a positive lower limit to the fitness function (Wagner and Krall 1993). For example, the three-valued fitness function just considered (equation 3.17) satisfies the condition: no fitness values are less than $1 - s_2 > 0$. And no fitness values are less than $1 - s > 0$ in the one-locus case from chapter 1. In contrast, the multiplicative fitness function used to find the multilocus mutation/selection equilibrium (equation 3.13) has no positive lower limit, since there we set no upper limit on $k$. Thus, no error catastrophe is possible on that fitness function.

To understand this result, consider that the error threshold corresponds to the mutation rate at which mutation overwhelms purifying selection. Or looking at the situation slightly differently, the error threshold requires an upper limit to the strength of purifying selection against deleterious mutations. Insisting that all fitness values be greater than some positive value meets that requirement. On the other hand, absent a nonzero lower limit on the fitness function, there can be no mutation rate that overwhelms the selective pressure to maintain at least some higher-fitness haplotypes in the population. Equation (3.13) has no such limit. Rather, there always exists a haplotype with enough deleterious mutations $k$ such that $(1 - s)^k$ is less than any arbitrarily small, positive value. Thus, no matter how high the downward mutational pressure becomes, the population will always evolve to a point on the fitness function at which it experiences a just-equal upward selective pressure. This ensures the existence of a mutation/selection equilibrium at all mutation rates.

We now return to the biological possibility of back mutations. Of course, in nature, genomes have finite lengths, and this means that as deleterious alleles accumulate, the probability of beneficial mutations increases. As in the one-locus treatment, this effect has only negligible consequences for mutation/selection equilibrium so long as selection is much stronger than mutation (mathematically, so long as $\mu \ll s$). However, back mutations can eventually halt the error catastrophe, as seen in power user challenge 3.10.

---

**Power User Challenge 3.10**

Explain why back mutations, absent in our treatment of the multilocus error catastrophe, eventually protect populations evolving on fitness functions bounded from below by some nonzero value. Assume that the haploid genome has finite $L$ biallelic loci, and any fitness function of the form $w_k > w_{k+1} > 0$, where $k$ is the number of deleterious mutations in the genome. Hint: What is the ratio of beneficial to deleterious mutations as a function of $k$? You should assume that the only beneficial mutations are the reversions of deleterious mutations (but see empirical aside 3.3). Reconcile your answer with our one-locus treatment of this issue developed in teachable moment 1.20.

Moreover, there are good empirical and theoretical reasons to believe that, beyond their own reversion, deleterious mutations can often introduce additional opportunities for beneficial mutations. Called *compensatory* or *second site mutations*, these are conditionally beneficial; that is, only beneficial in the presence of one or more deleterious alleles elsewhere in the genome. In this case, the beneficial mutation rate will increase even more quickly than predicted in power user challenge 3.10. These facts, reviewed in empirical aside 3.3, represent an additional limitation on the error catastrophe.

**Empirical Aside 3.3:** Beyond reversion: compensatory mutations

After a deleterious mutation has appeared in a genome, that mutation's reversion increases the probability of beneficial mutation. However, the mechanistic basis of a deleterious mutation can often be compensated for by mutations elsewhere in the genome, further increasing the beneficial mutation rate. For example, to be functional, proteins are often required to fold into a specific conformation, and mutations that reduce a protein's folding stability can be compensated by many stabilizing mutations elsewhere in the same protein. Similarly, many biologically important RNA molecules depend on so-called Watson–Crick pairings (between the nucleotides adenosine [A] and uracil [U], or between cytosine [C] and guanine [G]). Thus, mutations at one base can be easily compensated for by a second mutation at the corresponding base (e.g., an A→C mutation in an A:U base pair is compensated for by a U→G mutation.) And work in viruses (Escarmís et al. 1999; Silander et al. 2007), bacteria (Barrick et al. 2010), and eukaryotes (Estes and Lynch 2003) all find evidence of compensatory mutations. Indeed, in one virus it was estimated that on average, each deleterious mutation may create 10 or more compensatory sites in the genome (Poon and Chao 2005; Poon et al. 2005).

**Power User Challenge 3.11**

Do back mutations represent a form of epistasis? What about compensatory mutations? Explain.

### 3.1.6 Sequence Space and Fitness Landscapes

Our treatment of multilocus mutation/selection equilibrium and error catastrophes assumes that fitness is only a function of the number of deleterious mutations in a genome but independent of which sites (or loci; see teachable moment 3.8) are mutated. While such rotationally symmetric fitness functions are perhaps a reasonable heuristic when modeling the selective elimination of deleterious mutations, they are often inadequate when we turn to questions of adaptation. (With apologies to Leo Tolstoy's *Anna Karenina*, we might be comfortable assuming that the fitness effects of all deleterious mutations are alike, but suspect that each beneficial mutation is beneficial in its own way.) As promised in section 1.2.4, we now explore the theoretical complexities of the problem.

To build intuition, consider the HIV-1 genome, which is 9,749 base pairs long. Point mutation can convert any particular haplotype into one of $3 \times 9{,}749 = 29{,}247$ others. In the parlance of computer science, the *Hamming distance* between two haplotypes is the number of point mutations that distinguish them. Single mutations in the HIV-1 genome can only convert one haplotype into the 29,247 alternatives at Hamming distance 1. In contrast, there are vastly more conceivable HIV-1 genomes. More specifically, assuming only point mutations, this number is $4^{9{,}749}$, since we can specify any of four nucleotides at each position. (Or roughly $10^{5{,}900}$; for reference, the number of particles in the observable universe is estimated to "only" be $10^{80}$. Also, for reference, the human genome is more than a 100,000-fold longer than the HIV genome; there are thus roughly $10^{18{,}000{,}000{,}000}$ conceivable genomes the size of ours.) Of course, almost all of these combinations will be nonfunctional, but at this moment, we are only concerned with the scope of mutational possibilities.

How do we model the evolution of populations from low- to high-fitness haplotypes that are more than one Hamming unit apart? To begin, we introduce *sequence space*, a discrete, high-dimensional geometric space in which haplotypes are represented by points arranged so that mutational neighbors are spatially adjacent, and vice versa (Maynard Smith 1970). Assuming biallelic loci, sequence space is comprised of $L$ dimensions, in which each haplotype is represented by a vector of coordinates. There are thus $2^L$ points in this space, as required. These ideas are illustrated in figure 3.7.

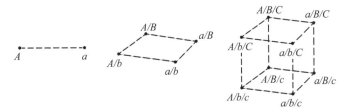

**Figure 3.7**
Sequence space for $L = 1$ (left), 2 (center), and 3 (right) biallelic loci. In each case, all $2^L$ distinct haplotypes, defined by all possible combinations of alleles, are associated with unique points in a discrete space in such a way that haplotypes Hamming distance 1 apart are adjacent. Note that parallel edges in sequence space represent the same mutation on all conceivable genetic backgrounds. Thus, sequence space for $L$ biallelic loci occupies $L$ dimensions.

> **Teachable Moment 3.9:** Strings and alphabets
>
> Also in the parlance of computer science, a haploid genome of $L$ sites is a *string* of symbols of that length, and the set of symbols allowed in at each position is called its *alphabet*. In the development presented here, we assume an alphabet of two, the wild type and mutant, as per the infinite sites model. The alphabet of DNA has four elements (the four nucleotides), while the protein alphabet has 20 amino acids. Few qualitative novelties in the population genetic theory emerge with alphabets larger than the two assumed in the text.

To understand how populations of organisms evolve through sequence space, we now introduce the *fitness landscape*: the projection of the fitness function over sequence space. This idea was first proposed by Wright (1932) as a visual aid for biologists, and it remains an important and influential concept in evolutionary genetics. (The interested reader is directed to de Visser and Krug [2014] and Fragata et al. [2019] for reviews of many recent applications. Interestingly, Wright defined a second fitness landscape over allele frequencies, which has close connections to Fisher's fundamental theorem of natural selection [Provide 1986]. We say nothing of that second conception here.) We described fitness functions such as that in equation (3.13) as rotationally symmetric: on these, a haplotype's fitness depends on its Hamming distance to the fittest haplotype, but because loci are interchangeable, not on its direction in sequence space.

> **Power User Challenge 3.12**
>
> Is it correct that all rotationally symmetric fitness functions lack epistasis? Why or why not? Is it correct that all fitness functions lacking epistasis are rotationally symmetric? Why or why not?

> **Teachable Moment 3.10:** Fitness seascapes
>
> Throughout this book, we assume that fitness values are constants. A refinement of the fitness landscape model respects the fact that values can change, as for example if fitness depends in part on interactions with the environment (including with other species). Recently, exciting progress has been made with principled models of changing fitness landscapes (Lindsey et al. 2013; Cvijović et al. 2015; Iram et al. 2020), sometimes called *fitness seascapes* (Mustonen and Lässig 2009).

The intuitive appeal of the fitness landscape is clear: the mathematics of population genetics are complex. In contrast, an evolving population can be regarded as a cloud of points under Wright's framework, whose radius and density reflect the amount of genetic variation at that moment. This cloud "climbs" the landscape in response to natural selection, while also generating new genetic diversity. So, if we could capture the contours of the fitness landscape, we could at least qualitatively describe the evolutionary fate of a population.

---

**Teachable Moment 3.11:** Epistasis on fitness landscapes

Epistasis refers to interactions between mutational effects. We encountered pairwise epistasis in section 3.1.3, but pairwise interactions can themselves vary with genetic background. Because sequence space explicitly represents each mutation on each conceivable genetic background (figure 3.7), it can accommodate all conceivable epistatic effects.

Another degrees-of-freedom argument (teachable moment 3.2) sharpens our understanding. Given $L$ biallelic loci in the genome, each locus requires a selection coefficient (e.g., equations [3.7b] and [3.7c] when $L=2$). But there are $2^L$ independent values on the fitness landscape. Because natural selection acts only on differences (or ratios) of fitness values, this leaves $2^L - L - 1$ degrees of freedom unaccounted for. These must correspond to independent epistatic terms (equation [3.8] when $L=2$).

Going further, we can think abstractly of the pairwise epistasis defined in equation (3.8) as our surprise at the fitness effect of the double mutant, given what we knew about the effects of the two mutations in isolation. We can generalize this idea by introducing $k$th-order epistasis: our surprise at the fitness effect of some set of $k$ mutations, given what we already knew about the effects of all subsets of those $k$ mutations. Given $L$ biallelic loci, there are $\binom{L}{k}$ distinct subsets of $k$ loci, and thus this number of distinct $k$th-order epistatic terms. Returning to the binomial theorem, this gives a total of $\sum_{k=2}^{L} \binom{L}{k} = 2^L - L - 1$ epistatic terms, as required. (The interested reader is directed to Poelwijk et al. [2016] to see several approaches for calculating these many epistatic terms.)

Finally, note that the foregoing counting argument applies equally to describing the genetic composition of a haploid, multilocus population. Given $L$ biallelic loci, there are now $2^L$ haplotype frequencies that, however must sum to 1. Thus, there are $2^L - 1$ degrees of freedom. We can represent these by that number of haplotype frequencies, or equally, convert them into $L$ allele frequencies and $2^L - L - 1$ linkage disequilibria. Continuing as in the previous paragraph, these will consist of $\binom{L}{2}$ pairwise disequilibria, $\binom{L}{3}$ three-way disequilibria, and in general $\binom{L}{k}$ $k$th-order linkage disequilibria. In total, these again sum to and

$$L + \sum_{k=2}^{L} \binom{L}{k} = 2^L - 1$$ degrees of freedom, as required.

---

Unless the fitness landscape is flat (i.e., unless all haplotypes have equal fitness), it will have at least one fitness maximum (or peak): a haplotype in which all mutations are deleterious. Landscapes can also have more than one peak; if their fitness values vary, we differentiate between the *global fitness peak* and *local fitness peaks*. Indeed, Wright's own interest in fitness landscapes was to articulate this possibility. He asked how populations might traverse the intervening *fitness valley* between peaks. We explore the population genetics of this problem in section 3.1.8.

---

**Power User Challenge 3.13**

What is the expected number of fitness peaks on a fitness landscape defined over $L$ biallelic loci if fitness values are randomly distributed? Hint: first find the probability that any particular haplotype is a fitness peak. What condition must be met?

---

**Power User Challenge 3.14**

What is the maximum number of fitness peaks possible on a fitness landscape defined over $L$ biallelic loci?

---

Importantly, while peaks separated by valleys are widespread on three-dimensional, physical landscapes, we should temper intuitions when thinking about biological fitness landscapes (Gavrilets 1997), which are of vastly higher dimensionality. In particular, although the reasoning in power user challenge 3.13 suggests that the number of fitness peaks on a landscape may be quite large, that conclusion assumes that fitness values are randomly distributed across the fitness landscape. A moment's reflection reveals that this is untenable: mutationally adjacent haplotypes are surely more similar in fitness than are mutationally more distant ones. On the other hand, many other important questions can still be approached in this framework, as we see next.

---

**Teachable Moment 3.12**

Another limitation to the reasoning in power user challenge 3.13 is that it assumes fitness values are continuous, which, in turn, implies that no two haplotypes can have numerically identical fitness. If on the other hand there are only a finite number of discrete fitness values possible, then mutationally adjacent haplotypes may have equal fitness values, reducing the expected number of peaks on the landscape. Indeed, a body of theory from physics called percolation theory can be used to show that under these circumstances, the number of peaks on the fitness landscape may be quite small (Gavrilets 1997). Moreover, as shown in chapter 2, natural selection becomes inefficient at distinguishing between fitness differences smaller than roughly the reciprocal of the population size. Environmental noise surely further weakens selection's sensitivity. Thus, even if true fitness values are continuously distributed, inefficiencies of natural selection may effectively "digitize" them.

---

### 3.1.7 Multilocus Adaptation on Fitness Landscapes under the Strong Selection/Weak Mutation Assumption

As noted above, haploid populations can be regarded as clouds of points in sequence space whose radius and density reflects the amount of genetic diversity they harbor. Unfortunately, quantitatively describing this cloud requires simultaneously tracking multiple, cosegregating lineages in the population. Moreover, while sequence space readily captures mutational adjacency, it does not easily allow the rigorous representation of genetic recombination.

We can, however, make progress by assuming that the time to fixation or loss of each beneficial mutation is vastly less than the waiting times between their appearance. This is called the *strong selection/weak mutation* (or *SSWM*) assumption (Gillespie 1984) and it allows us to represent populations as single points in sequence space. In this case we can also disregard genetic recombination, which is of no consequence in populations that carry at most one mutation at a time. (We relax both assumptions in section 3.2.4.1.)

Now imagine an evolving population of haploid, $L$ locus genomes that is currently Hamming distance $1 \leq k \leq L$ from the global fitness peak, meaning that the population is carrying alleles different from those found in the peak at $k$ of its $L$ loci. The number of *mutational trajectories* or temporal sequences of mutations to the peak is $k!$, since any of the $k$ different loci can mutate first, after which only $k-1$ differ, followed by $k-2$, and so on. (Our enumeration disregards trajectories that include mutational reversion and those on which more than one mutation fixes simultaneously. Those assumptions are relaxed in Weinreich [2010].)

Which of any of the $k!$ mutational trajectories between the current population and the global fitness peak are *selectively accessible*? That is, which have the property that each successive mutation will be favored by natural selection? The answer depends on the incidence of a particular form of epistasis called *sign epistasis*. A mutation exhibits sign epistasis if the sign of its fitness effect depends on genetic background, as illustrated in the comparison between the top two panels of figure 3.8. (In partial answer to power user challenge 3.11, compensatory mutations [empirical aside 3.3] exhibit sign epistasis.)

Note that the absence of sign epistasis does not mean the absence of epistasis entirely. It only means an invariance in the identity of the fitter allele at all loci; magnitudes of effect may still vary with genetic background. This more modest form of epistasis is called *magnitude epistasis* and is illustrated in the comparison between the left two panels of figure 3.8.

---

**Power User Challenge 3.15**

What is the relationship between sign and magnitude epistasis on the one hand, and positive and negative epistasis on the other? Hint: begin by considering beneficial and deleterious mutations separately.

---

To see the connection between sign epistasis and selectively accessible mutational trajectories to the global fitness peak, first imagine a fitness landscape lacking sign epistasis at any locus. Now consider a population Hamming distance $k$ from the global peak, meaning that at exactly $k$ loci, it is fixed for the allele not found in the peak haplotype. In the absence of sign epistasis, mutation to the allele found in the peak at any of these $k$ loci must be beneficial. Thus, the first mutation along all $k!$ mutational trajectories to the peak are beneficial. In the absence of sign epistasis, this reasoning applies equally at each haplotype along each of the $k!$ trajectories to the peak, meaning all such trajectories are selectively accessible.

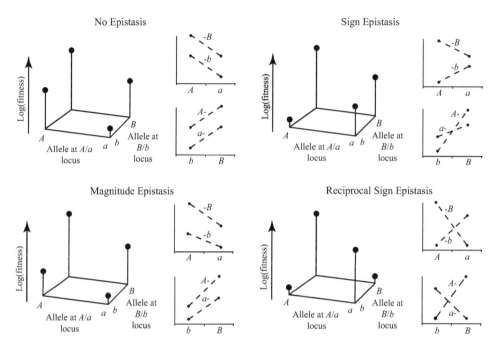

**Figure 3.8**
Sign epistasis and magnitude epistasis, and the number of selectively accessible trajectories on the fitness landscape. To the left in each panel, two-locus sequence space is represented on the $x$–$y$ plane and each haplotype's log-transformed Wrightian fitness is on the $z$-axis. To the right, two cross sections through the landscape are shown to illustrate the fitness effect of each mutation on each background. (Mutations at the $A/a$ locus on each background appear above; mutations at the $B/b$ locus on each background, below. Because fitness is log-transformed, dashed lines are parallel in the absence of epistasis.) In all panels, the $AB$ haplotype is the global fitness peak and we assume the population is first fixed for the $ab$ haplotype. Thus $k=2$ and there are $2!=2$ mutational trajectories to the global peak. In the absence of sign epistasis (left two panels), both trajectories are selectively accessible. Sign epistasis can render one (upper right) or both (lower right) of the two mutational trajectories selectively inaccessible. Figure modified from Weinreich et al. (2005).

---

**Power User Challenge 3.16**

Sign epistasis renders at least some of the $k!$ shortest mutational trajectories to the global fitness peak selectively inaccessible. Explain why.

---

**Power User Challenge 3.17**

The $k!$ mutational trajectories between a population Hamming distance $k$ from the global peak and the peak are exactly $k$ mutations long. As noted earlier, longer trajectories also exist, on which one or more loci are mutated more than once. In the absence of sign epistasis, can any of these longer trajectories be selectively accessible? Why or why not?

---

Importantly, sign epistasis *per se* is not sufficient for two (or more) peaks on the fitness landscape (e.g., the upper right panel of figure 3.8). Rather, two peaks on the fitness landscape must differ at least two loci, and further, in either haplotype, mutation at either locus to the allele in the other must be deleterious (see also power user challenge 3.14). The form of sign epistasis is called *reciprocal sign epistasis*, and is illustrated in the lower right panel of figure 3.8. Reciprocal sign epistasis reduces the number of selectively accessible trajectories from some haplotypes to the global peak to zero.

---

**Empirical Aside 3.4:** Sign epistasis in empirical fitness landscapes

We can fully characterize the number of fitness peaks and the incidence of sign epistasis with data for fitness (or a proxy) measured for all combinations of some set of mutations. Such *combinatorially complete datasets* grow exponentially in the number of loci examined; perhaps the largest such data set to date examined just 13 loci (Poelwijk et al. 2019). One survey of 16 such datasets found that 13 exhibit sign epistasis at least one locus, and 11 have more than one fitness peak (Weinreich, Lan, et al. 2013).

---

Maintaining our SSWM assumption, the identity of successively fixed mutations will be statistically independent. This allows us to compute the probability of realizing any particular selectively accessible trajectory as the product of the probabilities of its individual fixations. Mathematically writing $P_{i \to j}$ for the probability that the population evolves from haplotype $i$ to haplotype $j$, we have

$$P_{a \to b \to c \to \cdots \to y \to z} = P_{a \to b} P_{b \to c} \cdots P_{y \to z}, \tag{3.18}$$

where $b, c, \ldots, y$ are the mutationally adjacent haplotypes that define a particular trajectory from haplotype $a$ to haplotype $z$.

Now defining $M_i^+$ as the set of beneficial single-mutant neighbors of haplotype $i$, the probability that the next fixation from haplotype $i$ will be to haplotype $j$ is

$$P_{i \to j} = \frac{P_{\text{fix}}\left(x_0 = \frac{1}{2N}, N, s_{i \to j}\right)}{\sum_{k \in M_i^+} P_{\text{fix}}\left(x_0 = \frac{1}{2N}, N, s_{i \to k}\right)} \approx \frac{s_{i \to j}}{\sum_{k \in M_i^+} s_{i \to k}}. \tag{3.19}$$

Here, $s_{i \to j}$ is the selection coefficient between haplotypes $i$ and $j$ in a population of $2N$ haploids (see teachable moment 3.1), $P_{\text{fix}}\left(x_0 = \frac{1}{2N}, N, s_{i \to j}\right)$ is given by equation (2.24), and the approximation comes from our branching process assumption (equation 2.7). The denominator is the probability of fixation of any haplotype in $M_i^+$, so the ratio is the proportion of that total probability associated with fixation of haplotype $j$. (We used the same reasoning in teachable moment 2.21.) Results under this framework are illustrated in empirical aside 3.5.

**Empirical Aside 3.5:** What are probabilities of realization for mutational trajectories in nature?

TEM-1 β-lactamase is an enzyme that protects bacteria from antibiotics such as penicillin. All combinations of five point mutations in the gene known to jointly increase drug resistance approximately $10^5$-fold were constructed and characterized in *E. coli* (Weinreich et al. 2006). Sign epistasis was detected in four of the five mutations, resulting in only 18 of 5! = 120 selectively accessible mutational trajectories from the wild type. Further, a very sharp bias in probabilities of realization among the remaining 18 was reported, as shown in figure 3.9.

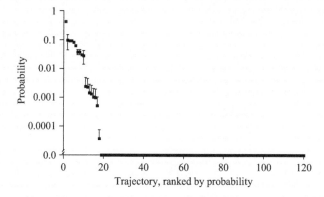

**Figure 3.9**
Probabilities of realizing all conceivable mutational trajectories in TEM-1 β-lactamase that maximize drug resistance via natural selection in *E. coli*, ordered from most- to least-likely. Nonzero values computed with equations (3.18) and (3.19). Zero values correspond to trajectories along which $M_i^+$ is empty for at least one constituent haplotype, an impossibility along all selectively accessible trajectories. Error bars correspond to s.e.m. over drug resistance measurement error (data from Weinreich et al. 2006).

### 3.1.8   Stochastic Tunneling

Our emphasis on selectively accessible trajectories in the previous section allowed us to sidestep the problem of evolution across a fitness valley. Mathematically, we implicitly assumed that $M_i^+$ was never empty (although see teachable moment 3.12). Of course, if the population arrives at the global fitness peak, evolution by natural selection must stop. But as noted, in principle, sign epistasis can also give rise to local fitness peaks, in which case further evolution depends on the population's ability to transit an intervening fitness valley. Indeed, theoretical interest in the population genetics of valley transits goes back to Wright himself. He assumed that this process could only happen in small populations, in which purifying selection is weak (figure 2.9, right). This, in turn, motivated his *shifting balance theory*, which holds that population subdivision is an essential requirement for population adaptation on multipeaked fitness landscapes.

We now quantitatively explore this process. Consider the simplest, two-locus, symmetric fitness valley (figure 3.10), which again illustrates the essential role of reciprocal sign epistasis: both the $a \rightarrow A$ and $b \rightarrow B$ mutations are deleterious in individuals with the $ab$ and $AB$ haplotypes, while the same mutations are beneficial in the $aB$ and $Ab$ haplotypes. How long (on expectation) will it take for a population to evolve from the local fitness peak ($ab$) to the escape haplotype ($AB$)?

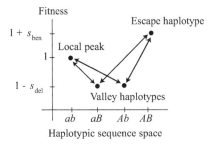

**Figure 3.10**
Two-locus, two-peaked fitness landscape with symmetric fitness valley. (Both $s_{ben}$ and $s_{del}$ are positive, and $s_{del}$ is not larger than 1.) The $ab$ haplotype is a local fitness peak, escape from which is accomplished when the population evolves to the $AB$ haplotype. Together, these define two valley haplotypes $aB$ and $Ab$, here of equal fitness. Note that the $x$-axis does not represent Hamming distance, which is instead represented by arrows. Figure modified from Weinreich and Chao (2005).

In chapter 2, we observed that deleterious mutations can reach fixation with nonnegligible probability if the compound parameter $Ns_{del}$ is not much larger than 1. This was Wright's intuition: sufficiently small populations can transit the fitness valley by sequentially fixing first one (deleterious) mutation and then a second, now-beneficial one. If $\mu$ is the per generation, per locus probability of mutation, then the per generation probability of mutation at either locus in a haploid population of $2N$ individuals is 2 loci $\times 2N$ individuals $\times \mu = 4N\mu$. Thus, the per generation probability of fixation of either $Ab$ or $aB$ is $4N\mu P_{fix}(x_0 = \frac{1}{2N}, N, -s_{del})$, where $P_{fix}$ is given by equation (2.24). From there, only one locus remains to be mutated, meaning that the per generation probability of the fixation of the $AB$ haplotype is

$$2N\mu P_{fix}(x_0 = \frac{1}{2N}, N, s_{ben} + s_{del}).$$

Finally, because the expected waiting time for an event in generations is equal to the reciprocal of its per generation probability (see again the derivation of equation [2.36] from [2.34]), the expected waiting time to transit the fitness valley in figure 3.10 by the sequential fixation of mutations is

$$E[t_{seq}] = \cfrac{1}{4N\mu P_{fix}\left(x_0 = \cfrac{1}{2N}, N, -s_{del}\right)} + \cfrac{1}{2N\mu P_{fix}\left(x_0 = \cfrac{1}{2N}, N, s_{ben} + s_{del}\right)}, \qquad (3.20)$$

where the subscript on $t$ signifies the fact that two mutations fixed sequentially. Equation (3.20) is shown by the dotted (mustard) line in figure 3.11. Note that once population size grows beyond $\sim\frac{1}{s_{del}}$, this time becomes tremendously long. This result motivates the attention to population subdivision, as per Wright's shifting balance theory.

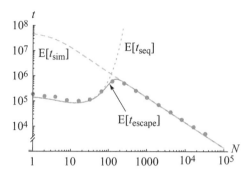

**Figure 3.11**
Time for populations to transit the fitness valley shown in figure 3.10 as a function of population size. Expectations assuming the sequential (equation 3.20), or simultaneous (equation 3.21) processes in isolation, and allowing either process (equation 3.22) are shown by dotted (mustard), dashed (green), and solid (blue) lines, respectively. Simulation results (filled circles) are averages across 1,000 replicate realizations. Here, $\mu = 10^{-5}$, $s_{del} = 10^{-2}$, and $s_{ben} = 10^{-1}$. Figure reproduced with permission from Weinreich and Chao (2005). © 2005 The Society for the Study of Evolution. All rights reserved.

However, a second population genetic process for crossing the fitness valley is also possible. Recall that at the one-locus mutation/selection equilibrium, deleterious mutations segregate at frequency $(1 - \tilde{p}) = \dfrac{\mu}{s_{del}}$. Thus, a haploid population of size $2N$ will on expectation be carrying $2 \text{ loci} \times 2N \text{ individuals} \times \dfrac{\mu}{s_{del}} = \dfrac{4N\mu}{s_{del}}$ $aB$ or $Ab$ haplotypes. Mutation at the remaining locus in these haplotypes will give rise to an offspring carrying the $AB$ haplotype, which happens with per generation probability $\mu \left( \dfrac{4N\mu}{s_{del}} \right) = \dfrac{4N\mu^2}{s_{del}}$. Such a haplotype is mostly competing with $ab$ haplotypes, over which it has fitness advantage $s_{ben}$, and will thus fix with approximate probability $P_{fix}(x_0 = \dfrac{1}{2N}, N, s_{ben})$. This gives

$$E[t_{sim}] = \dfrac{s_{del}}{4N\mu^2 P_{fix}\left( x_0 = \dfrac{1}{2N}, N, s_{ben} \right)}, \tag{3.21}$$

where the subscript on $t$ signifies the fact that the two mutations fix simultaneously. Equation (3.21) is shown by the dashed (green) line in figure 3.11. Once population size is larger than $\sim \dfrac{1}{s_{ben}}$, $E[t_{sim}]$ declines with population size because the number of low-fitness valley haplotypes increases with $N$. The sequential process is also sometimes called *stochastic tunneling* because the population seems to "tunnel" through a barrier between high-fitness haplotypes. (Our treatment assumes no recombination; Weissman et al. [2010] allow it.)

Finally, we compute the time to escape the local fitness peak and reach the escape haplotype via either process as

$$\mathrm{E}[t_{\mathrm{escape}}] = \frac{1}{\dfrac{1}{\mathrm{E}[t_{\mathrm{seq}}]} + \dfrac{1}{\mathrm{E}[t_{\mathrm{sim}}]}}. \tag{3.22}$$

Here, we reason that the rate to escape is the sum of the rates via each process, and the waiting time for the compound process is again the reciprocal of the rate. This is the solid (blue) line in figure 3.11, and is largely consistent with results from computer simulations (shown by the filled circles in the figure).

In summary, two regimes for crossing fitness valleys exist: small populations escape the local fitness peak via the sequential fixation of two mutations, while in large populations, the two mutations fix simultaneously. We conclude that Wright's shifting balance theory may be important if populations are at least periodically small, but large asexual populations are also able to quickly cross valleys on the fitness landscape.

---

**Power User Challenge 3.18**

Why does equation (3.22) modestly underestimate the correct value in the sequential fixation regime compared to simulations? (The first five points in figure 3.11 are all above $\mathrm{E}[t_{\mathrm{escape}}]$.) And why does it also do so in very large populations? (Ditto the last three points in the figure.)

---

**Power User Challenge 3.19**

What is $\mathrm{E}[t_{\mathrm{escape}}]$ after relaxing the symmetry in figure 3.10. You should write $s_A$ and $s_B$ for the selection coefficients acting against the $Ab$ and $aB$ haplotypes, respectively.

---

## 3.2   Stochastic Multilocus Theory

We now turn to stochastic effects in multilocus models. Our first challenge will be to extend coalescent theory to model the genealogy of a sample of multilocus haplotypes. Interestingly, while shared parentage at one locus causes lineages to merge (looking backward in time), recombination between loci causes them to split. This insight yields ready access to the equilibrium correlation between allelic identities at two loci as a function of the population size and recombination rate. Next, we investigate the influence of selection at one locus on the evolution of other, genetically linked loci. This represents a source of stochasticity not contemplated in chapter 2: the uncertainty over the genetic background on which a new selected mutation appears. This process bears strong parallels to random genetic drift, but important differences also exist.

### 3.2.1 Multilocus Coalescent Theory

In chapter 2, we modeled one-locus genealogies defined by some sample of (haploid) alleles under the Wright–Fisher model. We now extend those ideas by modeling two-locus genealogies with recombination. At one limit, recall that in the absence of recombination, the entire ensemble of alleles across a haploid genome is faithfully cotransmitted to offspring. Consequently, asexual populations (and also nonrecombining portions of genomes in sexual populations) will realize a single, shared genealogical history at all loci, which can be modeled with our chapter 2 theory. The other limit is free recombination between loci. In this regime, we can essentially (see power user challenge 3.20) regard each locus as representing an independent, replicate genealogy drawn from the same population, and chapter 2 theory will again suffice for each.

---

**Power User Challenge 3.20**

One generation of random mating brings one-locus diploid genotypes to Hardy–Weinberg frequencies (section 1.3.1). In contrast, recombination only drives linkage disequilibrium to zero exponentially (equation 3.5). Thus, even free recombination between loci ($r = \frac{1}{2}$) does not render the evolutionary history of two loci entirely independent statistically in finite time. Yet, we are content to approximately equate free recombination with $D = 0$. Find the expected number of generations between recombination events between loci when $r = \frac{1}{2}$ to see why.

---

**Teachable Moment 3.13**

Genealogies defined by freely recombining loci represent replicate coalescent realizations drawn from a single population's history. This has proven critical to many of the methods introduced in chapter 4, because whole-genome data sets can capture many thousands of such loci, vastly reducing statistical uncertainties in parameter estimation. Recall for example that the variance in pairwise coalescence times under Wright–Fisher assumptions is $4N^2$.

---

#### 3.2.1.1 The ancestral recombination graph

Now consider the genealogies defined by a sample of $n$ haplotypes carrying two loci with a recombination rate $0 < r < \frac{1}{2}$ between them. In this case, we might expect that some but not all coalescences will be shared between loci. Figure 3.12 illustrates the topology of one possible genealogy allowing recombination in a sample of $n = 3$, two-locus haplotypes. First, focusing on each locus in isolation yields two *marginal genealogies* (top right and left), so-called because they are genealogies at a locus averaged over all other loci in the sample. Their difference (lineages 1 and 2 are the first to coalesce on the left but not on the right) can be reconciled by a recombination event between lineages in the past. This history is illustrated in detail (bottom left) and in a more abstracted form (bottom right). This illustrates a point already noted: recombination causes lineages to split in

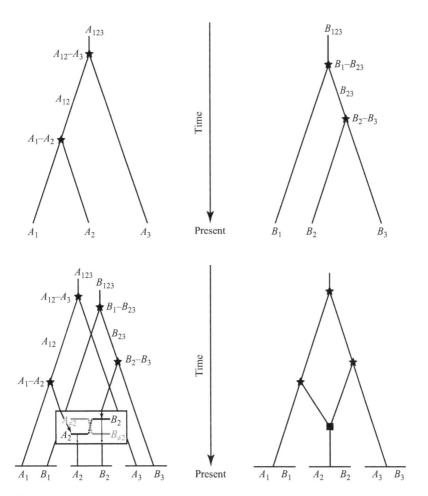

**Figure 3.12**
ARGs capture the genealogical consequences of coalescence and recombination Three two-locus haplotypes are sampled in the present generation with alleles at the two loci labeled $A_1B_1$, $A_2B_2$, and $A_3B_3$, respectively. We are agnostic on whether these alleles are genetically identical or different; our interest is only in the genealogies they define. Top: marginal genealogies for the lineages at the $A$ (left) and $B$ loci (right). Stars represent coalescent events, and coalesced lineages are named for their constituent lineages. Note that coalescent times in the two alleles' genealogies can differ (although their expectations are identical since in each generation, their constituent lineages are independently sampling parents in the same population). Bottom left: looking backward in time, the first (most recent) event is recombination between $A$ and $B$ loci which splits the $A_2$ and $B_2$ lineages. We are largely agnostic on the lineages of the other product of recombination (orange); although as indicated, we do know that they are not $A_2$ and $B_2$. Thereafter (still moving backward in time), the $A_2$ and $B_2$ genealogies are uncoupled. Next comes the coalescence between the $A_1$ and $A_2$ lineages (star), after which they share a lineage labeled $A_{12}$, which is identical to (i.e., genetically linked with) the $B_1$ lineage. Next, we find the $B_2$–$B_3$ coalescence (yielding lineage $B_{23}$, linked to the $A_3$ lineage), followed by the $B_{23}$–$B_1$ coalescence (still on the $A_3$ lineage, yielding $B_{123}$). Finally, comes the $A_{12}$–$A_3$ coalescence (on the $B_{123}$ lineage). Bottom right: a simplified representation the same ARG, where the black square represents recombination; lineage names also omitted for clarity.

two (looking backward in time), whereas coalescent events cause two lineages to merge into one. A genealogical representation that captures both coalescence and recombination events is called an *ancestral recombination graph* (*ARG*). Importantly, the ARG illustrated in figure 3.12 is quite simple, consisting of just $n=3$ samples at two loci. As either quantity increases, so too will the complexity.

---

**Power User Challenge 3.21**

Imagine a sample of $n = 2$ two-locus haplotypes, with alleles labeled $A_1B_1$ and $A_2B_2$. First draw the marginal genealogies for each locus. Now draw the ARG if the first event in the genealogy (looking backward in time) is a coalescence. If the per meiosis probability of recombination is $r$ and the sample comes from a diploid population of $N$ individuals honoring all Wright–Fisher assumptions, what is the probability of this outcome? Hint: see teachable moment 2.21. Finally, draw the ARG assuming that the first event in the genealogy is a recombination event and that the next two events are coalescence events.

---

**3.2.1.2 Recombination/drift equilibrium**   Quite obviously, correlations between genealogies at two loci go down as the recombination rate between them goes up. On the other hand, correlations go up as effective population size goes down, since smaller populations have shorter coalescent times during which recombination may act. The balance between these two processes will depend on their relative strengths: $r$ and $\dfrac{1}{2N}$, respectively. (As elsewhere, in moving from deterministic to stochastic theory, $r$ is now the per meiosis probability of recombination between loci rather than its rate.)

Sharpening the question, we seek the equilibrium correlation between alleles at loci between which the recombination rate is $r$ in a diploid population of size $N$. Using fairly elementary results from probability theory, one can show (Tenesa et al. 2007)

$$\tilde{r}_P^2 = \frac{1}{1+4Nr}, \tag{3.23}$$

where $r_P^2$ is the square of the Pearson correlation coefficient between loci introduced in teachable moment 3.3. (The same result can be found via a recurrence equation analogous to that which led to equation [2.31]; Sved 1971.) Equation (3.23) has a very satisfying interpretation. In partial answer to power user challenge 3.21, it is the probability that the two lineages coalesce before one allele on either lineage recombines onto some third lineage: $\dfrac{\dfrac{1}{2N}}{\dfrac{1}{2N}+2r} = \dfrac{1}{1+4Nr}$. Thus equation (3.23) tells us that the squared correlation coefficient between loci can be regarded equally as the probability that any two loci with a recombination rate $r$ between them share a single MRCA, or equivalently, as the fraction of loci at that genetic map distance that do so.

Note the formal parallels between equation (3.23) and the mutation/drift and migration/drift equilibria explored in chapter 2 (equations [2.32] and [2.33], respectively). But in place of heterozygosity at a single locus, the analogous quantity here is the correlation between alleles at two loci. Our biological interpretation of this result is also analogous: if the recombination rate is much smaller than the reciprocal of population size, random genetic drift will dominate recombination, and the genealogies of the two loci will be largely congruent. On the other hand, if the recombination rate is much larger than the reciprocal of population size, recombination will dominate drift and genealogies for the two will be largely uncorrelated.

## 3.2.2  Selective Sweeps and Genetic Draft

The mutations in the previous section were all assumed to be selectively neutral. Suppose instead that one is beneficial. (The consequence of linked deleterious mutations will be explored in section 3.2.3.) The fixation of a beneficial mutation will eliminate genetic variation at that site, as well as at all those genetically linked to it. This is called a *selective sweep*, since natural selection "sweeps" heterozygosity out of the population. The beneficial mutation is called a *driver mutation*, and selectively neutral mutations whose frequencies are perturbed by genetic linkage to a driver are called *passenger* or *hitchhiking mutations*. Recombination localizes the effects of the sweep to regions that are genetically "close" to the driver, but we begin with the simpler, no-recombination case.

### 3.2.2.1  Recurrent selective sweeps without recombination

In the absence of genetic recombination, the fixation of a beneficial mutation will eliminate genetic variation everywhere else in the genome, after which mutation begins to restore it. We explore the relaxation process in power user challenge 3.23, but for now ask a simpler question: averaged over recurrent selective sweeps, what is expected coalescent time in a nonrecombining genome? Our treatment follows Gillespie (2000).

Imagine a two-site, diploid, nonrecombining population of $N$ individuals (or $2N$ haploids) with nonoverlapping generations. Let all mutations at the first site be beneficial and suppose they become established with a per generation probability $\rho$. (Equivalently, $\frac{1}{\rho}$ is the expected interval between establishment events.) We make three comments on $\rho$. First, we disregard the stochastic vagaries of the beneficial mutations' establishment. Second, unlike the two-allele models of selection considered earlier, we assume the driver site can sustain an unlimited succession of beneficial mutations. Finally, we assume SSWM conditions (introduced in section 3.1.7); that is, that fixation is much faster than the interval between fixations. This substantially simplifies things because it means that at most, there will be one beneficial mutation in the population at any one time. Mathematically, SSWM reads $t_{\text{fix}} << \frac{1}{\rho}$; we relax this assumption in sections 3.2.4.1 and 4.4.4.

---

**Power User Challenge 3.22a**

If beneficial mutations are established with a per generation probability $\rho$, what is the approximate per generation probability of a beneficial mutation? Assume branching assumption conditions: $s \ll 1$ and $Ns \gg 1$.

Because of our strong selection/weak mutation assumption, we can regard the population's evolution as proceeding by the alternation of two distinct phases. The first are long intervals between selective sweeps. Since the expected interval between sweeps is $1/\rho$ and the time for fixation is $t_{\text{fix}}$, the population spends a proportion $\dfrac{1/\rho - t_{\text{fix}}}{1/\rho} = 1 - \rho t_{\text{fix}}$ of the time in this phase. During this time, the per generation probability of pairwise coalescence at the neutral site is $\dfrac{1}{2N}$, as in chapter 2. But these long intervals will be punctuated by rare and rapid selective fixations, which occupy a proportion $\rho t_{\text{fix}}$ of the time. In the absence of recombination, coalescence at the neutral site is assured during these intervals. Thus, during a sweep, the average per generation probability of pairwise coalescence is $\dfrac{1}{t_{\text{fix}}}$.

Putting these ideas together, we find that in the absence of genetic recombination, recurrent selective sweeps with a per generation probability $\rho$ give us

Probability (pairwise coalescence in the previous generation)

$$= (1 - \rho t_{\text{fix}}) \frac{1}{2N} + \rho t_{\text{fix}} \frac{1}{t_{\text{fix}}} \approx \frac{1}{2N} + \rho, \qquad (3.24)$$

where the approximation follows from our strong selection/weak mutation assumption. Thus, the expected time to pairwise coalescence now reads

$$\mathrm{E}[t_2] \approx \frac{1}{\dfrac{1}{2N} + \rho} = \frac{2N}{1 + 2N\rho}. \qquad (3.25)$$

In English, selective sweeps introduce a second source of stochasticity in coalescent times at neutral sites: the chance genetic linkage to driver mutations. Like random genetic drift, this second stochastic process—sometimes called *genetic draft* (Gillespie 2000)—also eliminates selectively neutral variation from the population. But unlike drift, there is a lower limit to the strength of draft. In the absence of recombination, coalescent times can now be no longer than the expected waiting time between selective sweeps, regardless of how large a population's (census) size becomes. This is illustrated in the left panel of figure 3.13. For small $N$, equation (3.25) is equal to the Wright–Fisher analog, $E[t_2] = 2N$. But as $N$ grows, its denominator eventually becomes roughly $2N\rho$, and $E[t_2]$ converges to $\dfrac{1}{\rho}$.

As first noted by Maynard Smith and Haigh (1974), this line of reasoning offers a resolution to Lewontin's paradox (empirical aside 2.4). Recall that the paradox emerges because under our one-locus theory, heterozygosity approaches 100% as population size grows, whereas heterozygosity in nature is rarely larger than 20%. In contrast, genetic draft introduces a census size-independent upper limit on pairwise coalescence times, thus also predicting an upper bound on heterozygosity, as illustrated in the right panel of figure 3.13.

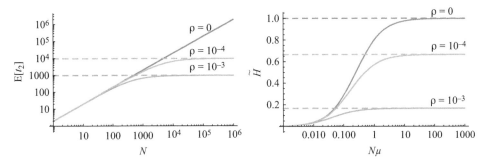

**Figure 3.13**

Recurrent selective sweeps shorten expected pairwise coalescent times and lower equilibrium heterozygosity in the absence of recombination. $\rho$ is the per generation probability of a selective sweep. Left: pairwise coalescent times as a function of census size (equation 3.25) plateaus at $\frac{1}{\rho}$. Right: correspondingly, equilibrium allelic heterozygosity at mutation/drift equilibrium (equation 2.32) in the presence of recurrent selective sweeps exhibits a much narrower dynamic range. ($\tilde{H}$ computed using the method described in teachable moment 2.21, replacing the Wright–Fisher expression for the per generation probability of coalescence with equation [3.24]. Results for $\rho = 0$ are identical to those in figure 2.12.)

---

**Power User Challenge 3.22b**

If beneficial mutations are established with a per generation probability $\rho$, approximately what is the per generation probability of a beneficial mutation, still assuming $s \ll 1$ but now relaxing the assumption $Ns \gg 1$? Explain how this provides another resolution to Lewontin's paradox distinct from that in figure 3.13.

---

**Teachable Moment 3.14:** Genetic draft and effective population size

Under the infinite sites models, equilibrium per site heterozygosity is given by $\tilde{\pi} = 2\mu E[t_2]$ (equation 2.41), where $E[t_2]$ is the expected pairwise coalescent time in the population. Under the Wright–Fisher model, $E[t_2] = 2N$, and in other demographic models this becomes $2N_e$, where $N_e$ is the coalescent effective population size (section 2.5). This rescaling allows us to apply Wright–Fisher theory to demographic models that violate its assumption of a well-mixed population.

Genetic draft also perturbs $E[t_2]$, suggesting the possibility that a similarly rescaled population size might allow us to again apply Wright–Fisher theory to populations undergoing selective sweeps. The path forward would be as in section 2.5: equate an expression for $E[t_2]$ under draft (e.g., equation 3.25) with its Wright–Fisher analog and solve for $N_e$.

However, as emphasized by Neher (2013), that approach is inappropriate for at least two reasons. First, our Wright–Fisher treatment implicitly disallows coalescent between more than two lineages in a generation. In contrast, each copy of a sufficiently strongly beneficial driver mutation must reasonably be allowed to sometimes yield three or more offspring, descendants of which may be present in our sample. Second, under drift, all lineages have the same probability of having anomalously good (or bad) luck, whereas under draft, lineages carrying a driver mutation persistently enjoy a higher-than-average probability of good luck. Put another way, the realized stochastic perturbations due to draft are heritable; under drift they are not.

Both of these violations of Wright–Fisher assumptions undermine the strategy of modeling selective sweeps as neutral processes with rescaled population sizes. More abstractly, they undermine our ability to model genealogies and mutations independently, as we did in chapter 2. Expected pairwise coalescent times under draft can reliably predict equilibrium per site heterozygosity (which depends only on the infinite sites assumption). However, the distortions in larger (i.e., deeper) genealogical structure induced by multiple coalescences and heritability in stochastic perturbations cannot be captured by Wright–Fisher theory at any effective population size. More complex coalescence models are required (reviewed in Neher [2013]), but these are beyond our scope.

### 3.2.2.2   Coalescent time immediately after a selective sweep with recombination   The situation becomes more complicated when we allow recombination: pairwise coalescent times must somehow also depend on the recombination rate with the driver mutation, and thus will vary across the genome. More specifically, a selective sweep will only result in pairwise coalescence at a passenger locus if neither lineage recombines away from the driver during the sweep. We follow chapter 13 of Coop (2020) to understand this position-dependent effect on coalescent times.

Imagine that a beneficial mutation has just appeared at a site in genetic linkage with (i.e., on a haplotype carrying) a neutral passenger mutation. If established, this beneficial mutation will begin to increase in frequency, simultaneously causing the frequency of the genetically linked passenger mutation to rise. We seek the expected coalescent time for two copies of the passenger mutation in the generation in which the driver reaches fixation. If neither passenger's lineage recombines away from the driver mutation during its sweep, their coalescence is assured sometime in the previous $t_{\text{fix}}$ generations, as per the no-recombination case. On the other hand, if a recombination event between the driver and passenger sites occurs in at least one of the two lineages, the two passenger mutations' expected coalescent time is $2N$ generations, as per chapter 2.

Recall that $r$ is the per meiosis probability of recombination between sites, and that half the products of meiosis are recombinants, so $\frac{r}{2}$ is the per meiosis probability that one copy of the neutral, linked mutation will recombine away from the driver. The per generation, per lineage probability of not recombining away from the sweeping driver is thus $1-\frac{r}{2}$ and the probability that neither recombined away during the sweep is $\left[\left(1-\frac{r}{2}\right)^{t_{\text{fix}}}\right]^2 \approx e^{-rt_{\text{fix}}}$. Putting these ideas together, we find that immediately after a selective sweep, the expected pairwise coalescent time at a neutral passenger site is

$$\begin{aligned}
\mathrm{E}[t_2^*(r)] &= e^{-rt_{\text{fix}}}t_{\text{fix}} + (1-e^{-rt_{\text{fix}}})2N \\
&\approx 2N(1-e^{-rt_{\text{fix}}}).
\end{aligned} \tag{3.26a}$$

The asterisk on $t_2$ signifies the fact that this is the coalescent time in the generation the driver reaches fixation. The approximation follows if we assume that selective fixation happens much more quickly than neutral coalescence, or mathematically, if $t_{\text{fix}} \ll 4N$. Another way that we can think about $\mathrm{E}[t_2^*(r)]$ is as the number of lineages in the population not swept to coalescence by the driver mutation.

This expression quantifies the intuition articulated in the first paragraph of this section, that coalescent times will vary in a recombining genome, and is illustrated in the top panel of figure 3.14. In the generation in which a driver mutation fixed, expected coalescent times are essentially zero very close to it ($r \approx 0$), since the probability of recombination during the sweep ($rt_{\text{fix}}$) is so low. As the recombination rate with the driver mutation increases, the probability of recombination during its sweep increases, and with it, expected coalescent times. Finally, once $r$ between the driver and passenger sites becomes larger than roughly $\dfrac{1}{t_{\text{fix}}} \approx \dfrac{s}{2\ln(2N)}$

(this is the reciprocal of equation [1.10]), the sweep's effect on coalescent times becomes negligible. Or in the language of teachable moment 2.17, the genetic "reach" of a sweep scales as compound parameter $\sim \dfrac{s}{r}$. ($2\ln(2N)$ is a constant if we imagine computing equation [3.26a] across a single genome.) In English, if the recombination rate $r$ between driver and passenger sites is much larger than the driver's selective advantage $s$, genetic draft will not affect coalescent times at the passenger site.

To this point, our theory has been developed in terms of recombination rate $r$. Recall that a pair of sites with recombination rate (or probability) 1% are said to be 1 cM apart. DNA sequence data also give *physical distances* between sites. Physical distances $l$ are measured in megabases of DNA (1 Mb $= 10^6$ bases). Dividing recombination rate $r$ between sites by their physical distance $l$ yields the physical recombination rate $c$ (in units of $\dfrac{\text{cM}}{\text{Mb}}$), namely the per generation, per megabase rate of recombination. Or equivalently, $r = cl$.

---

**Empirical Aside 3.6:** What are physical recombination rates in nature?

In a survey of mammals (Dumont and Payseur 2008), the mean physical recombination rate was approximately $0.822 \dfrac{\text{cM}}{\text{Mb}} = 8.22 \times 10^{-9}$ recombination events per base pair per generations. In humans, this figure is approximately $1 \dfrac{\text{cM}}{\text{Mb}}$, half that in rats and mice, and roughly $0.2 \dfrac{\text{cM}}{\text{Mb}}$ in two marsupial species. A broader survey of eukaryotes (Stapley et al. 2017) found comparable values in plants; although values $40 \dfrac{\text{cM}}{\text{Mb}}$ or higher were reported in a smaller number of species.

There is also considerable variation within genomes. For example, the human genome is characterized by hotspots of 1–2 kb within which recombination rates are elevated 5–10-fold relative to the mean (Myers et al. 2005), and similar patterns are seen in many other species. Heritable, sex- and population-specific variation in recombination rate are also common. See Stapley et al. (2017) for a recent review.

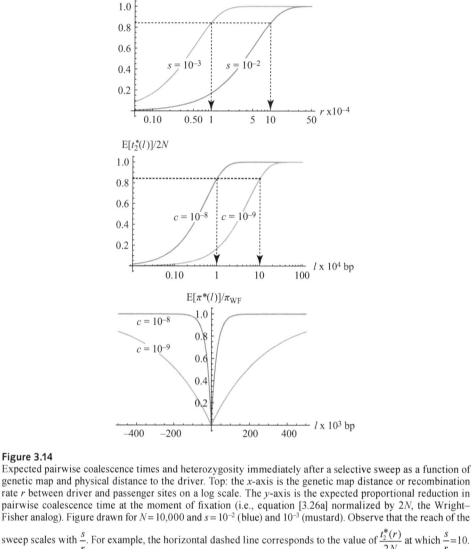

**Figure 3.14**
Expected pairwise coalescence times and heterozygosity immediately after a selective sweep as a function of genetic map and physical distance to the driver. Top: the $x$-axis is the genetic map distance or recombination rate $r$ between driver and passenger sites on a log scale. The $y$-axis is the expected proportional reduction in pairwise coalescence time at the moment of fixation (i.e., equation [3.26a] normalized by $2N$, the Wright–Fisher analog). Figure drawn for $N = 10,000$ and $s = 10^{-2}$ (blue) and $10^{-3}$ (mustard). Observe that the reach of the sweep scales with $\frac{s}{r}$. For example, the horizontal dashed line corresponds to the value of $\frac{t_2^*(r)}{2N}$ at which $\frac{s}{r}=10$. Thus, it crosses the mustard and blue lines at a recombination rate equal to their selection coefficients divided by 10. Center: the $x$-axis is physical distance $l$ between driver and passenger sites on a log scale. The $y$-axis is again the proportional reduction in expected pairwise coalescence time at the moment of fixation (now equation [3.26b], again normalized by $2N$). Figure drawn for $N = 10,000$, $s = 10^{-3}$, and $c = 10^{-8}$ (blue) and $10^{-9}$ (mustard) per base pair. The reach of the sweep now scales with $\frac{s}{c}$. For example, the horizontal dashed line corresponds to the value of $\frac{t_2^*(l)}{2N}$ at which $\frac{s}{lc}=10$ and thus crosses the blue and mustard lines at physical distances equal to $\frac{s}{c}$ divided by 10. Bottom: the $x$-axis is again physical distance $l$ between driver and passenger sites (now on a linear scale) and the $y$-axis is the expected proportional reduction in per site heterozygosity (equation [3.26c], again divided by the one-locus Wright–Fisher expectation, equation [2.42]). This panel can be regarded as representing a 1 Mb ($10^6$ base) piece of DNA centered on the driver mutation. Parameters $N$, $s$, and $c$ as in the middle panel of this figure.

Calibrating recombination rate in physical distance yields

$$\mathrm{E}[t_2^*(l)] \approx 2N(1 - e^{-clt_{\mathrm{fix}}}), \qquad (3.26b)$$

illustrated in the middle panel of figure 3.14. In this parameterization, the physical reach of the sweep scales as $\sim \dfrac{s}{c}$ Mb.

Finally, substituting equation (3.26b) into $\tilde{\pi} = 2\mu\mathrm{E}[t_2]$ (equation 2.41) yields

$$\mathrm{E}[\pi^*(l)] \approx 4N\mu(1 - e^{-clt_{\mathrm{fix}}}) \qquad (3.26c)$$

for the per site heterozygosity in the generation that the driver mutation fixes. Equation (3.26c) (again normalized by its Wright–Fisher analog) is shown in the bottom panel of figure 3.14. This illustrates a characteristic trough in heterozygosity induced by a selective sweep centered on the driver locus. As expected, the width of the trough scales as $\sim \dfrac{s}{c}$ Mb. Many mapping methods rely on this effect to locate sites of biological interest, and we return to these ideas in chapter 4. Of more immediate interest to us, in many species there is variation in physical recombination rates within genomes (empirical aside 3.6), the implications of which we develop next.

---

**Power User Challenge 3.23:** The trough in heterozygosity in the bottom panel of figure 3.14 is transient; what is its relaxation time?

Equation 3.26c gives $\mathrm{E}[\pi^*(l)]$, the expected per site heterozygosity at a neutral locus $l$ nucleotides from a driver mutation that has just fixed in the population. What is $\mathrm{E}[\pi_t(l)]$, its expected heterozygosity $t$ generations later? Finding this answer is only slightly more difficult than was finding $N(t) = N(0)e^{rt}$ in section 1.1.1. First, use the strategy in equation (2.31) to write an expression for $\mathrm{E}[\Delta\pi_t(l)]$. After letting $\Delta t \to 0$, integrate your answer with respect to time. Here is the only wrinkle: you will need a function $Q_t(l)$ of $\pi_t(l)$ that lets you rewrite your differential equation in the form $\dfrac{dQ_t(l)}{dt} = BQ_t(l)$ where $B$ is some constant. (Close examination of teachable moments 2.16 and 2.19 will illustrate the characteristics your $Q$ needs.) Finally, use equation 3.26c to find the constant of integration. Hint: since $\pi^*(0) = 0$, finding $\mathrm{E}[\pi_t(0)]$ will build intuition for the more general question.

---

#### 3.2.2.3 Recurrent selective sweeps with recombination

The attentive reader may be wondering why setting $c = 0$ in equation (3.26b) doesn't recover equation (3.25). The answer is that they are answers to different questions. Equation 3.25 is the expected pairwise coalescence time across a nonrecombining genome in the presence of recurrent selective sweeps. Equation (3.26b) is the same quantity at a site in a recombining genome the moment a driver at physical distance $l$ Mb away reaches fixation.

What is the expected pairwise coalescence time in recombining genomes subject to recurrent selective sweeps? Some progress is easy: if recurrent beneficial mutations occur with probability $\rho$ at a site $l$ Mb away, we have

Probability (pairwise coalescence in the previous generation)

$$=\frac{1}{2N}+\rho e^{-clt_{\text{fix}}},$$

which immediately gives

$$\mathrm{E}[t_2(l)]=\frac{2N}{1+2N\rho e^{-clt_{\text{fix}}}}.$$

This is simply our no-recombination result (equation 3.25), modified to honor the fact that recombination means that coalescence is not assured by the selective sweep.

But this result is still dependent on $l$. We build intuition by assuming that driver mutations are uniformly distributed physically along the chromosome, or equivalently that the per generation probability that a beneficial mutation will become established ($\rho$) is independent of position in the genome. Thus, the expected distance between any arbitrary neutral site in the genome and the next beneficial mutation to become established is constant. To further simplify, imagine that each such possible beneficial mutation has the same selection coefficient $s$, so the expected duration of all fixation events is also constant.

Under these two conditions, the average probability that any particular selectively neutral site has recently been swept to fixation by genetic draft depends only on the local physical recombination rate $c$. Neutral sites in genomic regions where $c$ is low will be within the reach of more driver mutations, suggesting shorter pairwise coalescence times and lower heterozygosity in those regions. Conversely, those in regions where $c$ is high will be influenced by fewer sweeps, and should thus experience longer coalescence times and higher heterozygosity. In summary, heterozygosity within genomes is predicted to be positively correlated with local physical recombination rate. (The interested reader is directed to chapter 13 in Coop [2020] for a quantitative treatment of this reasoning.)

As noted in empirical aside 3.6, there is considerable variation in physical recombination rates within and among organisms. Indeed, another potential resolution to Lewontin's paradox is that physical recombination rates tend to be lower in species with larger census sizes (Buffalo 2021). But support for the prediction that pairwise coalescence time is positively correlated with recombination rate is much stronger than the implications for Lewontin's paradox provided by interspecific comparisons. One of the most exciting empirical findings in the last decade of the twentieth century was the observation that within species, the level of neutral polymorphism in the genome is positively correlated with local recombination rate (empirical aside 3.7), exactly as predicted by the reasoning in the previous paragraph.

**Empirical Aside 3.7:** Heterozygosity is often positively correlated with recombination rate, but divergence is not

This pattern was first observed in *Drosophila melanogaster* (Aguade et al. 1989; Begun and Aquadro 1992) is also evident in humans (Keinan and Reich 2010) and several other animals. Interestingly, it seems to be absent in plants (reviewed in Leffler et al. 2012).

However, caution is warranted in concluding that genetic draft is the mechanism responsible for the pattern, for two reasons. We treat the first one here, and the second in the next section. First, theory predicts that heterozygosity at mutation/drift equilibrium should also be positively correlated with mutation rate. Thus, if mutation rate were positively correlated with recombination rate, we would also expect the relationship described in empirical aside 3.7. For example, suppose that recombination *per se* were mutagenic, or alternatively, that neutral mutations were more common in regions of low recombination owing to the sorts of sites found there.

We can control for this effect by noting two facts. First, the rate of the molecular clock is also predicted to scale with the neutral mutation rate (empirical aside 2.1). Second, that scaling is unaffected by genetic draft (teachable moment 3.15). Thus, genetic divergence between species provides an immediate control for a mutation rate-dependent explanation of this effect. Indeed, while genetic divergence does exhibit a weak dependence on recombination rate in some species (e.g., Kulathinal et al. 2008), this effect seems unable to fully account for the correlation between heterozygosity and recombination rate described in empirical aside 3.7.

---

**Teachable Moment 3.15:** In the absence of recombination, genetic draft has no influence on the molecular clock

Imagine a neutral mutation $A$ at frequency $x$, which is also its probability of eventual fixation (equation 2.21). Now imagine a driver mutation destined to fixation appears in the population at another site. With probability $x$, the driver will be linked to $A$, in which case $A$ will fix. Thus, the probability of fixation for $A$ due to draft is equal to that due to drift in the absence of recombination (although the former process is much slower than the latter). This point was first noted by Birky and Walsh (1988) and we return to it in chapter 4.

---

### 3.2.3  Background Selection

The selective elimination of recurrent deleterious mutations will also reduce equilibrium heterozygosity. This is called *background selection*; the reasoning developed in the last section shows that the reach of background selection will also be inversely related to local recombination rate. Consequently, background selection offers another hypothesis for the pattern described in empirical aside 3.7 (Charlesworth et al. 1993).

While a rigorous mathematical treatment of background selection is beyond our scope, we can develop some intuition. (The interested reader is directed to Cvijović et al. [2018], chapter 3 in Walsh and Lynch [2018], and citations in both publications to learn more.) To begin, imagine a diploid, nonrecombining population of $N$ individuals (or equivalently, of $2N$ haploids) with nonoverlapping generations. Suppose it is evolving on a multiplicative, rotationally symmetric fitness function ($w_k = (1-s)^k$; equation [3.13]). Then lineages carrying $k$ deleterious mutations have deleterious selection coefficient $w_k - w_0 = (1-s)^k - 1 \approx -sk$, the absolute value of which is (by definition) their per generation probability of selective elimination. Thus, if $s$ is at all "large," such lineages will contribute very little to heterozygosity, and almost all lineages will find their coalescences among the mutation-free haplotypes in the population. Recall from section 3.1.4 that the equilibrium frequency of that class under this fitness function is $e^{-\frac{U}{s}}$, giving $E[t_2] \approx 2Ne^{-\frac{U}{s}}$ for this model. (In fact, this quantitative reasoning breaks down for surprisingly large values of $s$ [Cvijović et al. 2018]: even lineages destined for rapid selective loss can contribute to heterozygosity before they are eliminated. But the qualitative lesson stands: background selection shortens expected pairwise coalescence times.)

As it did in the case of genetic draft, recombination will again increase the pairwise coalescence times relative to no-recombination expectations in models of background selection. More specifically, recombination can move selectively neutral variation found on lineages carrying one or more deleterious mutations onto selectively unloaded backgrounds before they are eliminated by purifying selection. Thus, we expect a greater reduction in pairwise coalescence times and heterozygosity at sites closer to the target of purifying selection.

Going further, we again assume a physically uniform distribution of sites at which deleterious mutations can occur. In this case, background selection also predicts that equilibrium heterozygosity will be positively correlated with local, physical recombination rate (empirical aside 3.7). Also, as before, the rate of the molecular clock is predicted to be unaffected by background selection, allowing an identical control for recombination-mediated effects on mutation rate. Thus, recurrent selective sweeps and background selection are both consistent with the pattern described in empirical aside 3.7. Methods to distinguish these two hypotheses remain an active area of investigation (Comeron 2017).

### 3.2.4  The Hill–Robertson Effect

We have seen how pairwise coalescence times at a neutral locus (and thus, heterozygosity) is reduced by recurrent selection elsewhere in the genome. That happens whether selection coefficients are positive or negative, and the effect is attenuated by genetic recombination. We now turn to the situation of two genetically linked loci, both under selection. We will find that genetic linkage between loci reduces the efficiency of natural selection in the presence of random genetic drift. This is called the *Hill–Robertson effect* (Felsenstein 1974), though recently, Otto (2020) suggested the mechanistically more explicit term *selective interference*. As in the case of genetic draft, genetic recombination attenuates this effect, since it reduces the impact of the stochastic uncertainty over the genetic background in which a new mutation appears.

The Hill–Robertson effect also bears some similarities to the deterministic two locus treatment developed in section 3.1.3, where recombination likewise rendered selection more efficient. But importantly, that effect was dependent on negative epistasis, whereas the Hill–Robertson effect requires no epistasis. Nevertheless, a deeper understanding of the present problem can be found in that earlier treatment: recall that negative epistasis induced negative linkage disequilibrium, reducing fitness variance, and thus, making selection less efficient. Recombination reduced linkage disequilibrium (in absolute value) by recreating the underrepresented haplotypes, increasing both fitness variance and the efficiency of selection (see figure 3.3 left).

Thus, the Hill–Robertson effect seems to imply that selection and drift alone might induce negative linkage disequilibrium. That it should do so may be surprising, and we follow Barton and Otto (2005) to see why. In partial answer to power user challenge 3.2, in expectation, mutation *per se* introduces no linkage disequilibrium. To see this, first imagine a population of $N$ diploid, two-locus organisms in which allele $A$ is at frequency $x$ at the first locus, and the second locus is fixed for the $b$ allele. The corresponding haplotype frequencies and linkage disequilibrium are shown in the first row of table 3.2 Now suppose mutation introduces a single copy of the $B$ allele at the second locus. As in teachable moment 3.15, the $B$ allele will appear on an $Ab$ or $ab$ haplotype with probabilities given by their respective frequencies: $x$ and $1-x$. Haplotype frequencies and linkage disequilibria in each of these cases are shown in second and third lines of the same table. Finally, the expected linkage disequilibrium after the appearance of the $B$ allele is the mean of the values under these two cases, weighted by their respective probabilities, or

$$\mathrm{E}[D] = x\frac{1-x}{2N} + (1-x)\left(-\frac{x}{2N}\right) = 0. \tag{3.27}$$

If the expected linkage disequilibrium after mutation is zero, the mechanism of the Hill–Robertson effect must lie in the variability around that mean. And indeed, although the probabilities of the two outcomes shown in table 3.2 precisely offset the corresponding values of $D$, the selective consequences of these two possibilities are quite different. This follows for the reason already given: the efficiency of natural selection depends on the sign of linkage disequilibrium. More specifically, if the mutation happens to make $D$ positive, natural selection will tend to more quickly push the alleles at both loci to fixation (or loss, depending on the sign of their selection coefficient). On the other hand, if by chance $D$ is negative, selection will be slowed. Thus, natural selection enriches the population for segregating lineages in which $D < 0$, by slowing their transit to fixation or loss compared to those in which $D > 0$. This is the essence of the Hill–Robertson effect.

Two further points are important. First, while the preceding argument was developed for a single copy of a new mutation ($b \rightarrow B$), it applies equally at all frequencies of both alleles (Barton and Otto 2005). Second, the Hill–Robertson effect is not

**Table 3.2**
Expected haplotype frequencies and linkage disequilibria before and just after $b \rightarrow B$ mutation[a]

| | $p_{AB}$ | $p_{Ab}$ | $p_{aB}$ | $p_{ab}$ | $D$ |
|---|---|---|---|---|---|
| Before $b \rightarrow B$ mutation | 0 | $x$ | 0 | $1-x$ | 0 |
| If $b \rightarrow B$ mutation occurs on an $Ab$ haplotype | $\dfrac{1}{2N}$ | $x-\dfrac{1}{2N}$ | 0 | $1-x$ | $\dfrac{1-x}{2N}$ |
| If $b \rightarrow B$ mutation occurs on an $Ab$ haplotype | 0 | $x$ | $\dfrac{1}{2N}$ | $1-x-\dfrac{1}{2N}$ | $-\dfrac{x}{2N}$ |

[a] Assuming $A$ allele is present at frequency $x$ and $B$ is absent before the mutation.

restricted to "small" populations. It is true that the granularity of possible linkage disequilibria values between two loci increases with population size, suggesting that realized deviations due to variance around $E[D] = 0$ decline correspondingly. However, realized deviations increase again with the number of loci, meaning that the Hill–Robertson effect may be important in perhaps surprisingly large populations (Iles et al. 2003). Population subdivision also influences the strength of drift, thus further increasing the likely importance of the Hill–Robertson effect.

---

**Power User Challenge 3.24**

Is there a one-locus diploid analog to the Hill–Robertson effect? If so, how does it work? Hint: see power user challenge 3.5.

---

Thus far, our discussion of the Hill–Robertson effect has been agnostic on the sign of selection coefficients. We now focus on two specific cases that have been studied since the early 1930s. First, the Hill–Robertson effect causes recombination to accelerate adaptation when beneficial mutations are plentiful. This is called the *Fisher–Muller effect*: established, cosegregating beneficial mutations will of necessity compete with one another. In the absence of genetic recombination, some such mutations will be stochastically lost. Note that this is above and beyond the stochasticity of establishment. Rather, they will be lost because they had the bad luck to establish in negative linkage disequilibrium with other, already-established beneficial mutations. The other is the possibility that, in the presence of recurrent deleterious mutations, the most-fit haplotype in the population can be lost to drift. Here again, the risk is that deleterious mutations will appear in negative linkage disequilibrium, making them harder for natural selection to purge. This is called *Muller's ratchet*, a form of drift load (see teachable moment 1.19) that, like the Fisher–Muller effect, can be offset by genetic recombination. We now examine each case in greater detail.

**3.2.4.1 The Fisher–Muller effect**   The Fisher–Muller effect occurs when beneficial mutations are lost to competition with other beneficial mutations. We quantify this effect in terms of the *velocity of adaptation* ($v$), defined as the fitness increase ($\Delta w$) per unit time ($\Delta t$). Mathematically,

$$v = \frac{\Delta w}{\Delta t}.$$

Our benchmark is the velocity in the absence of the Fisher–Muller effect, that is, under the SSWM assumptions introduced in section 3.1.7. To simplify further, assume that all beneficial mutations have the same selection coefficient $s$. Under SSWM, velocity $v$ is the fitness gain per fixation event ($\Delta w = s$), divided by the expected time between fixations. This latter quantity ($\Delta t$) is the reciprocal of the per generation probability of establishment (and thus, fixation) of a beneficial mutation. Power user challenge 3.22a can be solved by appeal to the fact that, under SSWM assumptions, the per generation probability of establishment of a beneficial mutation is the product of the population wide, per generation beneficial mutation rate ($2N\mu$) and the per beneficial mutation probability of fixation ($\approx 2s$ under branching process assumptions). Algebraically then, $\rho \approx 4N\mu s$, and as elsewhere, the corresponding time between fixation events $\Delta t = \dfrac{1}{\rho} \approx \dfrac{1}{4N\mu s}$, in which case the velocity of adaptation is

$$v = \frac{\Delta w}{\Delta t} \approx \frac{s}{\dfrac{1}{4N\mu s}} = 4N\mu s^2. \tag{3.28}$$

In English, velocity $v$ is proportional to both population size and the beneficial mutation rate: both quantities contribute directly to the population-wide rate of production of beneficial mutations. And $v$ is proportional to the square of the selection coefficient. This is also easily understood: $s$ is the fitness gain per fixation event and half the probability that the next beneficial mutation will fix. (If selection coefficients follow some probability distribution, equation (3.28) applies equally after replacing $s^2$ with $E[s^2]$.) These ideas are illustrated in figure 3.15.

But as $N$ or $\mu$ increase (or $s$ declines), the time between fixation events will eventually become comparable to the duration of fixation events, in which case successive beneficial mutations cannot be modeled in isolation. This is the domain of the Fisher–Muller effect: in the absence of genetic recombination, beneficial mutations can now be lost to both drift and to draft.

Theoreticians differentiate between two forms of the Fisher–Muller effect (Desai and Fisher 2007). The *multiple mutations model* assumes that all beneficial mutations have the same selection coefficient $s > 0$. In this case, only beneficial mutations

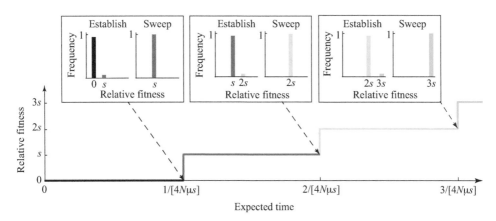

**Figure 3.15**
Schematic illustration of adaptation under SSWM assumptions. The key feature of this regime is that the time to complete each establishment/fixation event (shown in boxes) is vastly shorter than the time between events. Consequently, fitness increases in an approximately stepwise manner: every $\dfrac{1}{4N\mu s}$ generations (in expectation), a beneficial mutation with fitness advantage $s$ is established and fixes. The velocity of adaptation is the slope of this stepwise process averaged over many establishment/fixation events.

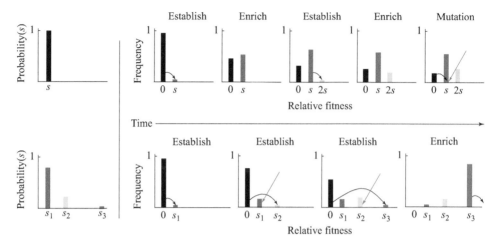

**Figure 3.16**
Schematic illustration of the Fisher–Muller effect. Top row: the multiple mutations model. Here, all beneficial mutations share the same selection coefficient $s$ (top-left panel). The behavior of this model is illustrated in the top-right panels. In the first and third of these panels, beneficial mutations establish novel, highest-fitness classes (shown in purple and blue, respectively). In the second and fourth panels, natural selection is seen to enrich these highest-fitness classes, contributing to population adaptation. But in the fifth panel, a beneficial mutation occurring on the black genetic background contributes a novel lineage (shown in red) to an intermediate fitness class. Even if this mutation escapes genetic drift, it is wasted because it makes no contribution to population adaptation, now being driven by the still-fitter blue class. Bottom row: the clonal interference model. Here, beneficial selection coefficients are drawn from some probability distribution (bottom-left panel). The behavior of this model is illustrated in the bottom-right panels (drawn after figure 2a in Neher 2013). In the first panel, a lineage with selection coefficient $s_1$ is established. But in the next, a second lineage with selection coefficient $s_2 > s_1$ is established, at which point the first beneficial mutation has been wasted (red arrow). Similarly, in the third panel, a yet-fitter lineage is established with selection coefficient $s_3 > s_2$, thereby wasting the second beneficial mutation.

occurring in individuals in the currently established, most-fit class can contribute to adaptation. Beneficial mutations occurring in members of less-fit classes are wasted, since at best, they only contribute to already-established more-fit classes. The multiple mutations model is illustrated in the top row of figure 3.16; the red arrow indicates a beneficial mutation lost to draft.

In contrast, the *clonal interference model* assumes that beneficial selection coefficients $s > 0$ are drawn from some probability distribution. In this case, lineages established with mutations of modest effect can be outcompeted (and so, wasted) if subsequent lineages establish with mutations of larger effect. In this case, adaptation is dominated by that subset of mutations of sufficiently large effect that their expected time to fixation is less than the waiting time to the establishment of mutations of still larger effect. The clonal interference model is illustrated in the bottom row of figure 3.16; again, red arrows indicate beneficial mutations lost to draft.

**Empirical Aside 3.8:** Evidence for the Fisher–Muller effect in laboratory populations of microbes

Data on the relationship between population size and the velocity of adaptation in natural populations are limited, since confounding factors of ecology and genetics complicate interpretation. But many informative experiments have been performed using replicate laboratory populations of microbes reared at different population sizes and mutation rates in fixed environments. Here, conclusions are unambiguous: in nonrecombining viruses, bacteria, and yeast, the pace of adaptation grows more slowly than linearly with population size (de Visser et al. 1999; Miralles et al. 1999; Desai et al. 2007). Moreover, this effect is ameliorated when recombination is allowed (Colgrave 2002; Goddard et al. 2005; Cooper 2007). More recently, competition between lineages carrying distinct beneficial mutations has been directly demonstrated with whole-genome sequence data from experimental populations of asexual yeast (Lang et al. 2013). This pattern also vanishes when recombination is allowed (McDonald et al. 2016).

The foregoing verbal arguments make clear that genetic linkage slows the velocity of adaptation in both the clonal interference and multiple mutations cases compared to that predicted under SSWM conditions. Quantitative progress on both forms of the Fisher–Muller effect have been made (Gerrish and Lenski 1998; Desai and Fisher 2007; Fogle et al. 2008; Park et al. 2010), but results are quite technical. As seen in sections 3.2.2 and 3.2.3, however, genetic recombination localizes draft physically (e.g., figure 3.14), and thus will weaken the Hill–Robertson effect (see also the center panel of figure 3.4). More specifically, pairs of loci that are genetically far enough to remain in approximate linkage equilibrium will respond to natural selection roughly independently. But if two simultaneously sweeping mutations are close, their probabilities of fixation will be perturbed.

The rigorous mathematics of the influence of recombination rate on the velocity of adaptation are also beyond our scope (the interested reader is directed to Neher et al. [2010], Weissman and Barton [2012], Neher [2013], Park and Krug [2013], and Weissman and Hallatschek [2014]), but some heuristics are possible. Following Neher (2013) we ask about the size of a sweeping beneficial mutation's Hill–Robertson "footprint"; that is, for how many generations will how many megabases of the genome be affected by a sweep? In section 3.2.2.2, we saw that the width of the trough of heterozygosity during a selective sweep scales as $\sim \dfrac{1}{ct_{\text{fix}}}$, where $c$ is the per generation, per base pair probabilty of recombination, and $t_{\text{fix}}$ is the duration of the sweep. Roughly speaking, that means that for a second beneficial mutation to escape the Fisher–Muller effect, it must occur further than this from some first, already established beneficial mutation. Thus, the Hill–Robertson effect causes each selective sweep to render $\sim \dfrac{1}{ct_{\text{fix}}}$ base pairs inhospitable to subsequent beneficial mutations for $t_{\text{fix}}$ generations, or $\sim t_{\text{fix}} \dfrac{1}{ct_{\text{fix}}} = \dfrac{1}{c}$ base pair-generations.

That this result it is independent of selection coefficient is reminiscent of the Haldane–Muller principle (section 1.2.2); although here, it applies to beneficial rather than deleterious mutations. More strongly beneficial mutations transiently have a wider physical reach across the genome, but their fixation is faster. Conversely, more weakly beneficial mutations have a narrower physical reach, but exert that influence for longer because their fixation is slower. What the mathematics tells us is that these two effects approximately cancel each other out. Thus genomes (or more precisely, regions of genomes) with higher physical recombination rates will be more effective at capturing beneficial mutations, regardless of selection coefficients. Put another way, the rate of selective sweeps in a sexual organism may be more strongly influenced by the total number of recombination events per generation (given by an organism's genetic map length, empirical aside 3.2) than by selection coefficients. We return to this point in section 3.3.1.

**3.2.4.2 Muller's ratchet**   Whereas the Fisher–Muller effect contemplates competition between beneficial mutations in the absence of recombination, Muller's ratchet considers the analogous question for deleterious mutations. To build intuition, imagine a nonrecombining, two locus population carrying a deleterious mutation at one locus. If a deleterious mutation now occurs at the second locus, expected linkage disequilibrium $D$ between them will be zero (equation 3.27). But the Hill–Robertson effect means that both deleterious mutations will be purged more quickly in realizations in which they are on the same haplotype ($D > 0$) and more slowly otherwise. Or equivalently, stochasticity means that populations will on average carry negative linkage disequilibrium among deleterious mutations.

Moving beyond this two locus picture, recall from section 3.1.4 that on a rotationally symmetric fitness landscape, an asexual population at mutation/selection equilibrium assumes a Poisson distribution of fitness classes (equation [3.15] and figure 3.5). The Hill–Robertson effect compresses that distribution toward the middle by enriching for intermediate fitness classes at the expense of the more extreme classes. Or equivalently, random genetic drift now joins mutation in opposing purifying selection. The magnitude of this effect grows as population size or selection coefficients decline, and also as mutation rate increases (see Goyal et al. [2012] for a rigorous, quantitative treatment of this balance). But if strong enough, the Hill–Robertson effect will result in extinction of the fittest haplotype, corresponding to one "click" of Muller's ratchet. And even if the just-lost, previously most-fit haplotype is transiently recreated by mutation (and establishes), the Hill–Robertson effect will again cause its extinction. Hence "ratchet": under these circumstances (i.e., suitably small selection coefficient and population size, and suitably large mutation rate), loss of the most-fit haplotype is irreversible in the long run.

Moreover, on that landscape, asexual populations susceptible to one click of Muller's ratchet will soon become susceptible to a second click. Recall first that in models of nonoverlapping generation, the selection coefficient between competing haplotypes is one less than the ratio of their fitness values, and further, that under equation (3.14), selection coefficients depend not on the absolute number of mutations in the two haplotypes, but only on the number of mutational differences between them. Mathematically, $\frac{w_{k+1}}{w_k} - 1 = \frac{(1-s)^{k+1}}{(1-s)^k} - 1 = -s$ is the selection coefficient between any two haplotypes that differ by one deleterious mutation. Thus disregarding drift, after the first click of the ratchet, the population will soon equilibrate to haplotype frequencies

$$\tilde{p}_k = \begin{cases} 0 & \text{if } k < 1 \\ \dfrac{e^{-U/s}\left(\dfrac{U}{s}\right)^{k-1}}{(k-1)!} & \text{otherwise} \end{cases},$$

where $k \geq 0$ as before. This expression is identical to equation (3.15), except that mutation-free haplotypes are absent and the frequency of haplotypes carrying $k+1$ deleterious mutations is now what the frequency of individuals with $k$ mutations was before the click of the ratchet.

Now however, the Hill–Robertson effect will again compress that distribution toward the center, making a second clock of the ratchet equally likely. After the second click, the population will again soon equilibrate, this time to haplotype frequencies

$$\tilde{p}_k = \begin{cases} 0 & \text{if } k < 2 \\ \dfrac{e^{-U/s}\left(\dfrac{U}{s}\right)^{k-2}}{(k-2)!} & \text{otherwise} \end{cases},$$

and so on. In other words, our theory predicts that in the absence of epistasis, asexual populations beyond the threshold of Muller's ratchet will rapidly deteriorate genetically. (The situation allowing epistasis, as well as back mutation, is explored in teachable moment 3.27.)

---

**Power User Challenge 3.25**

Our description of Muller's ratchet illustrates that equation (3.15) is not the unique solution to equation (3.14). Follow the path in teachable moment 3.7 to prove that

$$\tilde{p}_k = \begin{cases} 0 & \text{if } k < j \\ \dfrac{e^{-U/s}\left(\dfrac{U}{s}\right)^{k-j}}{(k-j)!} & \text{otherwise} \end{cases}$$

also satisfies that equation for all values of $j$, $k \geq 0$.

---

Finally, our treatment of Muller's ratchet assumes soft selection (see section 1.2.5). But Muller's ratchet can in principle drive population mean fitness so low that population size itself begins to decline. Should this happen, purifying selection will be weakened, increasing the pace of Muller's ratchet. This, in turn, will increase the rate of decline in population size, further heightening the risk of a subsequent click of the ratchet. This feedback loop between declining population size and a faster ratchet, termed *mutational meltdown* (Lynch et al. 1993), can plausibly lead to population extinction.

---

**Empirical Aside 3.9:** Muller's ratchet and the degeneration of Y chromosomes

In many mammals, sex at birth is determined by sex chromosomes, called X and Y. Females have two recombining X chromosomes and males have one X and one Y. Although the Y chromosome is thought to be derived from the X, the two do not recombine. The Y chromosome is also much reduced genetically, carrying fewer genes per Mb than any other. Why? Muller's ratchet provides one obvious explanation (Charlesworth 1978).

---

**Power User Challenge 3.26**

Can the error catastrophe ever cause a mutational meltdown? Can mutational meltdown cause an error catastrophe? Explain your answer to each question.

---

**Power User Challenge 3.27**

Our treatment of Muller's ratchet assumes no epistasis between deleterious mutations. Describe a population's sensitivity to the ratchet if epistasis is negative on a rotationally symmetric fitness landscape. Suppose instead that it is positive. Next, again assuming no epistasis, describe a population's sensitivity to the ratchet if we allow back mutation. Equivalently, our treatment of the ratchet assumes the genome has an infinite number of loci; suppose instead that it does not. Hint: consider the framing introduced in power user challenge 3.10.

---

As always, sexual recombination disrupts the Hill–Robertson effect, here stopping Muller's ratchet. Because linkage disequilibrium is negative, recombination converts haplotypes carrying an intermediate number of deleterious alleles into those carrying extreme numbers (figure 3.3, left). The resulting increase in fitness variance in the population can protect the most fit haplotype from loss.

## 3.3  Modifier Theory

We now have quantitative models that describe an evolving population's genetic composition over time in response to five processes: natural selection, mutation, migration, sexual recombination, and stochasticity. To this point, the intensities of these processes have been parameterized as constants (see teachable moment 1.1). But suppose the strength of some of these were also genetically determined, and thus in principle, subject to evolutionary change. For example, mutations in genes encoding DNA polymerases and proofreading enzymes can affect the carrier's mutation rate, raising the question of whether and how natural selection will respond. Similarly, the recombination rate is under genetic control (e.g., empirical aside 3.6). And in addition to parameter values, many structural features of our models (e.g., haploidy vs. diploidy, sexual vs. asexual reproduction, random vs. assortative mating) are at least in part genetically determined, raising the same kinds of questions. *Modifier alleles* (or *modifiers* for short) are genetic variants that influence parameter values or other features of population genetic models, and *modifier theory* is the domain of population genetics concerned with their evolution.

Before going further, we define *direct selection*. Direct selection acts on mutations that directly affect their carrier's physiology or behavior and, thus, their fitness. It is the only form of selection we have encountered so far, and is well-illustrated by mutations that protect bacteria from antibiotics. Importantly, modifier alleles can also be directly selected. For example, imagine a virus whose time to reproduction is dominated by the time to duplicate its genome. Suppose further that the speed of genome duplication is inversely related to the accuracy of replication. In this case, a genetic variant with a higher mutation rate would enjoy a shorter generation time, and thus, higher fitness. Similarly, genetic recombination could contribute to successful chromosomal segregation at meiosis, improving the fitness of the resulting gametes, quite apart from its influence on linkage disequilibrium.

Critically, however, modifier alleles can also respond to *indirect selection*, selection mediated by statistical associations to directly selected alleles. For example, the fate of an allele that increases mutation rate (called a *mutator*) will be influenced by whatever directly selected beneficial and deleterious mutations it produces elsewhere in its carrier's genome. If a mutator produces a beneficial mutation, the mutator's frequency will increase in expectation via genetic hitchhiking. At the same time, mutators will be subject to increased background selection due to the higher deleterious mutation rate they confer. These statistical associations between the mutator allele and directly selected beneficial and deleterious mutations elsewhere in the genome are what drives indirect selection on the mutator.

Statistical associations between a modifier allele and one or more directly selected alleles thus lie at the heart of indirect selection. Consequently, the strength of indirect selection will be inversely related to the persistence of such associations. Genetic linkage provides one mechanism, especially effective in asexuals. But other mechanisms are possible. For example, modifiers of migration rate can experience indirect selection mediated by persistent statistical association between locally beneficial alleles and local environment. This point also illustrates the fact that modifier theory is inherently a multilocus problem.

---

**Power User Challenge 3.28**

Modifier alleles are not the first case of indirect selection we have encountered. Name another. Hint: the strength of indirect selection is always dependent on the persistence of statistical linkage to a directly selected allele.

A more abstract way of understanding the distinction between directly and indirectly selected alleles is that the former affect carriers' mean value of some trait, and thereby their carriers' expected fitness. Selection thus regards all copies of a directly selected allele equivalently. This point was implicit in our treatment of natural selection in chapters 1 and 2. In contrast, modifier alleles affect not only means but also the statistical distribution of fitness values realized among carriers. For example, an asexual diploid heterozygous at two loci can only produce two gametes rather than the four available to a sexual diploid with the same genotype. Consequently, natural selection will regard different copies of a modifier allele differently, if they cooccur with directly selected mutations of different fitness effects.

A great many modifier loci have been explored, each heritably influencing one or more population genetic model parameters or features. We now examine the two best-studied; see also Karlin and McGregor (1974) and Otto (2013). For each, we focus only on mechanisms of indirect selection, overlooking the possibility of additional, direct selective effects. In practice, the effects of both kinds of selection have to be considered to understand a modifier allele's evolutionary fate.

### 3.3.1  The Evolution of Sex

Sexual reproduction is almost ubiquitous among eukaryotes, and genetic recombination is also widespread among microbes (empirical aside 3.1). Yet many organisms reproduce clonally, and the question of the evolutionary advantage of sex was perhaps first posed by Weisman (1887). The ubiquity of sexual reproduction is all the more remarkable when one appreciates the fact that in anisogamous diploid species (see empirical aside 1.10), sexual reproduction *per se* brings a two-fold reduction in fitness. This so-called *two-fold cost of (anisogamous) sex* reflects the fact that the metabolic investment females make in each gamete (an egg) is immensely larger than is that made by males, whose gametes (sperm) are much smaller. Yet their genetic representation in the resulting diploid is equal. Thus, a modifier allele that causes females to reproduce *parthenogenically*—that is, by clonal reproduction from individual eggs—would be found at twice the frequency among progeny than would the wild type (sexual) allele. Many nongenetic costs of sexual reproduction have also been identified (e.g., of mate finding), only sharpening the question of why sexual reproduction is so widespread.

The origin of sexual reproduction—that is, the sequence of physiological, genetic, reproductive, and behavioral innovations required to convert a clonally reproducing population into a sexually reproducing one—are outside of the scope of this book; the interested student will find reviews of the topic in Birdsell and Wills (2003) and Goodenough and Heitman (2014). We instead focus more narrowly on understanding the persistence of sexual reproduction. For example, the reasoning in the previous paragraph might lead us to wonder why parthenogenic females aren't more widespread. Yet they only represent perhaps 0.1% of eukaryotes, and tend to be evolutionary dead ends (Simon et al. 2003). And at a finer scale, what can be said about the evolution of recombination rates in nature?

In fact, we already know quite a bit. The key insight is that in populations carrying negative linkage disequilibrium, recombination makes natural selection more efficient (section 3.1.3). Moreover, both negative epistasis and the Hill–Robertson effect can induce negative linkage disequilibrium. Thus, for example, imagine two populations identical in every respect, except that one reproduces sexually and the other asexually. Assuming further that epistasis is negative among all beneficial mutations, linkage disequilibrium will also be negative on expectation. In this case, the sexually reproducing population will enjoy a greater velocity of adaptation than will the asexual (center panel, figure 3.4), by virtue of the fact that recombination will increase the population's variance for fitness (figure 3.3, left). If these two populations are competing for the same pool of resources, we might then expect the sexual population to eventually displace the asexual one.

This is an example of an *optimality* or *group selection* argument, in which natural selection is assumed to favor the genetic variant that optimizes fitness in a group or population of organisms. The key feature of group selection models is that they envision competition between closed populations (e.g., sexual vs. asexual), rather than between individuals. Put another way, populations in these models are never polymorphic for the modifier. As a result, the evolutionary success or failure of a modifier allele in this framing is dependent only on its long-term fitness effects. It is protected from adverse short-term fitness consequences.

Importantly, adverse short-term fitness consequence of modifiers can be substantial. Returning to the case of negative linkage disequilibrium induced by negative epistasis (or equivalently, by the Hill–Robertson effect), the left panel in figure 3.3 demonstrates that recombination will more often disrupt single mutant $Ab$ and $aB$ haplotypes than it creates them. The resulting increase in fitness variance is what favors closed groups of sexual organisms. However, the average fitness of recombinants (mathematically, $\frac{1}{2}(W_{AB}+W_{ab})$ is less than that of nonrecombinant ($W_{Ab}=W_{aB}$), precisely because epistasis is negative. Thus, an individual, newly sexual mutant competing against a population of asexuals may be quickly lost within a population: while some of its offspring will enjoy positive indirect selection ($W_{AB}>W_{Ab}=W_{aB}$), others will suffer a greater impact due to purifying indirect selection ($W_{ab}<W_{Ab}=W_{aB}$).

---

**Teachable Moment 3.16:** Short- and long-term fitness consequences of modifier alleles, the reduction principle, and group selection

A tension between short- and long-term fitness consequences is inherent in modifier theory. Natural selection is a process of differential survival: fitter lineages spread through a population at the expense of less-fit lineages. By definition, modifier alleles influence fitness variance among carriers. Thus, for some copies of a modifier to experience indirect positive selection, others must experience indirect purifying selection. Critically, on the premise that most directly selected mutations are deleterious (empirical aside 1.3), modifier alleles will much more quickly experience the latter than the former, demanding attention to their evolution at two timescales.

In the present case, recombination facilitates more efficient natural selection via Fisher's fundamental theorem. Consequently, modifier alleles that cause their carriers to reproduce sexually (or that simply increase recombination rate in an already-sexually-reproducing population) have long-term benefits because their carriers will enjoy a higher velocity of adaptation. But they do so explicitly via the selective elimination of their genetic siblings, a short-term cost.

The short-term costs of modifiers are the focus of the *Reduction Principle* (Altenberg et al. 2017). The Reduction Principle states that in populations already residing at a maximum on the fitness landscape, natural selection will favor modifier alleles that reduce offspring variance. In other words, by assuming that the population has equilibrated to an unvarying environment, the Reduction Principle disregards the long-term fitness opportunities of modifier alleles. At the other extreme, group selection models only consider these long-term consequences: the short-term costs are hidden within the groups, whose extinction is assumed to be impossible. Modifier theory is concerned with both effects.

Nevertheless, an extensive theoretical literature has demonstrated that when populations are away from fitness maxima (i.e., if fitness values vary in time or space, as they surely must), indirect selection may favor modifier alleles for sexual reproduction, as well as those that increase the recombination rate in sexual organisms (reviewed in Otto [2009], Hartfield and Keightley [2012], Dapper and Payseur [2017] and Singhal et al. [2019]). Experimental evidence for selection on modifiers of sex and of recombination is summarized in empirical aside 3.10.

**Empirical Aside 3.10:** Does selection act on recombination rate?

That recombination rates vary widely within and between genomes (empirical aside 3.6) begs the question of evolutionary mechanism. Several lines of evidence support the hypothesis that recombination rate is under natural selection. First, recombination is positively correlated with lifetime reproductive output in humans (Kong et al. 2004). Additionally, the difference in the genome-wide average recombination rate between two populations of *Drosophila pseudoobscura* is much larger than expected on the basis simply of random genetic drift (Samuk et al. 2020). Finally, consistent with its role in alleviating the Fisher–Muller effect, recombination rate has been seen to increase in laboratory populations of exposed to directional selection (Korol and Iliadi 1994; Becks and Agrawal 2012). Similarly, recombination rates are higher in domesticated species than their progenitors, and more compellingly, in populations exposed to environmental stress (reviewed in Coop and Przeworski [2007] and Ritz et al. [2017]).

**Power User Challenge 3.29**

Which if any of the evidence in support of the hypothesis that selection has acted on the recombination rate offered in empirical aside 3.10 is consistent with the action of indirect selection? Explain.

### 3.3.2  The Evolution of Mutation Rate

Mutation rate modifiers similarly experience opposing short- and long-term fitness consequences. Because most mutations are deleterious, mutator alleles will most often reduce the fitness of their carriers. But mutators may also occasionally introduce beneficial mutations, allowing them to hitchhike to high frequency. In asexuals, the resulting statistical association between mutator and driver mutations is perfect, and mutators have been shown to have hitchhiked to fixation in human cancers (Loeb [2011], within which cells propagate asexually), asexual human pathogens (Suárez et al. 1992; LeClerc et al. 1996), and laboratory populations of asexual microbes (reviewed in Raynes and Sniegowski [2014]). Of course, hitchhiking (and background selection) will be disrupted by genetic recombination, weakening indirect selection on mutators.

That mutators might hitchhike to high frequency or even to fixation assumes a ready supply of beneficial mutations. But suppose instead that the population is at or near a maximum on the fitness landscape. In this case, almost all mutations will be deleterious. The Reduction Principle means that selection should now act to reduce mutation rate; where will it equilibrate? Of course, there are thermodynamic limits: replication can never be perfect, but we can say more.

To begin we ask a simpler question: how many offspring does each deleterious mutation cost a lineage? To simplify, assume deleterious mutations all have selection coefficient $s$. Then in expectation, deleterious mutations will persist for $\sim\frac{1}{s}$ generations before selective elimination (teachable moment 3.6), during each of which time, its lineage will have $s$ fewer offspring. Thus, each deleterious mutation corresponds to $\sim s\frac{1}{s}=1$ fewer offspring. (This is the Haldane–Muller principle restated.) Now consider an asexual population with genome-wide deleterious mutation rate $U$. Suppose further that a subpopulation carries an *antimutator* allele whose deleterious mutation rate is reduced by an amount $\Delta U$. Because this is an asexual population, each deleterious mutation avoided in the antimutator subpopulation corresponds to one addition offspring, so the antimutator's indirect selection coefficient is $\Delta U$. On the other hand, the drift-barrier hypothesis (empirical aside 2.3) tells us that this downward pressure on mutation rate will cease when $\Delta U$ is roughly $\frac{1}{N}$.

---

**Power User Challenge 3.30**

Still assuming the absence of beneficial mutations, what is the selection coefficient acting on the antimutator if there is free recombination between the antimutator and all deleterious mutations it creates? Hint: see power user challenge 3.20.

---

### 3.3.3  Other Modifiers

Many other models have been explored. For example, selection coefficients themselves can also be under genetic control. Consider molecular chaperones, a class of proteins that assist in the proper functional folding of their so-called client proteins. As such, they can mask the deleterious effects of destabilizing missense mutations in their clients, reducing their selection coefficient. Thus, genes for molecular chaperones reduce their carriers' sensitivity to some deleterious mutations, an effect called *mutational robustness.* As above, modifiers of mutational robustness are a two-edged sword. Lineages with higher mutational robustness will enjoy a short-term advantage, because their mutant offspring will suffer a smaller fitness deficit than will the offspring of genetically less robust parents (de Visser et al. 2003). On the other hand, in the long run, mutational robustness can render a population less able to adapt and more susceptible to Muller's ratchet, at least in asexual populations (LaBar and Adami 2017).

The population genetics of many other modifier alleles have been explored, and the interested reader is directed to Crow (1991) on modifiers of segregation distortion, Otto and Gerstein (2008) on modifiers of haplo/diploidy, and Bagheri (2006) on modifiers of dominance. Each one of these similarly respond to both the short- and long-term fitness consequences described in teachable moment 3.16.

## 3.4   Chapter Summary

- Multilocus model complexity grows exponentially in the number of loci. Consequently, multilocus theory is less fully developed than is single-locus theory.

- Linkage disequilibrium describes the statistical associations of alleles at different loci in a population. Natural selection, mutation, population subdivision, and random genetic drift can all induce linkage disequilibrium, while genetic recombination reduces it.

- Adaptation in multilocus models can be visualized on the fitness landscape, in which mutationally adjacent haplotypes are spatially adjacent. Sign epistasis can induce multiple maxima on the fitness landscape, complicating the process of adaptation.

- Natural selection reduces expected coalescence times. This induces a correlated reduction in neutral genetic diversity along chromosomes, and in the efficiency of natural selection. Genetic recombination attenuates both effects.

- Many parameters and other features of population genetic models are themselves under genetic control, and thus mutationally labile. Modifier theory is the domain of population genetics that considers their evolution.

# 4

## The Data

The first three chapters of this book have been devoted to developing a framework that predicts how a population's genetic composition will evolve over time in response to five processes: natural selection, mutation, sexual reproduction, migration, and stochastic effects. In this final chapter, we invert that apparatus and explore approaches to detecting the influence of those same processes in natural populations from gene and genome sequence data. For example, given DNA sequence data at a protein-coding gene from a sample of organisms in nature, what does our theory let us say about the influence of selection at that locus? And, how do whole-genome sequence data tell us about the demographic history of a population over the last 10 or 10,000 generations?

This chapter is thus perhaps the payoff for many readers, whose primary interest may be in inferring the evolutionary history of some specific biological population. The explosive growth of gene and genome sequence data made possible by technological advances since the start of the present century, together with ongoing improvement in computational software and hardware, make this a tremendously dynamic field of inquiry. Consequently, an exhaustive survey of this work is doomed to rapid obsolescence. Nor will we discuss the statistical complexities of estimation that emerge from the imprecisions of measurement (teachable moment 1.12). Not because those aren't critical challenges, but rather because that topic is well beyond the scope of this book. Rather, we seek only to develop clear intuitions for many of the current methods of population genetic analysis, in order to broadly illustrate opportunities for inference. In so doing, we also provide tastes of some recent results. Many citations to the relevant literature are provided for the interested reader, who will also find surveys of recent work in books by Nielsen and Slatkin (2013), Hahn (2019), Coop (2020), and Hartl (2020).

Before beginning, we note that data from our own species motivate much of the work described in this chapter. The human genome project was launched in 1990, driven ultimately by interest in the genetic basis of human disease. Roughly 10 years after the first complete human genome sequences (Venter et al. 2001; Collins et al. 2003), over 1,000 human genome sequences were published (Genomes Project Consortium 2012). Less than a decade after that, almost 150,000 human *exomes* (sequence data covering the roughly 1% of the human genome that encodes proteins) were collected in a single database (Karczewski et al. 2020). The National Institutes of Health now plan to collate 1,000,000 complete genomes over the next 10 years, together with diverse health information from those individuals (NIH 2021a), in order to advance the goal of personalized genomic medicine (see also empirical aside 1.17).

Moreover, a great deal is known from the archeological record about history of our species (very briefly summarized in empirical aside 4.1). Critical for our purposes, recent technical advances now make practical the recovery of human *ancient DNA*, genome sequence data from physical remains of deceased individuals. As of this writing, several hundred of these have been published, ranging in age from less than 1,000 years to over 40,000 years. We also have whole genome sequence data from two of our closest, now extinct relatives: the Neanderthals (Prüfer et al. 2014; Prüfer et al. 2017) and Denisovans (Reich et al. 2010; Meyer et al. 2012). Such longitudinal, time-series data can give direct access to temporal changes in the genetic composition of our species that are complementary to the instantaneous snapshots offered by current-day sampling methods.

---

**Empirical Aside 4.1:** Human evolutionary history

Archeological, anthropological, and genetic evidence suggest that anatomically modern humans first appeared in Africa roughly 200,000 years ago. Migration by small populations from Africa into Eurasia occurred 40,000 to 60,000 years ago, leading to the global peopling of the world. This is sometimes called the out-of-Africa migration, and represents one of the strongest demographic signals in our genomes. Genetic evidence further demonstrates that these immigrants encountered and interbred with archaic populations, including the Neanderthals (known from fossil remains dating to at least 130,000 years ago) and Denisovans (identified only from DNA recovered from bone fragments dating to as much as 200,000 years ago). The interested reader is directed to an excellent review (Ahlquist et al. 2021) and two recent books (Rutherford 2017; Reich 2018) to learn more about our species' history as revealed by population genetics.

---

On the other hand, theory and methods are agnostic on the source of data, and the reader should appreciate that techniques of the sort described in this chapter are equally applicable to any species on Earth. See empirical asides 4.2 and 4.3 for more.

---

**Empirical Aside 4.2**

For the first 50 years of its life, theoretical population genetics asked many more questions than data were able to answer. But since the introduction of molecular biology into the field in the mid-1960s, the quality and quantity of data began to drive new theory (see the first two paragraphs of chapter 5 in Lewontin 1974 for a prescent assessment of what the subsequent decades would bring). And since the start of the genomic era, theory has been racing to keep up with the almost unlimited flood of genetic data we now enjoy. Most notably, the introduction of so-called "next generation sequencing" has induced a super exponential drop in the cost of gene and genome sequence data acquisition (NIH 2021b). Consequently, it is now practical for individual investigators to bring many of the methods described here to bear on any population or species of interest (e.g., da Fonseca et al. 2016).

Importantly, conclusions made with these data often suggest new avenues for deducing underlying biological mechanisms. An excellent example comes from the peppered moth *Biston betularia*. Recall from chapter 1 that the frequency of the *carbonaria* type in nature is correlated with air pollution levels over time (empirical aside 1.2 and power user challenge 1.8) and space (empirical aside 1.15). A population genetic survey of the causal mutation for the *carbonaria* type implies a single, recent origin (van't Hof et al. 2011). These correlative observations suggest that nineteenth-century industrialization and subsequent clean air regulations drove differential predation of the two types. But they do not demonstrate this fact.

Confirmation of that hypothesis comes from extensive and careful observation of bird predation on both forms of the moth on light and dark backgrounds by Kettlewell (1955, 1956) and Majerus (Cook et al. 2012). Many of population genetic methods described in this chapter similarly suggests mechanistic hypotheses, motivating exciting new avenues of biological inquiry.

**Empirical Aside 4.3:** The population genetics of dogs

Domestic dogs represent another superb model organism in which to apply the methods described in this chapter. Their domestication occurred perhaps 15,000 years ago, with most modern breeds coming into existence less than 200 years ago. Thus, we have tremendously detailed extrinsic information on the recent demographic history in those populations. While we have comparably detailed historical records for some human populations, dog breeds have the added advantage of being the product of extremely intense artificial selection for specific traits. In addition, we have access to genome sequences from village dogs, largely unselected, semi-feral, conspecific human commensals, as well as to ancient dog DNA. See recent reviews by Boyko (2011) and Freedman and Wayne (2017) for more.

## 4.1 Measuring Genetic Parameters

One of the most important findings in the first three chapters is that model behavior often depend on the relative strengths of two forces. Quantitatively, boundaries between biological regimes are given by compound parameters: products or ratios of individual parameters (see also teachable moment 2.17). For example, assuming the Wright–Fisher model of random genetic drift, the threshold between polymorphism and monomorphism lies roughly when the product of population size and mutation rate (mathematically, $N\mu$) is 1. Consequently, our ability to estimate a population's size from its heterozygosity depends on an independent estimate of its mutation rate. Similarly, immediately after a selective sweep, the width of the resulting trough in heterozygosity scales as the ratio of the per base recombination rate to the selection coefficient (written as $\frac{c}{s}$). Thus, our ability to estimate a driver mutation's selection coefficient from heterozygosity in its vicinity depends on an independent estimate of recombination rate.

Fortunately, of the five forces with which population genetics is concerned, mutation and recombination are properties of individuals; the other three are properties of populations. More specifically, because pedigrees (section 1.4.4.2) let us control the number of reproductive events separating two individuals, they allow direct access to per generation mutation and recombination rates. For this reason, it is simpler to measure these two, and so this is where we begin.

### 4.1.1  Mutation Rate

In 1928, H. J. Muller introduced *mutation-accumulation* (*MA*) lines, a pedigree-based method for measuring mutation rates (Muller 1928) that is still in widespread use today. MA lines are replicate, laboratory-reared populations of organisms that experience frequent population bottlenecks. Ideally these bottlenecks cause the next generation to be derived from a single randomly sampled individual. We know from chapter 2 that such a sampling regime comes very close to eliminating the influence of natural selection: all nonlethal mutations have an equal probability of being propagated. Thus, MA lines capture an almost random sample of new mutations.

Further assuming that the starting organisms are well-adapted to their environment, the *Bateman–Mukai method* uses fitness assays of replicate MA lines, (i.e., lines founded with the same genotype) to estimate both genome-wide deleterious mutation rate ($U$) and mean deleterious selection coefficient ($\bar{s}$) from the mean per bottleneck decline in fitness across lines ($\overline{\Delta W}$), and the per bottleneck change in fitness variance across replicate lines ($\Delta \mathrm{Var}(W)$). The first can be estimated by computing the mean total fitness decline across replicate lines and dividing by the number of bottlenecks. Adopting the common convention of signaling an empirical estimate of some quantity with the caret symbol ($\wedge$), this is written as $\widehat{\overline{\Delta W}}$. The second quantity equals the variance in fitness across replicates at the end of the experiment, again divided by the number of bottlenecks. This follows because fitness variance before the first bottleneck is zero, since all lines are initialized with genetically identical organisms. Mathematically, $\Delta \mathrm{Var}(W) = \mathrm{Var}(W)$, the empirical estimate for which is written as $\widehat{\mathrm{Var}(W)}$.

We now model $\overline{\Delta W}$ and $\Delta \mathrm{Var}(W)$ in an MA experiment. Our treatment follows Halligan and Keightley (2009). First, under our assumption that population bottlenecks entirely neutralize natural selection, the rate of fixation is equal to the per bottleneck deleterious mutation rate (see empirical aside 2.1). Thus, the expected per bottleneck rate at which fitness declines is the product of the genome-wide per bottleneck deleterious mutation rate ($U$) and the mean deleterious selection coefficient ($\bar{s}$), written as

$$\mathrm{E}[\overline{\Delta W}] = U\bar{s}. \tag{4.1}$$

We can also model the expected change in fitness variance across replicate lines in terms of $U$ and $\bar{s}$. Equation (1.13a) lets us write the per mutation increase in fitness variance as $\dfrac{1 \times \overline{s^2} + (n-1) \times 0^2}{n} = \dfrac{\overline{s^2}}{n}$, where $n$ is the number of replicate MA lines. Multiplying by the expected number of mutations per bottleneck across $n$ lines ($nU$) gives $E[\Delta \mathrm{Var}(W)] = U\overline{s^2}$. Next, equation (1.13c) lets us write $\overline{s^2} = \mathrm{Var}(s) + \overline{s}^2$, and assuming all deleterious mutations have the same selection coefficient (i.e., $\mathrm{Var}(s) = 0$) yields $\overline{s^2} = \overline{s}^2$, and thus,

$$E[\Delta \mathrm{Var}(W)] = U\overline{s}^2. \tag{4.2}$$

Given empirical estimates $\widehat{\Delta W}$ and $\widehat{\mathrm{Var}(W)}$ from an MA experiment as described in the paragraph before last, we estimate the genome-wide mutation rate as

$$\hat{U} = \frac{\widehat{\Delta W}^2}{\widehat{\mathrm{Var}(W)}} \tag{4.3}$$

and mean deleterious selection coefficient by

$$\hat{\overline{s}} = \frac{\widehat{\mathrm{Var}(W)}}{\widehat{\Delta W}}. \tag{4.4}$$

(Recall that that we substitute $\widehat{\mathrm{Var}(W)}$ for $\widehat{\Delta \mathrm{Var}(W)}$, since the fitness variance among replicates is zero at the beginning of the experiment.) The Bateman–Mukai approach can also be used to estimate mean dominance coefficient among deleterious mutation in diploids (Halligan and Keightley 2009).

---

**Teachable Moment 4.1:** Inverting predictive theory to make inferences about nature

The foregoing example allows us to illustrate two points articulated in the first two paragraphs of this chapter. First, methods of data analysis begin by modeling expected values for some empirically accessible quantity in terms of population genetic parameters of interest (e.g., equations 4.1 and 4.2). These equations are then inverted to yield expressions for those parameters in terms of empirically accessible quantities (e.g., equations 4.3 and 4.4).

Of course, one can never hope to measure any quantity in nature without some experimental uncertainty. (The attentive reader will have noticed that we ignored the stochasticity implied by the expectation operator $E[\cdot]$ in equations [4.1] and [4.2].) And no inferential analysis is complete without an exploration of the implications of such uncertainty. Well-developed theory exists to address this issue but is, however, outside of our scope. Readers are encouraged to study any undergraduate textbook on statistical inference to learn more.

---

Unfortunately, the Bateman–Mukai approach suffers from at least two weaknesses. First, it only estimates the rate of measurably deleterious, nonlethal mutations, which are just a fraction of the total. Second, it only provides a genome-wide total; it provides no information on variation in mutation rate across the genome.

Fortunately, the explosive growth in genome sequencing capacity (empirical aside 4.2) has given new life to the mutation accumulation experiment. It is now routine to simply count all the nonlethal mutations that have accumulated in an MA line by comparing the genome sequences of the starting and final genotypes. The comparison of complete genome sequences from human parent/offspring trios (Jónsson et al. 2017) is also a form of MA, and these methods underlie the data in empirical aside 1.4.

Quite a different, population genetics-based approach to estimating mutation rates, first proposed by Messer (2009), was recently applied to humans (Carlson et al. 2018). Those authors began with a data set of 3,560 whole human genome sequences. They focused only the roughly $3.6 \times 10^7$ singleton point mutations in the data. (Point mutations are often called *single-nucleotide polymorphisms* (SNPs). These authors focused only on singleton SNPs; that is, SNPs seen only once in the data set.) Because these mutations are at approximate frequency $\frac{1}{2 \times 3560} \approx 1.4 \times 10^{-4}$ in the population, their fate will as yet have been nearly unaffected by natural selection. (The 2 in the denominator reflects the fact that humans are diploid.) This represents a vastly larger number of mutations than can practically be observed in an MA framework. Beyond providing a very accurate estimate of average mutation rate, the authors found that mutation rates in humans exhibit a dependence on the particular nucleotides found within a few base pairs of the mutation. Using the same approach, Harris and Pritchard (2017) demonstrated variation in mutation rate among human populations.

### 4.1.2 Recombination Rate

As was the case for mutation rate measurements, there are again both pedigree- and population genetics-based approaches to estimating recombination rates between loci. Morgan and his students invented the pedigree-based method (teachable moment 3.4), which yield genetic map distances $r$, in cM. With the advent of genome sequencing, pedigree analyses now also give physical recombination rates $c = \frac{r}{l}$, in $\frac{cM}{Mb}$. Unfortunately, the comparatively small number of meioses captured in a traditional pedigree limits the resolution of this approach.

*Single-sperm genotyping* represents a more powerful, pedigree-based technique. Here, recombination frequencies are characterized by examining individual sperm cells collected from a single male. The vast number of recombination events represented in such a sample (e.g., $\sim 10^8$ cells in humans) allow recombination rates to be measured at much finer resolution. As with the finest-resolution mutation rate measurement described above, many context-dependencies on the recombination rate exist (Bell et al. 2020). An important limitation to this approach, however, is that it only yields access to recombination rates in males, whereas recombination rates often differ between the sexes (empirical aside 3.6).

Population-based approaches for estimating the recombination rate are also possible, using linkage disequilibria observed between pairs of loci. Their statistical power again reflects the fact that data from a sample of $n$ haplotypes capture meioses over roughly the past $E[t_{MRCA}] = 4N_e \left(1 - \frac{1}{n}\right) \approx 4N_e$ generations if $n$ is at all large. Recall that at recombination/drift equilibrium, the expected squared Pearson correlation coefficient between loci is $\tilde{r}_P^2 = \dfrac{1}{1+4Nr}$ (equation 3.23). Thus, inverting this result yields an estimate of the recombination rate between loci, given observed values of $r_P^2$ and a population size estimate.

A related approach takes advantage of the fact that the variance in the Hamming or mutational distances between haplotypes is inversely correlated to the absolute value of linkage disequilibrium $D$. To illustrate, consider a population of two-locus haplotypes written as $AB$, $Ab$, $aB$, and $ab$ in which all allele frequencies are $\dfrac{1}{2}$. Regardless of linkage disequilibrium, the average pairwise Hamming distance is given by the sum of the per site heterozygosities, here equal to 1. (To see this, think about the situation for just one site.) Now suppose that there is no linkage disequilibrium (i.e., $D=0$), so haplotype frequencies are $p_{AB}=p_{Ab}=p_{aB}=p_{ab}=\dfrac{1}{4}$. In this case, the variance in Hamming distances is $\dfrac{1}{4}(0-1)^2+\dfrac{1}{2}(1-1)^2+\dfrac{1}{4}(2-1)^2=0.5$. At the other limit, assume a maximally extreme $D=-\dfrac{1}{2}$. Now haplotype frequencies will be $p_{AB}=p_{ab}=0$ and $p_{Ab}=p_{aB}=\dfrac{1}{2}$ and the variance becomes $\dfrac{1}{2}(0-1)^2+\dfrac{1}{2}(2-1)^2=1$. (Variance is also 1 if $D=+\dfrac{1}{2}$.)

The statistics for inferring recombination rates from Hamming distances were first developed in Hudson (1987) and Wakeley and Hey (1997), and resulting inferences have been extensively validated using comparably high-resolution single-sperm genotyping data. The chief difficulty in all population-based estimates of the recombination rate is that natural selection, mutation, demography and stochasticity can also cause linkage disequilibrium, substantially complicating the inferential process. (The interested reader is directed to Stumpf and McVean [2003], Clark et al. [2010], chapter 4 in Walsh and Lynch [2018], and Peñalba and Wolf [2020] for many of the technical details.) Moreover, because these approaches yield population averages, they are unable to detect variation among individuals (or between sexes) in the recombination rate. Nevertheless, their immense statistical power, coupled with the ever-larger data sets now available, make these methods tremendously informative, and they underlie many of the estimates provided in empirical aside 3.6.

## 4.2  Measuring Effective Population Size

As repeatedly noted in the first three chapters of this book, a population's demography (its size and structure) influences its population genetic behavior. Thus, for example, equilibrium heterozygosity depends on effective population size, and population subdivision can cause demic allele frequencies to diverge by random genetic drift. And as we will see, gene and genome sequence data from an expanding population can bear similarities to those from one in which a selective sweep has recently completed. Thus, it is of utmost importance to be able to estimate the demographic history of natural populations.

In chapter 2, we saw that in the Wright–Fisher model, the strength of random genetic drift scales with the reciprocal of the population's census size. Any other demographic model's effective population size quantifies its sensitivity to drift relative to that benchmark. For example, in section 2.5, we found that an unequal sex ratio increases drift relative to that in the Wright–Fisher model, meaning that unequal sex ratio reduces effective population size below census size. Thus, in this section, we ask how to estimate a population's effective size. We take up the estimation of other demographic features in section 4.3.

### 4.2.1  Direct Observation of Reproductive Variance

The most direct method to measuring effective population size relies on observations of the distribution of offspring number, census size estimates, and sex ratios. This approach is especially common in questions of conservation biology and agricultural breeding. See Frankham (1995) for a theoretically rigorous survey of this sort of work, which finds that the variance effective population size may typically be on the order of 10% of census size. A more genetically based analog is to record changes in neutral SNP (or allele) frequencies over time. Recall from chapter 2 that the Wright–Fisher model assumes allele frequencies are binomially distributed, and quantifies generation-to-generation variance in terms of effective population size via equation (2.18). Inverting that theory provides access to an estimate of effective population size on the basis of allele frequency variance over time. The interested reader will find technical details on the application of this approach in Wang et al. (2016) and citations therein. Note, however, that both these methods are limited to populations of short-lived organisms, since they require observations in at least two successive generations.

### 4.2.2  Equilibrium Effective Population Size Estimates

In chapters 2 and 3 we developed theory to predict levels of heterozygosity and linkage disequilibrium at mutation/drift and recombination/drift equilibrium, respectively. We now invert that theory to find estimates of effective population size from observed levels of each of those quantities. This represents an important motivation for directly estimating mutation and recombination rates (section 4.1). More abstractly, this anticipates a recurring feature of this chapter: any population's genetic variability in the present can only reflect past mutation or recombination events. Consequently, all methods of population genetic inference must ultimately rest on one of these two kinds of data.

For example, since $\tilde{\pi} = 4N\mu$ (equation 2.42), $\dfrac{\hat{\pi}}{4\hat{\mu}}$ is an estimate of effective population size. Given extrinsic estimates of $\mu$, this approach motivated many of the estimates of $\pi$ provided in empirical aside 2.5. A closely related approach focuses not on neutral heterozygosity but rather on heterozygosity of mildly deleterious mutations. Recall from chapter 2 that the efficiency with which natural selection purges deleterious mutations is inversely related to effective population size. As effective population size goes up, selection also becomes more efficient at preventing deleterious mutations from rising to high frequency (although the quantitative theory is quite technical; Kimura [1969]). Thus, mildly deleterious heterozygosity will be inversely related to effective population size (see also empirical aside 2.3).

The added complexity of this approach is that mildly deleterious heterozygosity will also depend on deleterious selection coefficients. Or more precisely, mildly deleterious heterozygosity will depend on mutation rate $\mu$ and the product $N_e s$. Nevertheless, writing $\bar{s}$ for the mean selection coefficient, bulk estimates of $N_e \bar{s}$ for missense mutations (see empirical aside 4.4) are possible (Eyre-Walker and Keightley 2007), yielding estimates of effective population size from empirical levels of missense heterozygosity (Galtier and Rousselle 2020).

---

**Empirical Aside 4.4:** The genetic code is redundant

Proteins are polymers of amino acids, small molecules whose sequence and identities along the protein confer its specific biochemical properties. In nature, almost all proteins are composed of 20 different amino acids (teachable moment 3.9), which, in turn, are encoded by *codons*, nucleotide triplets in the corresponding gene. However, as there are $4^3 = 64$ possible nucleotide triplets and just 20 amino acids (plus a stop codon that signals the end of a protein), the mapping from DNA to protein given by the *genetic code* is redundant. Consequently, most amino acids can be represented by more than one codon.

This redundancy gives rise to two kinds of mutation in protein-coding genes: *missense* or *nonsynonymous* mutations, which change the amino acid encoded by a codon, and *synonymous* mutations, which do not. (*Nonsense* mutations are a third kind, which convert amino acid-encoding codons to stop codons. Because these cause premature termination of the encoded protein, they are almost always tremendously deleterious to their carrier, and thus only rarely segregate in natural populations.)

As we shall see, the redundancy of the genetic code represents an immense inferential opportunity for population genetics. This follows because on first principles, one might assume that natural selection acts only on missense mutations (although see empirical aside 4.5). Because synonymous and nonsynonymous mutations can occur in the same gene, comparisons of their incidence in natural populations represent an opportunity to control for local differences in mutation rate, divergence times, and other parameters. We will return to this point several times in section 4.4.

---

**Empirical Aside 4.5:** Natural selection can act on synonymous mutations

The redundancy of the genetic code (empirical aside 4.4) means that many amino acids can be represented by more than one *synonymous codon*. Synonymous codons are often found in unequal frequency, a phenomenon called *codon usage bias*. The intensity of codon usage bias varies among species as well as between (and even, within) genes in the same species. Many genome-wide correlates with codon usage bias among species are known, including nucleotide composition and transfer RNA (tRNA) abundances. Similarly, at the gene level, correlation with length, position within the gene, expression level, and recombination rate have been described.

Two kinds of mechanisms to explain codon bias have been proposed: mutational bias and selective effects. Mutational bias (empirical aside 1.4) enriches for some nucleotides over others, and thus for some synonymous codons over others. (Indeed, mutational bias has even been shown to affect amino acid composition; e.g., Singer and Hickey 2000.) Mutation bias often explains interspecific differences in codon usage.

Plausible selective mechanisms that have been explored include translational accuracy and speed, messenger RNA structure and stability, and splicing accuracy and speed. These are perhaps more likely responsible for within-genome differences. See Hanson and Coller (2018) and Iriarte et al. (2021) to learn more.

Just as heterozygosity can be used to estimate effective population size if mutation rate is known, so too can linkage disequilibrium, given recombination rates. More specifically, we can again invert $\tilde{r}_{\mathrm{P}}^2 = \dfrac{1}{1+4Nr}$ (equation 3.23), now to find an expression for $N_{\mathrm{e}}$ written in terms of empirically determined values of Pearson's correlation coefficient $r_{\mathrm{P}}^2$ between alleles at two loci separated by known genetic map distance $r = cl$. Interestingly, estimates based on linkage disequilibrium capture effective population size in the more recent past than do those based on heterozygosity. This follows because, while the per base-pair recombination and mutation rates are often comparable (mathematically, $\mu \approx c$), linkage disequilibrium is almost always measured between loci separated by thousands or hundreds of thousands of base pairs $l$. In answer to power user challenge 3.23, the time to mutation/drift equilibrium scales with the reciprocal of the mutation rate. By the same reasoning, the time to recombination/drift equilibrium scales with the reciprocal of the recombination rate. Thus, values of mean $\widehat{r_P^2}$ between pairs of loci $l$ base pairs apart will reflect effective population size $\sim \dfrac{1}{cl}$ generations in the past (Hayes et al. 2003). In English, allelic correlations between pairs of more distant loci reflect more recent history than does heterozygosity, because such correlations come to equilibrium more quickly.

This approach was applied to SNPs in humans separated by between 5 and 100 kb by Tenesa et al. (2007). Again taking 1 cM $\approx$ 1 Mb (empirical aside 3.6), those distances correspond to $r \approx 5 \times 10^{-5}$ and $10^{-3}$ Morgans (i.e., recombination events per generation), respectively. Thus, these data have recorded effective population sizes between 500 and 10,000 generations in the past (taking $\dfrac{1}{2r}$ generations as the time to equilibrium; further, taking 20 years as a plausible generation time, this corresponds to 10,000 to 200,000 years). These data suggest that effective population size in humans was fairly constant before a massive expansion (see empirical aside 4.1), here estimated to have started 20,000 years ago.

**Power User Challenge 4.1**

Imagine a population of haploid nonrecombining organisms in which the left half the genome has a per site, per generation neutral mutation rate $\mu_{\mathrm{left}} = 10^{-5}$ while in the right half it is $\mu_{\text{right}} = 10^{-8}$. Write $\hat{\pi}_{\text{left}}$ and $\hat{\pi}_{\text{right}}$ for the observed, per site heterozygosities in the left and right halves of the genome, respectively. Over what timescales are $\hat{\pi}_{\text{left}}$ and $\hat{\pi}_{\text{right}}$ informative about historical effective population sizes in this population?

### 4.2.3 Nonequilibrium Mutation-Based Effective Population Size Estimates in Asexuals

As just noted, heterozygosity-based effective population size estimators that assume mutation/drift equilibrium reflect values from $\sim \frac{1}{\mu}$ generations in the past, where $\mu$ is the locus-wide neutral mutation rate. Taking a typical locus to be 1,000 bp and the per base mutation rate as $10^{-8}$, that corresponds to perhaps $10^5$ generations. And similarly, linkage disequilibrium-based analyses that assume recombination/drift equilibrium shed light on the biology perhaps $10^4$ generations in the past. We now relax our equilibrium assumptions to find methods that are sensitive to more recent effective population sizes.

Again, there are mutation- and recombination-based approaches. In this section, we explore population size inferences based on nonequilibrium patterns of SNPs in asexual haplotypes. This choice is both historical (asexually propagated human mitochondrial DNA were among the first sources of relevant data) and pedagogical (their analysis is simpler). On the other hand, as noted in teachable moment 3.13, the lack of recombination in these data means they represent just one genealogical realization from the underlying demographic history, and thus, parameter estimates based on these data have high statistical uncertainty. In section 4.2.4, we extend these methods to sexual populations. Finally, in section 4.2.5, we explore how to estimate population size from nonequilibrium patterns of linkage disequilibrium.

#### 4.2.3.1 The distribution of pairwise Hamming distances

Polymorphism data for human mitochondrial DNA (mtDNA) motivated the earliest nonequilibrium estimates of effective population size. Mitochondria are eukaryotic subcellular structures that carry their own genetic material. Because many cells have hundreds of copies of mtDNA, their DNA is more easily purified than nuclear DNA. Moreover, mitochondria are maternally inherited and pass through a very sharp bottleneck each generation in their transmission to the egg. Thus, their evolution is functionally asexual and haploid.

In 1987, Rebecca Cann and colleagues published a watershed study of polymorphism across roughly 10% of the mitochondrial genome in 147 humans sampled from five geographic populations. The broadest demographic conclusion of the paper was that the ancestor of all modern human mtDNA (the *mitochondrial Eve*) lived in Africa roughly 200,000 years ago. The authors also found evidence of repeated waves of migration from Africa subsequently entering non-African populations.

We will explore methods for inferring population structure and migration using genetic data in section 4.3. But for our present purposes, the most interesting aspect of those data is a sharp discordance between the empirical distribution of pairwise Hamming distances among human mtDNA and that expected at mutation/drift equilibrium (Rogers and Harpending 1992). Before examining the data, we develop an expectation.

The problem is reminiscent of the mutation/drift equilibrium heterozygosity developed in chapter 2 under the infinite alleles model (equation 2.32). Recall that there, we wrote a recurrence equation for expected heterozygosity at a locus in the $t+1$st generation as a function of heterozygosity in the $t$th generation (equation 2.31). However, the infinite alleles model simply partitions pairs of alleles into those that are identical and those that are not; it is mute on the question of Hamming distances between them.

To go further, we adopt the infinite sites model, and introduce $J_i(t)$ for the probability that a pair of haplotypes differ at $i$ sites in generation $t$. In the absence of recombination, we now have a family of recurrence equation for all Hamming distances $i$, $0 \le i \le \infty$ which read

$$\mathrm{E}[J_0(t+1)] \approx J_0(t)[1 - \frac{1}{N} - 2U] + \frac{1}{N}$$

$$\mathrm{E}[J_{i+1}(t+1)] \approx J_{i+1}(t)[1 - \frac{1}{N} - 2U] + 2UJ_i(t),$$

for the Wright–Fisher model (Rogers and Harpending 1992). Here, $U$ is the per generation probability of mutation somewhere in the locus. Although written for haploids (i.e., assuming that the probability that two haplotypes are identical by descent is $\frac{1}{N}$, as appropriate for mtDNA), these equations otherwise bear parallels to equation (2.31), and the interested student may enjoy working through their derivation. (Critically, under the infinite sites assumption, each mutation increases Hamming distance by one.)

The solution to these recurrence equations at mutation/drift equilibrium (i.e., the values of $J_i$ such that $J_i(t+1) = J_i(t)$ for all $i$) is

$$\tilde{J}_i \approx \frac{1}{1+2NU} \left( \frac{2NU}{1+2NU} \right)^i \tag{4.5}$$

(Watterson 1975). The algebra to reach this expression is more involved than was the infinite alleles case, but happily, its form—a geometric distribution—suggests a biological interpretation. We recognize the term inside parentheses as equation (2.32): the infinite alleles probability that two alleles differ in a haploid population. Now simultaneously applying this infinite alleles reasoning to each site in the locus, this term raised to the power of $i$ is the probability that two alleles differ at exactly $i$ sites. And the first term on the right is one minus equation (2.32), or the probability that the two alleles defined by the remaining sites are identical.

To apply equation (4.5) to the mtDNA data from Cann et al. (1987), we only require an empirical estimate for $2NU$. As in section 4.1.2, mean pairwise Hamming distance is the sum of per site heterozygosities, which at mutation/drift equilibrium in a haploid population honoring Wright–Fisher assumptions is $2NU$. In human mitochondria, this value was 9.5 (Cann et al. 1987).

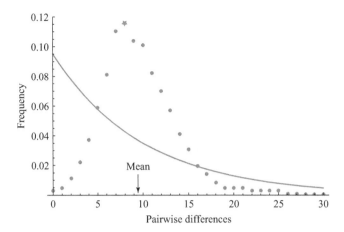

**Figure 4.1**
Pairwise Hamming distance distribution in human mitochondrial DNA. *x*-axis: Hamming distance (i.e., number of pairwise differences) between pairs of sequences; *y*-axis: frequency of observation. Circles: data from Cann et al. (1987); star: the most-often observed value. Solid line: theoretical expectation in population at mutation/drift equilibrium given by equation (4.5). The expectation was fit to the data by making their means equal; see text for more. (Figure after Rogers and Harpending 1992.)

Figure 4.1 presents the empirical distribution of Hamming distances between human mtDNA sequences (circles) together with the corresponding expectation at mutation/drift equilibrium (solid line). Clearly, the data are sharply incongruent with expectation, suggesting a violation of at least one of the assumptions underlying equation (4.5). Given the archeological and historical data mentioned earlier (empirical aside 4.1), a recent population expansion is a very plausible candidate for this discrepancy. Indeed, a recent expansion predicts just the pattern seen in the data. Rogers and Harpending (1992) provides a satisfying, analytic approximation of the situation, but here we simply use coalescent theory to develop our intuition.

At mutation/drift equilibrium, any pair of asexual haplotypes share a coalescence some $t_2$ generations in the past, and under the infinite sites model, their expected Hamming distance over $L$ sites is $2E[t_2]\mu L = 2E[t_2]U$ (this is equation 2.41 multiplied by the number of sites). Thus, in expectation, the distribution of Hamming distances in a population should be identical to its distribution of pairwise coalescent times, rescaled by $2U$. (This is true only in expectation as we are overlooking stochasticity in the mutational process.) Indeed, this provides a second interpretation for the form of equation (4.5). Namely, at mutation/drift equilibrium, the expected distribution of pairwise Hamming distances is geometrically distributed in the absence of recombination, precisely because the distribution of pairwise coalescent times is geometrically distributed (equation 2.35).

In contrast, population expansion causes a great many coalescent events in a very short period of time. For example, imagine a one-generation expansion in which $N_{small}$ lineages give rise to $N_{large}$ lineages. Such a genealogy will necessarily experience $N_{large} - N_{small} - 1$ coalescent events in that generation, and less abrupt population expansion similarly give rise to a transient compression in coalescent times compared to equilibrium expectation. Correspondingly, shortly after an expansion, a

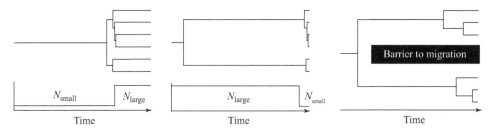

**Figure 4.2**
The influence of population size and structure on coalescent genealogies. Left: after a population expansion, expected coalescent times are scaled upward. Middle: a population contraction rescales expected coalescent times downward. Right: population subdivision prevents coalescent events between demes.

great many pairs of lineages will share very recent coalescent times, and thus (in expectation), very small pairwise Hamming distances. This situation is illustrated in the left panel of figure 4.2. As mutations then accumulate, the distribution of pairwise Hamming distances will move to the right. Indeed, it will do so at a rate given by the mutation rate, since under the infinite sites model, new mutations always increase Hamming distances. Eventually, new coalescent events will cause Hamming distances between some pairs of lineages to shorten again, and on a timescale given by $N_{large}$, the distribution will return to the equilibrium predicted by equation (4.5).

Going still further, before re-equilibration, the most "typical" (or modal) Hamming distance observed in the data gives the modal time elapsed since the population expansion, expressed in units of mutation rate. For example, in the human mitochondrial data shown in figure 4.1, the most-observed Hamming distance is 8 (see star in figure). As seen in the paragraph-before-last, the expected Hamming distance between any two haplotypes is $2E[t_2]U$. So for these data, we crudely estimate that the population expansion occurred $\hat{t}_2 = \dfrac{8}{2\hat{U}} = \dfrac{4}{\hat{U}}$ years ago, where $U$ is written per year. Given human mtDNA per base mutation rate $\hat{\mu} \approx 3 \times 10^{-8}$ per year (Cabrera 2021), and that $L \approx 1,500$ in the data (Rogers and Harpending, 1992), the relevant mutation rate is $\hat{U} \approx 4.5 \times 10^{-5}$ per year, yielding a back-of-the envelope estimate of a dramatic population expansion $\approx 89,000$ years ago. Subsequent analysis of complete mtDNA sequence (i.e., $L \approx 13,000$) from a larger and geographically broader sample of humans moved the date of population expansion forward to $\approx 50,000$ years ago (Atkinson et al. 2007). Analogous questions of human demographic history have also been explored using genetic variation on the haploid, asexual, male-transmitted Y chromosome (Shen et al. 2000).

**4.2.3.2 Skyline plots**   Although these insights are exciting, the power of this approach is weakened by a subtle nonindependence among pairwise Hamming distances (Felsenstein 1992). This follows because a sample of $n$ haplotypes defines just $n-1$ coalescent events, whereas there are $\binom{n}{2} = \dfrac{n(n-1)}{2}$ comparisons underlying results such as those in figure 4.1. The consequences of this nonindependence are most easily seen in the pairs of haplotypes that find their coalescence at the MRCA of the sample. Disregarding stochasticity in the mutational process, their pairwise Hamming distances will be proportional to the sum of $n-1$, geometrically distributed coalescent times. This compounding of stochastic uncertainty weakens inferential power. Moreover, more than $\frac{1}{3}$ of all pairs in any sample will pass through this deepest coalescence (Felsenstein 1992), amplifying the impact of this summed stochastic uncertainty.

---

**Power User Challenge 4.2**

Given a sample of $n$ haplotypes from a population, how many of the $\binom{n}{2}$ pairwise Hamming distance estimates between them do not reflect the sum of at least two geometrically distributed random variables?

---

Ultimately, it is the density of coalescent events through time that tells us about the temporal trajectory of population sizes (figure 4.2). Periods of faster coalescences imply smaller population sizes, while slower intervals signal larger population sizes. A more direct (and efficient in the statistician's sense, Felsenstein 1992) approach to reconstructing the historical trajectory of population size would thus be to explicitly and independently estimate the $n-1$ intervals between coalescent events (written as $t_n, t_{n-1}, \ldots, t_2$ in our earlier notation; e.g., right-hand panel in figure 2.13). Given empirical estimates of these inter-coalescent times $\hat{t}_i$, we could then invert $E[t_i] = \dfrac{2N}{\binom{i}{2}}$

generations (equation 2.38) to estimate the succession of $\hat{N}_i = \hat{t}_i i(i-1)$ in each interval.

This is exactly what so-called *skyline plots* present (Pybus et al. [2000] reviewed in Ho and Shapiro [2011]): each $\hat{N}_i$ plotted over its corresponding time interval $(\hat{t}_i, \hat{t}_{i-1})$. (The name reflects the fact that the discontinuous jumps between successive $\hat{N}_i$ are reminiscent of a city skyline.) Genealogies (and thus the $\hat{t}_i$) defined by a sample of haplotypes can easily be estimated using well-established methods of phylogeny (which, however, are outside our scope; see Ho and Shapiro [2011]). More importantly for us, phylogenetic methods yield inter-coalescence interval estimates in mutation counts, whereas the coalescent theory outlined in the previous paragraph assumes that the $\hat{t}_i$ are in generations. Under the infinite sites model (and the molecular clock; see empirical aside 2.1), division by an extrinsic mutation rate estimate yields the necessary conversion. (Absent this, the $y$-axis in a skyline plot represents $\hat{N}_i \mu$ instead of $\hat{N}_i$.) And similarly, to read the $x$-axis in time instead of generations requires that after converting the $\hat{t}_i$ to generations, they must also be multiplied by an extrinsic estimate of generation time.

Skyline plots have yielded many compelling empirical insights. For example, its original application found strong evidence for an exponentially growing population in HIV-1 (Pybus et al. 2000), as might be expected of a pandemic pathogen (although subtle demographic differences between strains were also noted). A subsequent study applied a refinement of the original method to samples of the hepatitis C virus (HCV) in Egypt (Drummond et al. 2005). Egypt currently has the world's highest prevalence of hepatitis C, likely caused by inadequate needle sterilization during the treatment of a devastating parasitic infection called schistosomiasis in the mid-twentieth century. Consistent with this hypothesis, a skyline plot of Egyptian HCV sequences demonstrates an approximately 100-fold increase in population size beginning roughly in 1950 (Drummond et al. 2005).

### 4.2.4 Nonequilibrium Mutation-Based Effective Population Size Estimates in Sexuals

A serious deficiency of both of the preceding approaches is their requirement that data come from nonrecombining regions of the genomes. The distribution of pairwise Hamming distances in a Wright–Fisher population (equation 4.5) assumes a nonrecombining segment of the genome. The phylogenetic reconstruction of genealogies required for skyline analyses also assumes no recombination. As noted earlier, inferences drawn by such methods are thus based on just the one genealogy realized by the sample. Recall from chapter 2 that both the time at and order in which lineages coalesce are subject to stochastic variance, above and beyond our uncertainty about the underlying demographic history. In addition, asexual loci contain no information about the demographic history of a population for dates preceding its MRCA (e.g., before the mitochondrial Eve or Y chromosomal Adam).

In contrast, recombination decouples realized genealogies across the genome (see teachable moment 3.13). Thus, genetic variation in sexually recombining regions of the genome reflects many independent replicate coalescent realizations, all drawn from the same (albeit, unknown) population history. (They also represent vastly more data; e.g., the nuclear genome in humans is roughly 10,000-fold larger than is the mitochondrial genome, thereby also vastly reducing uncertainty introduced by the mutational process.) And while the genealogical MRCA of the population still represents an unavoidable inferential time horizon, variance among realized genealogies across a recombining genome means that considering many loci is expected to yield temporally deeper insights (although with increasing uncertainty) than would any single locus. We return to this point in power user challenge 4.3 and teachable moment 4.3.

### 4.2.4.1 The site frequency spectrum

Recall from chapter 2 that each SNP observed in a sample of $n$ haplotypes partitions the sample into fractions of size $j$, $1 \leq j \leq n-1$, where $j$ haplotypes carry the derived nucleotide and $n-j$ of them carry the ancestral one. The SFS is defined as the fraction of derived SNPs observed $j$ times in a sample of $n$ haplotypes for all $1 \leq j \leq n$. Recall from chapter 3 that recombination has no effect on individual SNP frequencies, only on their correlations. This means that the expected SFS in both recombining and non-recombining regions of the genome will be as described in section 2.4.2.

Our chapter 2 treatment assumed the Wright–Fisher (i.e., constant-sized population) model. Suppose now that $N$ varies in time. For example, again imagine a population expansion from $N_{small}$ to $N_{large}$ individuals. After this, there will be a corresponding drop in probabilities of coalescence (figure 4.2, left). As in section 4.2.3.1, after roughly $E[t_{MRCA}]\ 4N_{large}$ generations, the genealogy will reequilibrate to our chapter 2 expectations. But its transient distortion will also transiently distort the expected SFS. To see how, recall that our derivation of expected SFS involved labeling each genealogical branch with an integer $i$ representing the number of descendent haplotypes in the sample (figure 2.14, left). After an expansion, branches leading to single haplotypes in a present-day sample (i.e., those labeled "1," also sometimes called *external branches*) will be anomalously long, as seen in the left panel of figure 4.2. Thus, after a population expansion, the SFS will be skewed to the left: we expect to see more SNPs at low frequencies than we would at equilibrium. Looking backward in time, this reflects a rush of coalescent events required to "fit" all the lineages defined by $N_{large}$ individuals into just $N_{small}$ ancestors. By the same reasoning, after a population contraction, deep branches in the genealogy, that is, those labeled with large integers will be anomalously long. This effect is illustrated in the center panel of figure 4.2, and will skew the SFS to the right: we correspondingly expect to see more SNPs at intermediate frequencies than we would at equilibrium.

---

**Power User Challenge 4.3**

The SFS has a finite time horizon: it cannot provide information about a population's demographic history older than the time of the population's MRCA. Is this time horizon deeper after a population expansion or population contraction, and why? You should assume that both demographic events occur at the same time in the past.

---

For example, Keinan et al. (2007) observed a deficit of low-frequency SNPs among approximately 70,000 SNPs examined in Americans of European ancestry relative to the equilibrium expectations given by equation (2.45). A similar pattern was seen among approximately 50,000 SNPs examined in two Asian populations. Both observations are consistent with a population bottleneck associated with the out-of-Africa migration (empirical aside 4.1).

We can also understand these effects looking forward in time (Hahn 2019). A population expansion increases the population-wide, per–site mutation rate from $N_{small}\mu$ to $N_{large}\mu$. Transiently, these additional mutations will necessarily be at low frequency (since they will all be recent), skewing the SFS to the left. Conversely, a population contraction means the elimination of many lineages, disproportionately affecting rare mutations, and thus skewing the SFS to the right. Critically, since both contractions and expansion perturb the SFS via their influence on rare mutations, our ability to detect recent events goes up with sample size.

**4.2.4.2  Tajima's *D* statistic**   The SFS observed in a sample of $n$ haplotypes has $n-2$ degrees of freedom (teachable moment 3.2): there are $n-1$ frequencies at which SNPs can be found, which of course must also sum to 1. It is thus an immense distillation of the underlying data, namely the nucleotide identity at each variable nucleotide, multiplied by $n$ samples. (Most notably, the SFS overlooks linkage disequilibrium among polymorphic sites in the data, a point we take up shortly.) Tajima's $D$ statistic (Tajima 1983) is a still further distillation: polymorphism data are now represented by just one number. (In fact, it is one of several such scalar distillations of the SFS; Achaz [2009].)

Recall from chapter 2 that under Wright–Fisher assumptions and the infinite sites model, the equilibrium per site heterozygosity (i.e., the probability that a site differs in two randomly sampled haploid individuals) is $\tilde{\pi} = 4N_e\mu = \theta$ (equation 2.42). Thus, $\hat{\pi}$ is an estimate of $\theta$. Under the same assumptions, we saw that at equilibrium, the expected number of variable sites in a sample of $n$ alleles is $\mathrm{E}[S_n] = 4N\mu\sum_{i=1}^{n-1}\frac{1}{i} = \theta\sum_{i=1}^{n-1}\frac{1}{i}$ (equation 2.43). Consequently, $\dfrac{\widehat{S_n}}{\sum_{i=1}^{n-1}\frac{1}{i}}$ represents a second estimator of $\theta$.

Tajima's $D$ is the difference between these two estimators, normalized in a way to account for sampling error. Mathematically,

$$D = \frac{\hat{\pi} - \dfrac{\widehat{S_n}}{\sum_{i=1}^{n-1}\frac{1}{i}}}{Z},$$

where $Z$ captures sampling error in $\hat{\pi}$ and $\widehat{S_n}$ (Tajima 1983). In expectation, $D$ will be zero, but demographic departures from Wright–Fisher assumptions can disproportionately inflate one estimate or the other.

To see how, recall that population expansions and contractions are disproportionately manifest at the left end of the SFS. More specifically, population expansion enriches for sites carrying rare SNPs, while they are depleted after a contraction. Now consider how we compute per site heterozygosity $\pi$. Conceptualized as the fraction of sites at which two random haplotypes differ, the most obvious approach is to compare all $\binom{n}{2}$ pairs of haplotypes in a sample of $n$ to find the average fraction of sites at which pairs differ. Mathematically, this reads

$$\hat{\pi} = \frac{\sum_{i=1}^{n}\sum_{k=i+1}^{n}\dfrac{\hat{d}_{i,k}}{L}}{\binom{n}{2}}, \tag{4.6}$$

where $\hat{d}_{i,k}$ is the observed Hamming distance between haplotypes $i$ and $k$, and $L$ is the number of sites sequenced. (Interested readers can convince themselves that the double summation visits all pairs of haplotypes exactly once.)

But another approach is to imagine cycling through all $S_n$ SNPs, tabulating the contributions to per site heterozygosity $\hat{\pi}$ from each. A SNP seen in $j$ of $n$

haplotypes increases $\pi$ in proportion to $j(n-j)$, since that number of the $\binom{n}{2}$ pairs of haplotypes differ at the site. (The remaining $\binom{j}{2}+\binom{n-j}{2}$ pairs of haplotypes are identical there.) More specifically, this approach to estimating $\pi$ reads

$$\hat{\pi}=\frac{\sum_{i=1}^{S_n}\dfrac{\hat{j}_i(n-\hat{j}_i)}{L}}{\binom{n}{2}},\tag{4.7}$$

where $\hat{j}_i$ is the number of haplotypes in the sample in which SNP $i$ is observed. This second formulation highlights the fact that SNPs at intermediate frequencies contribute more to $\pi$ than do those at low (or high) frequency, since $j(n-j)$ is maximized when $j\approx\dfrac{n}{2}$.

---

**Power User Challenge 4.4**

In chapter 1, we introduced the heterozygosity of a population as the probability that two randomly chosen alleles were different. Mathematically, this two-locus framing yielded $H=2p_A(1-p_A)$ (equation 1.23d), where $p_A$ is the frequency of the $A$ allele in the population. Equation (4.7) captures something very similar: the per site heterozygosity at a SNP that partitions $j$ haplotypes in a sample of $n$ haplotypes. Try to derive one from the other. Hint: you will not quite succeed, but the discrepancy reveals something interesting about the difference in the way these two problems are framed.

---

Now returning to the problem at hand, we see that for a given number of SNPs $S_n$, $\pi$ is depressed relative to Wright–Fisher expectations after a population expansion, which transiently enriches for new, low-frequency SNP, thereby causing Tajima's $D$ to be negative. For example, average Tajima's $D$ was significantly negative in a survey of 151 loci from Americans of European ancestry (Stajich and Hahn 2005), consistent with population expansion after the peopling of Europe $\approx 50,000$ years ago (empirical aside 4.1). Conversely, because a population contraction transiently eliminates low-frequency SNPs, it inflates $\pi$ compared to expectations based on $S_n$, driving Tajima's $D$ positive.

**4.2.4.3  The pairwise sequential Markovian coalescent**  The SFS and Tajima's $D$ are both sensitive to departures from Wright–Fisher equilibrium. But can we also estimate a sexual population's (nonequilibrium) effective population size trajectory over time? One strategy would be to construct an average skyline plot using data from many loci, reflecting many (nearly) independent coalescent realization drawn from the same (unknown) temporal demographic history.

This is exactly what the *pairwise sequential Markovian coalescent* (*PSMC*) of Li and Durbin (2011) does. Given a single, diploid genome, the method first identifies independently evolving blocks of DNA separated by historical recombination events using a sophisticated computational approach called the Hidden Markov model (HMM; Dutheil 2017). The technical details of their HMM are beyond our scope, but conceptually, it moves a "sliding window" of 20,000 base pairs from one end of the genome to the other, computing local heterozygosity $\pi$ in each. In the absence of recombination, successive windows should share some particular pairwise coalescent time $t_2$, and so in expectation, local heterozygosity will be $2\mu t_2$. But as repeatedly emphasized, recombination decouples genealogies, and the HMM recognizes such statistical decouplings by abrupt changes in local heterozygosity.

Next, the PSMC estimates coalescent time at each evolutionarily independent block of DNA, yielding a distribution of pairwise coalescent times $\hat{t}_2 = \dfrac{\hat{\pi}}{2\hat{\mu}}$ from along the length of the genome. Thus, the PSMC provides another window onto the density of coalescent events in a population through time. But whereas the skyline method estimates the time between each of the $n-1$ coalescent events with a sample of $n$ one-locus, nonrecombining haplotypes, the PSMC estimates pairwise coalescent times for many thousands of evolutionarily independent loci in a single diploid individual. A histogram of these times provides an estimate of the probability of coalescence as a function of time in the past. Since the per generation coalescent probability is $\dfrac{1}{2N}$ under Wright–Fisher assumptions, the reciprocal of these estimated probabilities can be regarded as twice the population's effective size at each time interval in its history.

The reader may now appreciate the method's name: pairwise, because it examines just two haplotypes, sequential, because it moves from one end of the genome to the other, Markovian, because of the HMM, and coalescent, because of its focus on $t_2$. Li and Durbin (2011) applied this method to two genomes of European, African, and Asian ancestry. In each of the six, the HMM identified over $10^5$ blocks of independently evolving DNA. As a consequence, the resulting histograms of pairwise coalescent times provided a very high-resolution picture of effective population size leading back more than $10^6$ years. Data from all six genomes record very similar histories before approximately $10^5$ years ago, including an approximately two-fold decline and recovery in population size between $1.5 \times 10^5$ and $1.5 \times 10^6$ years ago. All six genomes record a second decline in population size beginning roughly $10^5$ years ago, but interestingly, the drop was faster and deeper in the ancestry of the non-African genomes, again concordant with an out-of-Africa bottleneck (empirical aside 4.1).

Note that the temporal depth of signal in these data (reaching back roughly $10^5$ human generations) is greater than in skyline plots for single loci (for example, the approximately $10^2$ years of HIV and HCV history described in 4.2.3.2 corresponds to no more than $10^4$ viral generations.). This follows from the fact that the PSMC method samples many more (and many older) coalescent events than those underlying the skyline method. More recently, the *multiple sequential Markovian coalescent* (*MSMC*) was introduced by Schiffels and Durbin (2014). As the name suggests, this method analyzes more than two haploid genomes, focusing only on the most recent coalescent event among these genomes for each nonrecombining block. Consequently, the sensitivity of the MSMC method extends to within as few as 100 generations in the past.

---

**Power User Challenge 4.5**

Can the MSMC's increased sample size also increase its sensitivity to more ancient demographic events? Why or why not?

---

### 4.2.5 Nonequilibrium Recombination-Based Effective Population Size Estimates

Heterozygosity at a locus scales with coalescent times and, thus, effective population size. The PSMC and MSMC methods use the distribution of heterozygosities over a large number of independently evolving loci in a sample of two or more haploid genomes to estimate a population's effective size through time. This approach is of course dependent on recombination, which decouples evolutionary histories among loci. But it also overlooks the historical signal captured by recombination events themselves. That information is embedded in the population's haplotype structure; that is, by the patterns of linkage disequilibrium among sites. Perhaps the most direct access to these patterns is through the number and lengths of observed *runs of homozygosity* (or *ROH*), blocks of the genome that are identical in two haplotypes; often, the two found in a diploid organism.

We largely follow Buffalo et al. (2016) in exploring how ROH record historical effective population sizes. First recall from chapter 1 that a diploid is said to be autozygous at a locus if its two homologs are IBD (i.e., coalesce in some benchmark generation). For example, given a pedigree, the probability of IBD at a locus is $F_I = \dfrac{1}{2^{k-1}}$ (equation 1.30). Here, $F_I$ is an individual's inbreeding coefficient, and $k$ is the number of meioses separating its two homologs in the pedigree. Assuming nonoverlapping generations in the pedigree, $k = 2t$, where $t$ is the number of generations to the benchmark generation, yielding $F_I = \dfrac{1}{2^{2t-1}}$.

Now consider a single chromosome of genetic map length $R$ cM. Recall that 1 cM is the genetic map distance between loci in which recombination occurs with 1% probability per meiosis (section 3.1.1). Thus $\dfrac{R}{100}$ is the expected number of recombination events per meiosis on the chromosome, which define $\dfrac{R}{100} + 1$ independently segregating blocks of DNA. (We add one because, absent recombination, the chromosome segregates as one block.) In expectation, this rises to $2t\dfrac{R}{100}$ meioses after $t$ generations (during which time each of the two homologs pass through $t$ meioses), defining $2t\dfrac{R}{100} + 1$ independently segregating blocks of DNA. Finally, with $d$

chromosomes, this becomes $2t\dfrac{R}{100}+d$ blocks (again, absent recombination, each chromosome segregates as one block).

Now introducing $n_{\text{IBD}}$ for the number of tracts in a diploid genome that are IBD $t$ generations after a benchmark generation, we have

$$\text{E}[n_{\text{IBD}}]=F_{1}\left(2t\frac{R}{100}+d\right)=\frac{1}{2^{2t-1}}\left(2t\frac{R}{100}+1\right),\qquad(4.8)$$

since $F_{1}$ is both the probability of IBD at a locus and the fraction of independently segregating blocks that are IBD. The expected length of autozygous tracts ($r_{\text{IBD}}$, also in cM) is even more easily found:

$$\text{E}[r_{\text{IBD}}]=\frac{100}{2t}cM,\qquad(4.9)$$

since each of the $2t$ meioses introduces another recombination event, on average every 100 cM. (Because genetic map length is defined by the frequency of recombinants, recombination events are by definition uniformly distributed on this scale; Fisher [1954] and see teachable moment 3.4.) Equations (4.8) and (4.9) are shown in figure 4.3 as a function of $t$, the number of generations since the two homologous, independently segregating segments find their common ancestor.

This development of multilocus IBD theory assumes that time $t$ to coalescence between homologs is known from the pedigree. (The expectations in equations (4.8) and (4.9) are only over the stochastic uncertainty of recombination.) But random genetic drift means that the two homologs in any individual must eventually coalesce. Thus, inverting this predictive theory suggests a path to estimating coalescent times and hence, effective population sizes from data on IBD tracts. Longer and more frequent tracts imply more recent coalescence, and thus, smaller effective population sizes. Shorter and fewer tracts signal longer times to coalescence and larger populations.

Importantly, IBD tracts cannot be directly observed in genome sequence data, but rather are manifest by the mutational process as ROH. Clearly, our ability to infer

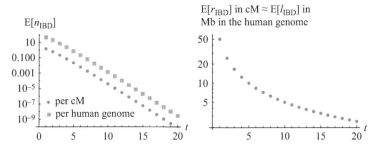

**Figure 4.3**
Expected IBD statistics as a function of number of generations $t$ since homologs had a common ancestor. Left: expected number of IBD tracts in a diploid given by equation (4.8) (blue: per 100 cM; mustard: per human genome, taking sex-averaged $R=3{,}500$ cM (Broman et al. 1998), and $d=22$ autosomal chromosomal pairs). Right: expected IBD tract length in cM given by equation (4.9), or equivalently, in Mb in the human genome, assuming 1 cM $\approx$ 1 Mb (empirical aside 3.6).

IBD tracts from observed ROH will be dependent on an adequate supply of mutations. This raises some challenging technical issues that we overlook, although power user challenge 4.6 provides some intuition. We instead equating IBD and ROH. We also omit development of the precise quantitative connection between ROH statistics and $N_e$. (The interested reader is directed to Powell et al. [2010], Huff et al. [2011], Harris and Nielsen [2013], MacLeod et al. [2013], Thompson [2013], and Shemirani et al. [2021] to learn more about both of these issues.)

---

**Power User Challenge 4.6**

How does mutation rate influence ROH in the absence of recombination? Imagine a nonrecombining haplotype $L$ bp in length with a per generation, per site mutation rate $\mu$. What is the expected ROH length for two haplotypes sampled at random from a diploid population of size $N$ adhering to all Wright–Fisher assumptions? Your answer should be in bp.

---

The conceptually more important issue is that the expected number of IBD (and ROH) tracts drops almost exponentially in time $t$ (left panel of figure 4.3). By five generations, this number is approximately 0.002 ROH per cM, meaning that only 0.72 ROH are expected across an entire human genome. By 10 generations, these numbers have dropped to $2 \times 10^{-6}$ cM and 0.001 ROH, respectively. This suggests that inferences based on ROH tracts will have an extremely short time horizon.

On the other hand, expected ROH tract length declines only linearly with time; for example, even after 200 generations, in expectation they will be 0.25 cM ≈ 250,000 bp long in humans. Thus, the problem is finding these informative but very rare events. Fortunately, by our assumption of random mating, expectations under our multilocus IBD theory apply equally to any two haploid genomes in the population, not only those pairs found in diploids. Thus, given genome sequence data from $n$ diploids, we can substantially improve sensitivity by looking for ROH between all

$$\binom{2n}{2} \approx 2n^2 \text{ pairs of haploid genomes (Palamara et al. 2012).}$$

The study of ROH has taught us great deal about the last 200 generations of demography in humans (Kirin et al. 2010; Palamara et al. 2012; Ralph and Coop 2013; Moreno-Estrada et al. 2014; Browning and Browning 2015; Ceballos et al. 2018; Saupe et al. 2021; Patterson et al. 2022), dogs (Boyko et al. 2010; Marsden et al. 2016; Sams and Boyko 2019; Mooney et al. 2021), and even the extinct woolly mammoth (Palkopoulou et al. 2015) and Neanderthals (Prüfer et al. 2014).

---

**Power User Challenge 4.7**

ROH in a genome record a population's demographic history over a much more recent interval than does the distribution of heterozygosities across evolutionarily independent segments in the same genome. Why?

---

---

**Teachable Moment 4.2:** Fisher's theory of junctions

Recombination only introduces genetic novelty if it occurs in a heterozygous region of a chromosome (table 3.1). R. A. Fisher's theory of junctions begins from the observation that once formed, such crossover events ("junctions") are themselves subject to random genetic drift, just as any other two-allele locus modeled in chapter 2. This insight yields another fertile avenue for the study of the evolution of runs of IBD, which has received renewed interest in the era of genome-scale sequence data. See Thompson (2018) for a recent introduction to these ideas.

---

**Teachable Moment 4.3:** Genealogical and genetic ancestors, and the many MRCAs of sexually reproducing individuals

In sexually reproducing organisms, the number of ancestors in any individual's genealogy doubles with each generation backward in time. Mathematically, the number of genealogical ancestors of any individual after $g$ generations is $2^g$. This number grows without bound, yet population sizes are finite, meaning that even in a randomly mating population, individuals very quickly begin to share genealogical ancestors. In a population of $N$ diploids, this happens when $2^g \approx 2N$, yielding $g \approx \log_2(2N) = \log_2(N) + 1$ generations (see Chang [1999] for a rigorous treatment of this issue). This consequence of inbreeding is sometimes called *pedigree collapse*.

Importantly, long before pedigree collapse, recombination means that the expected length and number of chromosomal segments inherited from any particular genealogical ancestor will already be very small (right panel in figure 4.3, now reading $t$ as the number of meioses back to the ancestor). This implies that only a very lucky few genealogical ancestors will also be genetic ancestors of any individual. Recombination further means that the identities of those lucky genetic ancestors will vary along the chromosome, or put another way, any two chromosomes in the present generation will represent a mosaic of MRCAs (bounded by the surviving junctions of teachable moment 4.2). This reasoning underlies the point made in teachable moment 3.13.

---

## 4.3    Describing Population Structure, Migration, and Admixture

Population genetic theory provides a rich avenue for estimating a population's effective size back to time horizons set by the coalescent, mating, and recombinational processes. We now turn to the other side of demography: describing population structure, migration, and *admixture* (matings between individuals from different populations) using genome sequence data. As before, these methods can be classified as relying on mutational or recombinational signals. Sections 4.3.1–4.3.3 explore the former group and we take up the latter in section 4.3.4.

### 4.3.1    Detecting Population Structure

The Wright–Fisher assumption of a well-mixed population is an idealized, mathematical convenience. In fact, nearly every species on Earth has genetic substructure. This already emerges as a consequence of random genetic drift in the presence of spatial substructure, and will be amplified by diversifying selection across the species' range. The first challenge is, therefore, to detect population structure.

We saw earlier that Tajima's $D$ statistic can detect both recent population expansions and contractions. It can also detect population subdivision and subsequent admixture. This follows because subdivision enforces deep coalescent times (figure 4.2, right), thereby enriching for SNPs of intermediate frequency. Recall that this causes Tajima's $D$ statistic to be positive. For example, Stajich and Hahn (2015) report that, averaged over 151 loci, this statistic was significantly greater than zero in Black Americans, consistent with mixed African and European ancestry. (Population contraction has the same influence on coalescence times, as seen in the middle panel of the same figure, and thus on the sign of Tajima's $D$.)

An even simpler approach to characterizing a population's genetic structure with genome-scale data is the *principal component analysis* (or *PCA),* first applied in this setting by Menozzi et al. (1978). This method rests on methods of linear algebra that are beyond our scope, but conceptually it is quite straightforward. First, recall sequence space (section 3.1.6), a discrete space in which mutationally adjacent haplotypes are spatially adjacent. Assuming $L$ SNPs in the data, each carrying one of just two nucleotides, any population can be visualized as a cloud of points in $L$ dimensions whose coordinates are given by a series of $L$ 0's and 1's, corresponding respectively to the wild type and mutant nucleotides at each SNP. Given genome-scale data for a sample of individuals, the PCA is a mathematical operation that that finds the projection of its cloud of points onto a low-dimensional surface that maximizes their spread. In other words, PCA finds the rotation of sequence space through which variance is apparently maximized. As a matter of practice, PCA is almost always onto a two-dimensional plane, permitting easy visualization. Of course, a great deal of genetic variation is obscured in the PCA projection, since many points (haplotypes) will appear to lie on top of one another.

Nevertheless, PCA has proven a very effective approach for detecting genetic structure in a population. For example, imagine two geographically isolated demes. Over time, random genetic drift (and perhaps also diversifying selection) will cause nucleotide frequencies to diverge between demes, manifest as some spatial differentiation between members of the two demes in sequence space. This spatial differentiation is often evident in the metapopulation's cloud of points in sequence space when viewed from the perspective that maximizes their spread; that is, in the PCA projection onto a plane.

One of the more remarkable applications of this method was given by Novembre et al. (2008), who presented a PCA using 197,146 SNPs in a sample of 1,387 Europeans, each of whose grandparents all came from a single European country. The authors first labeled each point with the grandparental country of origin. When the 1,387 points lying in a 197,146-dimensional space was projected onto a plane in such a way as to maximize variance, points clustered by grandparental country of origin. This suggests that in the generations leading to those grandparents, individuals tended to remain very close to their own place of birth. But more remarkably, clusters of points corresponding to each grandparental country of origin themselves clustered in a pattern that roughly but recognizably recapitulated the map of Europe. In other words, not only do individuals who share grandparental countries of origin cluster together, but at a larger genetic scale, grandparental countries of origin cluster in a way that recovers geographic distances.

The biological implication of this analysis is that among humans living in Europe, the probability of movement some distance during one's lifetime is inversely correlated with distance. Individuals tend not to leave their natal country, and if they do, they tend more often to move to nearby countries. This conclusion is of course not entirely surprising, but that its footprint should be evident in the PCA suggests that important information about a population's genetic structure survives the immense dimensionality reduction of the method.

More recently, a two-dimensional PCA analysis of individuals with significant European ancestry now living in Latin America was found to cluster largely with individuals now living in Iberia and Italy (Homburger et al. 2015). Again, this is not an entirely unexpected result given the history of southern European colonization in Latin America, but one that gives us confidence in such inferences, despite the immense dimensionality reduction.

A related approach to detecting the genetic structure of populations identifies subpopulations on the basis of variation in SNP frequencies. More specifically, these methods define subpopulations on the basis of shared allele frequencies, while simultaneously estimating each individual's proportional membership in each such group. The first such method introduced was called STRUCTURE (Pritchard et al. 2000), and many extensions have since been developed (reviewed in both Novembre [2016] and Lawson et al. [2018]). The technical details of this approach are not of interest here but, importantly, no methods estimate the number subgroups (generally written as $K$) defined by the data. Rather, results based on a range of values of $K$ are typically examined, and the investigators (and readers) can then judge how many subgroups seem to do the best job of describing the population's substructure. In an early application, setting $K=5$ for 1,056 globally sampled humans, STRUCTURE accurately partitioned individuals according to their continent of origin: Africa, West/Central Asia, East Asia, Oceania, and the Americas (Rosenberg et al. 2002). See also empirical aside 1.17.

### 4.3.2 Estimating Population Structure with $F_{ST}$

Given descriptive evidence of genetic substructure in a population, we might next wish to quantify its strength. Recall from chapter 1 that demic differences in allele frequency necessarily depress population-wide heterozygote frequencies, even if each deme honors all Wright–Fisher assumptions. This is the Wahlund effect, and reflects the fact that demic differences in allele frequency effectively enriches for assortative mating. Recall too that this depression in heterozygosity is quantified by estimates of $\hat{F}_{ST} = 1 - \dfrac{\hat{H}_{obs}}{2\hat{p}(1-\hat{p})}$, where $\hat{H}_{obs}$ and $\hat{p}$ are the observed, population-wide heterozygote and allele frequencies, respectively.

Reich et al. (2009) provide a nice example of the application of this theory. Those authors report much higher pairwise $F_{ST}$ values among culturally and linguistically recognized human subpopulations in India than in comparably distinct subpopulations in Europe. They suggest that the correspondingly larger allele frequency differentiation among Indian subpopulations may reflect both stronger genetic drift (owing to smaller founding populations) and more limited gene flow than occurred in Europe. Correspondingly, allele sharing between individuals within a group was substantially correlated with that group's mean $F_{ST}$ relative to all other groups.

A globally geographic connection to $F_{ST}$ values in humans was provided by Ramachandran et al. (2005). Building on ideas going back to Wright (1943) and Malécot (1948), those authors found a strong correlation between pairwise population $F_{ST}$ and geographic distances between populations, made stronger when geographic distances were computed assuming only the very shortest possible water crossings. More recently, the same approach was used to show that $F_{ST}$ builds up more slowly as geographic distances increase in Western European populations than in populations on either side of the Caucasus, Ural, or Himalayan Mountain ranges (Pagani et al. 2016).

These applications of $F_{ST}$ assume that the time over which allele frequencies have diverged is equal in each in each deme. A series of refinements to this same approach that test for more complex temporal patterns of population subdivision is also possible (Patterson et al. 2012; Peter 2016). And recently, an explicit connection between $F_{ST}$ and the PCA framework was presented (Peter 2022).

### 4.3.3  Estimating Migration with the Joint Site Frequency Spectrum

Another approach to detecting and quantifying population structure employs an extension of the SFS first encountered in chapter 2. Consider a sample of $n$ haplotypes taken from a metapopulation. Without regard to subpopulation of origin, recall from chapter 2 that each SNP partitions haplotypes into two groups, $j$ of which carry the derived nucleotide and $n-j$ of which carry the ancestral form. The SFS is a histogram displaying frequencies of SNPs that partition the population into each value $1 \leq j \leq n-1$. Under the infinite sites and Wright–Fisher assumptions (most notably here, a well-mixed population), its values are as predicted by equation (2.45) and shown in figure 2.15.

But each SNP also partitions haplotypes in each subpopulation. To illustrate, suppose there are two demes, $n_1$ of the haplotypes having been sampled from the first and $n_2$ of them from the second. (So $n_1+n_2=n$.) Now consider some particular derived SNP observed in the metapopulation. It may be seen any number $0 \leq j_1 \leq n_1$ times in the first subpopulation and any number $0 \leq j_2 \leq n_2$ in the second. This motivates a two-dimensional histogram, in which the frequency in the $(j_1, j_2)$th bin represents the fraction of derived SNPs seen $j_1$ times in the first sample and $j_2$ in the second. The layout of this two-dimensional histogram, called the *joint site frequency spectrum* (or *jSFS*), is shown in the left panel of figure 4.4. Note that the (one-dimensional) metapopulation SFS can be recovered from the two-dimensional jSFS by summing frequencies found in all bins $(j_1, j_2)$ in which $j_1+j_2=j$ for all $j$ larger than 1 and less than $n=n_1+n_2$. Also note that this construction generalizes to samples from more than two subpopulations.

The attentive reader may notice that, whereas SNPs in the metapopulation can only be observed $1 \leq j \leq n$, the bounds on subpopulation counts $j_1$ and $j_2$ are 0 and $n_1$

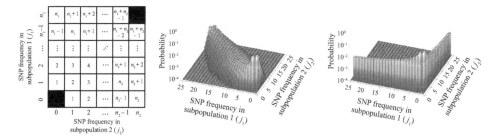

**Figure 4.4**
The joint site frequency spectrum (or jSFS) for two subpopulations. Left: construction of a two-subpopulation jSFS. Axes correspond to the number of samples in which a derived SNP is seen in each subpopulation, written as $j_1$ and $j_2$, respectively. Each SNP seen $0 \leq j_1 \leq n_1$ times in the first subpopulation and $0 \leq j_2 \leq n_2$ times in the second contributes to the $(j_1, j_2)$th cell in the resulting histogram. (Note that $n_1$ and $n_2$ need not be equal.) Each cell in the figure is also labeled by its $j_1 + j_2 = j$, and summing of counts across cells for each $0 < j < n_1 + n_2 = n$ yields the metapopulation SFS. Black cells in the lower left and upper right corners correspond to $j = j_1 + j_2 = 0$ and $j = j_1 + j_2 = n$, respectively; no SNPs in a population-wide sample of $n$ haplotypes can be observed 0 or $n$ times. Center and right: equilibrium jSFS assuming two well mixed and fully isolated subpopulations, respectively. Center: in a well-mixed population, the metapopulation-wide equilibrium SFS, given by equation (2.45), is binomially partitioned into the two samples in proportion to sample sizes; see text for details. Right: in the absence of migration between subpopulations, random genetic drift causes SNPs seen in each subpopulation to be lost from the other, causing the jSFS to equilibrate in the marginal cells corresponding to $j_1 = 0$ and $j_2 = 0$ with frequencies again given by equation (2.45). Center and right panels drawn for sample sizes $n_1 = n_2 = 24$.

or $n_2$. This reflects the fact that a SNP observed in one subpopulation may be absent or fixed in the other. Nevertheless, the (0,0)th and $(n_1, n_2)$th bins in the jSFS must still be zero: unless the SNP is observed at least once and no more than $n_1 + n_2 - 1 = n - 1$ times in the metapopulation sample of $n$ haplotypes, it cannot contribute to the jSFS. Put another way, as noted in the previous paragraph, the metapopulation SFS is embedded in the jSFS as the sums of frequencies in cells corresponding to each fixed value of SNP count $j = j_1 + j_2$. As per chapter 2, the metapopulation SFS counts are still bounded by 1 and $n - 1$.

Having defined the jSFS, we now turn to its interpretation. To begin, imagine a well-mixed population (i.e., with no subdivision) at mutation/drift equilibrium that honors all other Wright–Fisher assumptions. Our two "demic" samples can then be regarded as replicate samples from a single population. In this case, the probabilities of each possible joint SFS count $(j_1, j_2)$ can be found by binomially partitioning the population-wide SFS probabilities for each $j_1 + j_2 = j$ (given by equation [2.45]) with $p = \dfrac{n_1}{n_1 + n_2}$. This follows because each of $j$ copies of a SNP will in expectation be found in the first sample with this probability and the remainder must have been found in the other (see also equation [2.13]). The resulting jSFS is illustrated in the center panel of figure 4.4 for $n_1 = n_2 = 24$.

Now suppose that migration between subpopulations is halted but that each subpopulation continues to honor all Wright–Fisher assumptions. As time proceeds, polymorphic SNPs will be lost or fixed by drift in each subpopulation. But in the absence of migration, these losses and fixations will occur independently in each subpopulation. Of course, new SNPs will appear in each subpopulation by mutation but, by virtue of the infinite sites model, these will all be unique. Thus, over time, the now-subdivided population's jSFS will move to the margins corresponding to $j_2 = 0$ for the first population and $j_1 = 0$ for the second. Given enough time, these sub-populations will reach mutation/drift equilibrium, and their SFSs will equilibrate to the single-population values derived in chapter 2, as shown in the right panel of figure 4.4.

Finally, if migration between two previously isolated subpopulations resumes, SNPs seen in each will begin to contribute to polymorphism in the other. Thus, over time, counts in the jSFS will move back toward the center of the histogram, and in the limit of a well-mixed population, it will converge to the jSFS shown in the center panel of figure 4.4. More generally, the jSFS will find the migration/mutation/drift equilibrium: random genetic drift pushing SNP counts to the margins while migration and mutation pushes them into the interior. And indeed, migration rates can be estimated from an empirical jSFS using coalescent (Wakeley and Hey 1997) or diffusion theory (Gutenkunst et al. 2009). Gutenkunst et al. (2009) give an illustrative application of this approach, simultaneously estimating effective population sizes and migration rates both for the human out-of-Africa migration event and the settlement of the New World.

### 4.3.4  Estimating Admixture from Haplotype Structure

Thus far, we have only described SNP-based methods for characterizing population structure. Just as they did for estimates of effective population size, haplotype-based methods provide additional opportunities to detect recent admixture. For example, analyses of these data in our own species paint a fascinating complex picture of human history over the last 200,000 years (empirical aside 4.1).

To begin, recall that ROH tract lengths and counts can give insight into recent coalescent times and, thus, into a population's effective size over the last several hundred generations. Given genome sequence data for two or more putatively distinct populations, the exact same apparatus can be used to detect *introgression* (the genetic consequence of admixture) by one into the other. For example, returning to modern Latin American populations, Homburger et al. (2015) used this approach to date the earliest Native American/European admixture events, as well as later waves of admixture from African populations and additional European admixture. Similarly, working with human archaeological remains from Europe dated to ~45,000 years ago, Hajdinjak et al. (2021) demonstrated Neanderthal parentage in the preceding few generations. This is biologically important because, although Neanderthal introgression in humans is well-documented (e.g., several percent of the genome in most individuals of non-African ancestry is of Neanderthal origin; Sankararaman et al. 2014), we know less about where and when this admixture occurred.

A related approach uses the decay of linkage disequilibrium between pairs of introgressed SNPs to estimate dates of admixture. In chapter 3, we modeled the decay in linkage disequilibrium over time in a well-mixed population. Quantifying disequilibrium as $D = p_{AB}p_{ab} - p_{Ab}p_{aB}$, we saw that in the absence of natural selection,

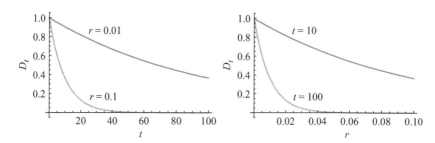

**Figure 4.5**
Decay of linkage disequilibrium $D$ as a function of time and of recombination rate. Results given by equation (3.5). Left: holding genetic map distance $r$ fixed, linkage disequilibrium decays as a function of time $t$, and more quickly for larger values of $r$. Right: holding time $t$ fixed, linkage disequilibrium decays as a function of genetic map distance $r$, and more quickly for larger values of $t$. Both: $y$-axis normalized by $D_0$.

$D_t = (1 - r)^t D_0 \approx D_0 e^{-rt}$ (equation 3.5), where $r$ is the per generation rate of recombination between loci and $t$ is the number of elapsed generations. Indeed, one could imagine fitting empirical observations of $D_t$ between some particular pair of loci over time $t$ to estimate $r$ between them: faster decay in $D$ would signal larger values of $r$, and vice versa. (In the language of teachable moment 1.1, $r$ is the dependent variable and $t$ is a parameter in this framing.)

But imagine now that admixture caused the simultaneous introgression of a great many SNPs into a population $t$ generations in the past. Given recombination rates $r$ between each pair of introgressed SNPs, equation (3.5) can equally be used to estimate $t$ from the many $D_t$ in the data: faster decay in $D$ as $r$ goes up would signal an older introgression, and vice versa. (Now $t$ is the dependent variable and $r$ is a parameter.) These two conceptions of equation (3.5) are illustrated in figure 4.5.

For example, Moorjani et al. (2011) plotted linkage disequilibrium against genetic map distance for all pairs of sub-Saharan–specific SNPs found in four present-day European populations. This approach revealed evidence for African introgression into Southern European populations approximately 55 generations ago, consistent with historical records of Arab migration into Europe at the end of the Roman Empire. (See also Antonio et al. [2019] for more on the population genetics of ancient Rome over the last 12,000 years.) More recent African admixture in present-day Palestinian populations was also detected, but no African admixture was detected in Northern European populations. Using the same approach, Sankararaman et al. (2016) demonstrated that Neanderthal admixture preceded Denisovan admixture in the ancestors of current-day Oceanian individuals.

**Teachable Moment 4.4:** Contrasting population genetic signals on the autosomes and sex chromosome can reveal sex-specific demography

Only females are diploid for the X chromosome in mammals (empirical aside 3.9), providing a window into sex-specific demographic history. There are again two basic approaches. First, in a population of $N$ diploids that honors all Wright–Fisher assumptions, autosomal effective population size is $2N$, while effective population size on the X chromosome is less: $2N_{female} + N_{male}$. Thus, the effective population size for autosomes is always greater than for the X by an amount that depends on sex ratio in the population. For example, Keinan et al. (2009) detected anomalously strong random genetic drift acting on the X chromosome during the out-of-Africa migration event of humans. This conclusion was based on comparisons of $F_{ST}$, the SFS, and $\pi$ for X-linked and autosomal loci, suggesting the possibility that this was a male-biased migration event. Sex-biased demography can also be detected in recombination-dependent statistics such as those for ROH, since recombination on the X chromosome is restricted to females (Buffalo et al. 2016).

## 4.4   Describing Natural Selection

We are finally ready to return to where we began in chapter 1. As first noted by Darwin and Wallace in 1858, heritable differences in rates of reproduction and death among organisms drive the evolution of the species through natural selection. We now invert the theoretical apparatus of population genetics developed in the first three chapters to detect signals of natural selection from genetic and genomic sequence data.

The first empirical evidence of natural selection acting on genetic sequences (dating to the 1960's) were gene-specific differences in the rate of the molecular clock (empirical aside 2.1; figure 2.7). By the 1980's, such data were complemented by gene-specific, within species polymorphism data. These single-gene, mutation-based approaches are the subjects of sections 4.4.1 and 4.4.2. In section 4.4.3 we explore genome-wide screens for natural selection, which comprise both mutation-based methods (sections 4.4.3.1–4.4.3.3) and recombination-based strategies (section 4.4.3.4).

### 4.4.1   Quantifying Natural Selection with the Molecular Clock

Recall from chapter 2 that each mutational difference between two species represents a fixation event in one of them at some point since their divergence. Recall too that under the neutral theory of molecular evolution (empirical aside 2.2), the expected fixation rate at a locus is its neutral mutation rate (equation 2.22). In chapter 2, we interpreted gene-specific differences in the rate of the molecular clock as reflecting differences in the neutral mutation rate. More specifically, we regarded the pace at which genes evolve as inversely dependent on their functional constraint, and thus the strength of purifying selection against mutations. We now explore these ideas mathematically.

---

**Teachable Moment 4.5:** Genes, orthologs, and paralogs

A gene is a locus that encodes a specific product; that is, a protein or an RNA. Organismal reproduction transmits genes from parent to offspring, and genes found in two organisms that trace their common ancestry to a coalescence event are called *orthologs*. For example, the β-globin locus, which can carry the *HbS* allele involved in sickle cell anemia (teachable moment 1.14) has a homolog in all tetrapods. Genes are also occasionally duplicated within a genome. Two copies of a gene that trace their common ancestry to a gene duplication event are called *paralogs*. For example, our own genomes carry genes for many paralogous globin proteins, involved in oxygen and carbon dioxide transport between the atmosphere and our mitochondria. We focus here on methods of analysis for orthologous genes, but many of these can also be used to study the evolutionary history of paralogs.

First, writing $0 \le f_0 \le 1$ for the probability that mutation in a gene is selectively neutral, and $\mu$ for the per site mutation rate in that gene, the above reasoning gives $\mu f_0$ for the neutral mutation rate, and thus the fixation rate in the gene. Introducing $d$ for the proportion of neutral sites that differ between orthologs, we have

$$E[d] = 2\mu f_0 t, \tag{4.10}$$

after $t$ generations of divergence, where the leading two again reflects the fact that fixation on either lineage contributes to divergence. (This $d$ is conceptually similar to but not identical to the $d_{ij}$ introduced in equation [4.6].)

Inverting equation (4.10) suggests a path to estimating the proportion of selectively neutral sites $f_0$ at a gene as a function of observed values of $d$. To control for time $t$ in the equation, we can compare divergence values at two or more genes in some particular species pair. Assuming no gene-specific variation in mutation rate, differences in observed divergence $d$ between two such genes imply differences in the proportion of neutral sites $f_0$. Faster diverging genes must be under a more modest burden of deleterious mutations (i.e., larger $f_0$), implying weaker functional constraint. Conversely, more slowly diverging genes (smaller $f_0$) must be under stronger functional constraints.

Importantly, this reasoning assumes the infinite sites model (mathematically, that $2\mu f_0 t \ll 1$), since the second and all subsequent mutations at a site will not increase $d$. (Indeed, reversion mutations lower it.) However, many interesting comparisons (e.g., those shown in figure 2.7) are across much deeper spans of time $t$ than this, requiring a refinement to our thinking. This is illustrated in teachable moment 4.6 and power user challenge 4.8.

---

**Teachable Moment 4.6:** The Jukes–Cantor model

We motivated the infinite sites model (section 2.4.2) by appealing to the fact that the per site mutation rate multiplied by the total time in a neutral genealogy is almost always much less than 1. But orthologs between species can diverge for arbitrarily many generations, demanding new theory. The Jukes–Cantor model is the simplest approach. We illustrate its derivation on the premise that all mutations are selectively neutral (i.e., $f_0 = 1$), but what follows can easily be adjusted for smaller values of $f_0$.

Given orthologous genes that are identical at a proportion $p_t$ of their sites in generation $t$, we first seek the proportion $p_{t+1}$ that will be identical in generation $t + 1$. Writing the per site, per generation probability of mutation (and so, also of fixation) $\mu$, the probability that neither lineage fixes a mutation in that interval is $(1-\mu)^2 \approx 1-2\mu$ if $\mu \ll 1$. In this case, the proportion $p_t$ of identical sites remain unchanged. Otherwise, with probability $2\mu$, there will be a fixation in one of the lineages. However, only a fraction $1-p_t$ of these will be at sites at which the orthologs currently differ. (Fixations at the remainder will not contribute to $p_t$.) Finally, assuming that all four nucleotides are equally frequent, there is a $\frac{2}{3}$ chance that this fixation will not revert the site back to the nucleotide seen in the other sequence. Putting these ideas together mathematically, we have the recurrence equation

$$p_{t+1} \approx p_t(1-2\mu)+\frac{2}{3}\mu(1-p_t).$$

We now move to continuous time, finding the differential equation $\dfrac{dp_t}{dt}=-\dfrac{8}{3}p_t\mu+\dfrac{2}{3}\mu$. This is then solved by the usual methods (see, for example, power user challenge 3.23) with the boundary condition that the sequences are identical at time $t=0$ (i.e., $p_{t=0}=1$) to give

$$p_t=\frac{1}{4}+\frac{3}{4}e^{-8\mu t/3}.$$

The Jukes–Cantor model thus predicts that the probability that two nucleotides are different after $t$ generations of divergence is

$$d_{\text{obs}}=1-p_t=\frac{3}{4}(1-e^{-8\mu t/3}). \tag{4.11}$$

This is shown by the solid blue line in figure 4.6. As $\mu t$ grows, this expression approaches $\frac{3}{4}$, corresponding to *mutational saturation* (horizontal, green dotted line). At saturation, the two sequences will be no more similar than if they were random. This limit is easily understood: after enough time has passed (this critical $t$ scales as $\frac{1}{\mu}$), the two sequences will be fully randomized relative to each other. Not surprisingly, the first-order Taylor approximation of equation (4.11) at small times $t$ (teachable moment 2.8) is $2\mu t$ (mustard dashed line), recovering the infinite sites result.

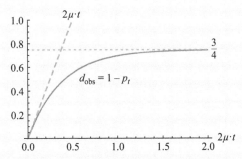

**Figure 4.6**
The probability that orthologous sites differ in two species as a function of generations since divergence $t$, normalized by the expected number of generations between neutral mutation $\left(\dfrac{1}{\mu}\right)$. Blue: assuming the Jukes–Cantor model (equation 4.11). Mustard: the infinite sites result (equation 4.10, assuming $f_0=1$). Green: mutational saturation, or equivalently, the probability that two random sites differ. Note that we have assumed an alphabet of four nucleotides at equal frequencies, and that mutation rates among all pairs of nucleotides are equal.

Finally, in expectation, the true number of fixations in $t$ generations is $2\mu t$. Introducing $d_{\text{true}}$ for this quantity, the exponent on e in equation (4.11) can be rewritten as $-4(2\mu t)/3 = -4d_{\text{obs}}/3$. Solving for $d_{\text{true}}$ yields

$$E[d_{\text{true}}] = -\frac{3}{4}\ln\left[1 - \frac{4}{3}d_{\text{obs}}\right]. \tag{4.12}$$

Thus, equation (4.12) lets us estimate the true proportion of sites that have been mutated as a function of the observed fraction under the Jukes–Cantor model. More complicated mutational models to account for larger alphabets, unequal nucleotide composition and variable mutation rates are possible, but only introduce quantitative changes. See power user challenge 4.8 and Yang (2014) to learn more.

---

**Power User Challenge 4.8**

The data in figure 2.7 illustrate amino acid differences rather than nucleotide differences. Derive the analogs to equations (4.11) and (4.12) for this case. You should assume that all amino acids are equally common and that mutation rates between all pairs of amino acids are equal. Next, as neither of these assumptions are true, discuss the impacts of their relaxation on time to mutational saturation.

---

The foregoing logic for analyzing pairs of orthologous genes can also be applied independently to nonsynonymous and synonymous fixations in a single protein coding gene (see empirical aside 4.4). Introducing $f_0^{\text{N}}$ and $f_0^{\text{S}}$, respectively, for the proportions of selectively neutral nonsynonymous and synonymous mutations, we have

$$E[d_N] = 2\mu f_0^{\text{N}} t \tag{4.13a}$$

$$E[d_S] = 2\mu f_0^{\text{S}} t. \tag{4.13b}$$

This is the theoretical basis for the widely used statistic $\hat{\omega} = \dfrac{\widehat{d_N}}{d_S}$, the observed ratio of nonsynonymous to synonymous fixations between species in a protein-coding gene. Mathematically,

$$E[\omega] = E\left[\frac{d_N}{d_S}\right] = E\left[\frac{2\mu f_0^{\text{N}} t}{2\mu f_0^{\text{S}} t}\right] = E\left[\frac{f_0^{\text{N}}}{f_0^{\text{S}}}\right].$$

In English, $\omega$ allows us to normalize the fraction of neutral nonsynonymous mutations ($f_0^{\text{N}}$) with reference to the fraction of neutral synonymous mutations in the same gene ($f_0^{\text{S}}$). Adding the very convenient (although rarely met; see empirical aside 4.5) assumption that all synonymous mutations are selectively neutral (i.e., that $f_0^{\text{S}} = 1$), $\hat{\omega}$ is often interpreted as the fraction of nonsynonymous mutations that are selectively neutral at the locus ($f_0^{\mathrm{N}}$). Thus, $\hat{\omega}$ represents the simplest measure of the distribution of fitness effects (DFE; see empirical aside 1.3) among nonsynonymous mutations in a protein-coding gene. To see this connection, recall that under the neutral theory of molecular evolution, the DFE for any gene has a fraction $f_0$ of its mass at selection coefficient $s = 0$, and the remainder at values of $s$ more negative than the opposite of the reciprocal of population size.

Importantly, under the neutral theory, observations of $\hat{\omega}$ greater than one would imply that purifying selection on synonymous mutations is stronger than on nonsynonymous mutations (algebraically, that $f_0^{\mathrm{S}} < f_0^{\mathrm{N}}$). While formally possible, the biologically more plausible interpretation for such results is that some proportion of fixed nonsynonymous mutations were driven by positive selection. Indeed, this point also complicates our interpretation of values less than one, which are consistent both with the neutral interpretation of the previous paragraph, and with a mixed model, in which some nonsynonymous mutations fix by drift and others by positive selection.

Nevertheless, taking advantage of the ready availability of sequence data since the turn of the present century, $\hat{\omega}$ has been used to great effect. For example, in comparisons between humans and our nearest living relative, the chimpanzee, Nielsen et al. (2005) find strong support for beneficial mutations in genes associated with sensory perception, immune response, cancer, and spermatogenesis. Importantly, in addition to functional constraints, many other correlates to molecular fixation rates are now also recognized, which must be accounted for (Soni and Eyre-Walker 2022).

### 4.4.2 The McDonald/Kreitman Test

It is tempting to interpret the observed ratio of nonsynonymous to synonymous fixations between orthologs of a protein-coding gene ($\hat{\omega}$) as an estimate of the fraction of nonsynonymous mutations that are selectively neutral in that gene ($f_0^{\mathrm{N}}$). The power of this approach stems from the fact that it offers an internal (gene-specific) control for both divergence times between orthologs and mutation rate. However, as just noted, this line of reasoning overlooks two critical biological facts. First, if synonymous mutations are not selectively neutral (see empirical aside 4.5), $\hat{\omega}$ will overestimate $f_0^{\mathrm{N}}$. Second, it is only crudely sensitive to beneficial mutations: while values of $\hat{\omega}$ larger than 1 imply their presence, values of $\hat{\omega}$ between 0 and 1 do not rule them out.

Simultaneously examining synonymous and nonsynonymous SNPs at the same gene allows us to address both of these limitations. Recall from chapter 2 that the expected number of selectively neutral SNPs in a sample of size $n$ is $\theta \sum_{i=1}^{n-1} \frac{1}{i}$ (equation 2.43; recall $\theta = 4N\mu$). This immediately yields

$$\mathrm{E}[S_n^{\mathrm{N}}] = f_0^{\mathrm{N}} \theta \sum_{i=1}^{n-1} \frac{1}{i} \tag{4.14a}$$

and

$$\mathrm{E}[S_n^{\mathrm{S}}] = f_0^{\mathrm{S}} \theta \sum_{i=1}^{n-1} \frac{1}{i} \tag{4.14b}$$

for the expected number of nonsynonymous and synonymous SNPs in the sample under the neutral model, respectively. Thus, the number of synonymous and nonsynonymous SNPs in a gene contains information about $f_0^N$ and $f_0^S$ that is independent of inferences based on the number of differences fixed between species.

This is the foundation of the McDonald/Kreitman test (1991), which arranges these four quantities as shown in table 4.1. Under the assumption that all mutations are either selectively neutral or immediately elimited by purifying selection (i.e., the neutral theory of molecular evolution), the ratios defined by each row (or equivalently, by each column) should be equal. (We note in passing that this prediction is applicable whenever sites can be partitioned into two or more functional classes, e.g., Akashi [1995].)

Rand and Kann's (1996) *Neutrality Index* (*NI*) offers a clear, biological intuition into deviations from expectation. Focusing on columns in table 4.1, it is defined as

$$NI = \frac{\frac{\hat{S}_n^N}{\hat{d}_N}}{\frac{\hat{S}_n^S}{\hat{d}_S}}$$

An *NI* greater than 1 for a protein-coding gene most likely signals that some nonsynonymous mutations are mildly deleterious. This is because, although deleterious mutations are quite effectively prevented from fixation, they can nevertheless segregate within populations (Kimura 1969), inflating the numerator of *NI*. Conversely, genes for which this quantity is less than 1 likely exhibit evidence of adaptation acting on nonsynonymous mutations. This follows because, if some fraction of beneficial nonsynonymous mutations were selectively fixed, $d_N$ will be inflated. Note that we while we do not assume synonymous mutations are selectively neutral, both interpretations assume that they are more weakly selected than nonsynonymous mutations. Moreover, both demography and very weak deleterious selection coefficients can also influence *NI* (Eyre-Walker and Keightley 2009; Messer and Petrov 2013).

Bustamante et al. (2005) applied this framework to 11,624 protein-coding genes in humans. They tabulated synonymous and nonsynonymous SNPs in sequence data from 39 humans, and counted fixations relative to chimpanzee orthologs. Summed across these genes, they identified 15,750 synonymous and 14,311 nonsynonymous SNPs, together with 34,099 synonymous and 20,467 nonsynonymous fixations.

**Table 4.1**
The McDonald/Kreitman table[a]

|  | Nonsynonymous | Synonymous |
| --- | --- | --- |
| SNPs within species | $E[S_n^N]=\theta f_0^N \sum_{i=1}^{n-1}\frac{1}{i}$ | $E[S_n^S]=\theta f_0^S \sum_{i=1}^{n-1}\frac{1}{i}$ |
| Fixed sites between species | $E[d_N]=2\mu f_0^N t$ | $E[d_S]=2\mu f_0^S t$ |

[a] Expectations from equations (4.13a), (4.13b), (4.14a), and (4.14b). $\theta=4N\mu$.

Together, this yields an *NI* value of approximately 1.51. Thus, averaged across all genes, the data are consistent with purifying selection acting on many nonsynonymous mutations. This follows because there are more nonsynonymous SNPs per nonsynonymous fixation (approximately 0.70) than there are synonymous SNPs per synonymous fixation (approximately 0.46). Going further, those authors also introduced a framework for estimating gene-specific selection coefficients acting on nonsynonymous mutations from gene-specific McDonald/Kreitman tables. Despite the genome-wide average influence of purifying selection, almost 10% of the genes were found to have undergone adaptation since the human/chimp divergence, exhibiting selection coefficients *Ns* ranging between 1 and 5.

### 4.4.3  Scanning Whole Genome Sequence Data for Natural Selection

In sections 4.2 and 4.3, we developed tools for characterizing a population's demographic history in terms of deviations from Wright–Fisher expectations using whole-genome sequence data. As noted, these fall into two groups: those based on mutations (quantified by heterozygosity, the SFS, and subpopulation-specific SNP-frequency differences) and on recombination events (using ROH and linkage disequilibrium statistics). Of course, natural selection can also induce deviations from Wright–Fisher expectations, again manifest in patterns of mutations and recombination events.

But before beginning, we note an important difference between demographic and selective population genetic effects. As first noted by Lewontin and Krakauer (1973), while demography is expected to be manifest genome-wide, selection is likely to act in a locus-specific manner. For example, a selective sweep is a genealogical bottleneck at the driver mutation. But recombination will cause its effect to be restricted to nucleotides in close physical linkage. In contrast, a population's genetic configuration following a demographic bottleneck will likely be seen across the genome. This insight will be central to our thinking.

#### 4.4.3.1  Detecting natural selection from heterozygosity   Under the Wright–Fisher model, equilibrium heterozygosity is proportional to the product of effective population size and the neutral mutation rate. This can be understood as the balance between the production of new mutations (at mutation rate $\mu$) and their elimination by random genetic drift (at rate $\frac{1}{2N_e}$, where $N_e$ is the effective population size). Natural selection increases the variance in reproductive success, thereby increasing the rate of loss of genetically linked, selectively neutral mutations. When selection is acting on a beneficial mutation, the resulting reduction in heterozygosity is called a selective sweep (section 3.2.2). When selection is acting on a deleterious mutation, it is called background selection (section 3.2.3).

Going further, when a beneficial mutation is driven to fixation by natural selection, the resulting selective sweep will transiently reduce heterozygosity across a region of the genome whose width scales as $\frac{s}{c}$ base pairs, where $s$ is the driver mutation's selective advantage and $c$ is the per base pair recombination rate (see figure 3.14). Thus, an obvious strategy for detecting positive selection using genome sequence data is to scan for reduced heterozygosity as an indicator of recently fixed mutations. For example, after controlling for local recombination rate (recall empirical aside 3.6), Sattath et al. (2011) found that heterozygosity is significantly reduced in the vicinity of recently fixed amino acid replacement mutations of the fruit fly *Drosophila simulans*. This is consistent with the idea that many recent nonsynonymous fixations may have driven selective sweeps. Also consistent, the pattern was not observed in the vicinity of synonymous fixations, at which selective sweeps are plausibly less likely.

Interestingly, this difference was not observed between nonsynonymous and synonymous mutations in humans (Hernandez et al. 2011), perhaps suggesting that selective sweeps were less common in our own species. On the other hand, Sattath et al.'s analysis overlooked the influence of background selection (Enard et al. 2014). On the contrary, those authors implicitly assumed that the density of deleterious mutations (and thus, the intensity of background selection) is uniform across the genome. While plausible in a gene-dense genome such as *Drosophila*, Enard et al. note that in the human genome, recently fixed amino acid replacement mutations happen to have occurred in regions of lower functional constraint. Conversely, recently fixed synonymous mutations lie in regions of greater constraint.

In other words, at least in humans, heterozygosity in the vicinity of replacement fixations is not depressed (selective sweeps notwithstanding), perhaps because they also lie in regions of reduced functional constraint, and thus a lessened burden of background selection. Conversely, stronger functional constraint in the vicinity of recent synonymous fixations could explain the comparatively depressed heterozygosity at those sites in humans. The interested reader is directed to Enard et al. (2014) to see hypotheses for these enrichments; our point is simply that attention to deleterious mutations (as well as to demographic history) is a critical step in these analyses (Bank et al. 2014).

**4.4.3.2 Detecting natural selection from the site frequency spectrum** The genealogical impact of natural selection on the SFS closely mirrors that of demographic processes illustrated in figure 4.2, left. For example, after a driver mutation induces a selective sweep, we should expect an excess of rare SNPs, exactly as after a population bottleneck, and for exactly the same reason. Namely, genetic diversity in the vicinity of the recently fixed driver mutation will only slowly be restored by new mutations, which by definition begin life at low frequency. Importantly, however, care is again required in interpretation: deleterious mutations are most likely to occur in genetic linkage with intermediate and high-frequency SNPs, also enriching the SFS for low-frequency SNPs. (The foregoing obviously applies equally to SFS-based summary statistics like Tajima's *D*.) Nevertheless, SFS distortions observed in genome-wide scans have yielded estimates of the distribution of selection coefficients (e.g., Boyko et al. 2008).

Balancing selection (teachable moment 1.23) will also distort the SFS, giving rise to an excess of intermediate-frequency alleles. This is as we saw for population subdivision (figure 4.2, right) and the reasoning is similar. Balancing selection by definition partitions the population's genealogy into two fractions defined by the two alleles. This increases the number of pairs of samples that have deep coalescences, and thus, the number of intermediate-frequency mutations relative to Wright–Fisher expectations. Importantly, genetic recombination will localize this effect to the immediate vicinity of the balanced polymorphism. (In contrast, this localization is likely weaker in a subdivided population, since recombination between haplotypes from different demes is uncommon.)

Importantly, the SFS at a gene also reflects the population's (unknown) demographic history, but the insight in Lewontin and Krakauer (1973)—that such effects will be manifest genome-wide—offers a strategy for overcoming this challenge. Namely, rather than comparing each gene's SFS to the expectation assuming Wright–Fisher equilibrium, those whose distributions are anomalously distorted relative to all others in the genome (so-called *outlier loci*) are recognized as likely targets of natural selection.

### 4.4.3.3 Detecting natural selection from deme-specific frequency differences

Recall from chapter 2 that random genetic drift causes allele frequencies in demes to diverge at a rate proportional to effective population size. Diversifying selection can accelerate this process, while balancing selection can slow it. Motivated by these ideas, Akey et al. (2002) computed $F_{ST}$ at 26,530 SNPs sampled from Black Americans of mixed African and European ancestry, Americans of predominantly European ancestry, and from East Asian individuals. Those results led the authors to identify 156 genes with anomalously large $F_{ST}$ values, implying the action of diversifying selection, and another 18 with anomalously small values, suggesting balancing selection. (Note again the importance of the Lewontin and Krakauer (1973) insight.) The authors also detected correlations in $F_{ST}$ values between loci at physical distances of ~200 kb, consistent with our expectation that the reach of selection will be attenuated by recombination (see figure 3.14).

Similarly, Keinan and Reich (2010) reported a strong positive correlation between intercontinental $F_{ST}$ values at SNPs in humans and local recombination rate. Assuming diversifying selection at this geographic scale (and uniformly distributed driver mutations), this finding is consistent with Hill–Robertson expectations (section 3.2.4), which predicts that higher recombination rates render natural selection more efficient.

### 4.4.3.4 Detecting natural selection from haplotype structure

Lastly, turning from mutation-based to recombination-based methods, recall that long, high-frequency ROH signal recent coalescence, or recent introgression from another population. Similarly, driver mutations should carry ROH consisting of genetically linked SNPs to high frequency during their selective sweeps. As before, the resulting ROH will also be anomalously long, because its rapid rise will limit the opportunity for recombination to decouple linked SNPs from the driver. In contrast, the much slower pace of random genetic drift means that a neutrally evolving ROH at high frequency will likely be much shorter. In other words, during a sweep, frequency and length of ROH are expected to be positively correlated, whereas in its absence, these two quantities should be negatively correlated. (This effect will also transiently distort the high frequency end of the SFS, but because that approach characterizes each

SNP's frequency independently, it overlooks the signal encoded in their correlations. This is exactly what long ROH capture.) Importantly, background selection appears to have no impact on ROH statistics (Enard et al. 2014).

A number of approaches for performing genome-wide scans for ongoing selective sweeps on the basis of long, high-frequency ROH have been developed (Sabeti et al. 2002; Voight et al. 2006; Wang et al. 2006). These have also been extended to explicitly capture the effects of geographically driven diversifying selection (Sabeti et al. 2007) and, recently, an approach for inferring the DFE from ROH data has been proposed (Ortega-Del Vecchyo et al. 2022).

A fascinating example of this work is found in mutations associated with lactase persistence in humans. Lactase is an enzyme required for the metabolism of lactose, a sugar found in milk. Lactase is expressed in all mammals during infancy, with expression declining rapidly after weaning. However, lactase expression persists into adulthood in some humans. This trait is especially common among individuals of Western European and East African ancestry, populations in which dairying was independently discovered.

The genetic basis of lactase persistence is known, and interestingly, the causal mutations differ in European and East African populations. This sort of *parallel evolution*—that is, the independent appearance of mutations conferring the same trait—is already strongly suggestive of a beneficial effect (see also empirical asides 1.2 and 1.15). Moreover, in both populations the ROH tracts carrying the causal lactase persistence SNPs are roughly 1,000-fold longer than those that carry the ancestral nucleotide (Tishkoff et al. 2007). Using the theory underlying equation (4.9), Bersaglieri et al. (2004) estimate the ages of these mutations as 5,000–10,000 years. And based on the theory represented by equation (1.16), the same authors estimate that they confer a 10% fitness advantage.

### 4.4.4  Hard and Soft Selective Sweeps

All the methods just described for detecting natural selection implicitly assume that biological adaptation proceeds via a succession of new mutations, each rendered beneficial by some recent environmental change. We now designate these as *hard selective sweeps* (Messer and Petrov 2013; Hermisson and Pennings 2017). Hard selective sweeps drive coalescence times to zero at the site of the driver mutation, thereby entirely ablating heterozygosity. Moving to the left and right along the genome, recombination will progressively attenuate the effect. This gives rise to the characteristic trough in heterozygosity centered on the driver seen in figure 3.14, as well as anomalously long, high-frequency ROH. This model is illustrated in the left column of figure 4.7.

However, with the explosive growth of gene and genome-sequence data, examples of adaptation associated with only modest reductions in heterozygosity are accumulating. One of the first recognized cases came from the study of drug resistance evolution in HIV. Pennings et al. (2014) examined sequence data from HIV populations in 17 patients before and after the emergence of resistance to a particular antiviral drug. Resistance to this drug is accomplished by a mutation that converts the lysine amino acid into an asparagine at position 103 of the virus' reverse

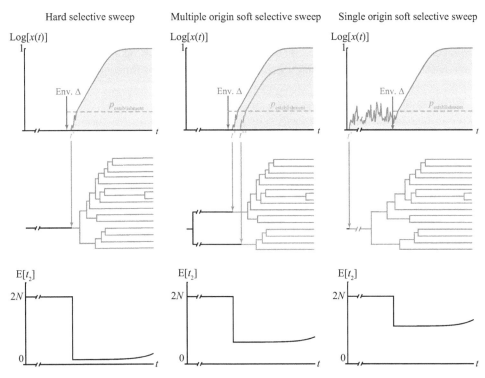

**Figure 4.7**
Hard and soft selective sweeps. Top row: log-transformed frequency of driver mutation(s) [written $x(t)$] as a function of time. The environment changes at the moment indicated in red by Env. Δ. Middle row: corresponding genealogies. Bottom row: typical pairwise coalescence times ($E[t_2]$) in regions near (i.e., at physical distances much closer than $\sim \frac{s}{c}$ Mb) to the driver mutation(s). Left: hard sweep. After an environmental change, a single, beneficial mutation destined to fix appears in generation $t^*$. While its frequency exhibits some stochastic perturbations before establishment, recall from chapter 2 that, thereafter, its time course is very nearly given by deterministic theory (compare with figure 1.2, right). As noted above, the corresponding genealogy is strongly reminiscent of that predicted after a population expansion (figure 4.2, left) and, correspondingly, pairwise coalescence times transiently drop to zero. Middle: multiple origin soft selective sweep. After an environmental change, beneficial mutations destined to fix appear in two different lineages in generations $t^*$ and $t^{**}$. Because the two lineages in which the beneficial mutations occur were already segregating before the environmental change, expected coalescent time near the driver will be longer than in the case of a hard selective sweep. Right: single-origin soft selective sweep. Here, a neutral mutation appears in generation $t^*$. Much later, it is rendered beneficial by an environmental change. Because the beneficial mutation was already present in multiple lineages before the environmental change, the drop in pairwise coalescence times near the driver will be even more modest than in the multiple origin selective sweep. Top and middle rows drawn after Hermisson and Pennings (2017) and Messer and Petrov (2013), respectively.

transcriptase enzyme. The drug-sensitive lysine in wild type HIV is encoded by the AAA codon, and owing to the redundancy of the genetic code (empirical aside 4.4), the resistant asparagine can be represented by either AAC or AAT.

In all 17 patients, resistance mutations were absent before the start of drug therapy. In six of them, HIV populations evolved resistance via the AAC codon, and those in another four did so via AAT. Moreover, in all 10 of these patients, HIV heterozygosity in a 1 kb region straddling the resulting asparagine was found to have been reduced by 71%. These populations thus exhibit all the hallmarks of a hard selective sweep, driven by an A→C mutation in six patients and an A→T in the other four.

However, in each of the remaining seven patients, both of the asparagine-encoding codons (AAC and AAT) were observed after resistance evolved. And in these individuals, HIV heterozygosity in the same 1 kb region straddling the evolved asparagine residue only dropped 15% during the course of resistance evolution. These results are incompatible with our picture of a hard selective sweep.

HIV has a high mutation rate (see empirical aside 1.4) and infected individuals carry immense numbers of the virus. Thus, the population-wide mutation rate within an infected patient is high, and it seems plausible that in each of these last seven individuals, both the AAC and AAT resistance mutations independently appeared and established after the start of drug therapy. This hypothesis would also explain the more modest drop in heterozygosity in the vicinity of the resistance mutation observed, since the two genetic backgrounds on which these mutations occurred are unlikely to have been identical.

Putting things in terms of genealogies, the evidence suggests that in these seven patients, coalescent times at the resistance mutation predated the start of drug therapy. Formally, we define a *soft selective sweep* (Messer and Petrov 2013; Hermisson and Pennings 2017) as an event in which natural selection responds to a recent environmental change by favoring beneficial variants whose coalescent times predate the environmental change. (The reader is cautioned not to confuse hard and soft selective sweeps with hard and soft selection, introduced in section 1.2.5.) In the present case, the soft sweep was apparently driven by the very rapid appearance (and establishment) of more than one mutation. This is sometimes called a *multiple origin* or *de novo* soft selective sweep, and is illustrated in the middle column of figure 4.7.

An environmental change might also render an already-segregating, neutral mutation beneficial. If natural selection then drives this mutation to fixation, its coalescent times will again predate the environmental change, thus, again fitting the definition of a soft selective sweep. This second possibility is sometimes called a *standing variation* or *single-origin* soft selective sweep, and is illustrated in the right column of figure 4.7. Here, adaptation is predicted to cause an even more modest drop in heterozygosity in the genetic vicinity of the driver mutation, owing to coalescent times that can long-precede the environmental change.

---

**Power User Challenge 4.9**

In the bottom row of figure 4.7, expected pairwise coalescence times have only begun to recover. Is that biologically accurate? Why or why not?

---

The relative importance of hard and soft selective sweeps in any population ultimately reflects the supply of beneficial mutations. Hard sweeps will be the rule when adaptation is mutation-limited (i.e., when the product of population size and the beneficial mutation rate is small), while soft sweeps will be more common in cases where this product is large. Quantitative theory exploring the probabilities of hard and soft sweeps (of both sorts), together with strategies for their detection from gene and genome sequence data are well developed, and the interested reader is directed to Messer and Petrov (2013), Jensen (2014) and Hermisson and Pennings (2017) to learn more.

## 4.5 Chapter Summary

- We can estimate population genetic parameter values by inverting our predictive theory and applying the resulting expressions to data from nature.

- Because all genetic variability in a population is the consequence of mutation or recombination, all methods of population genetic inference must ultimately rest on one of these two kinds of data.

- Advances in gene- and genome-sequencing and computational technologies since the turn of the present century are critical to many of the modes of inference now in common use.

- While data from humans are often most readily available, methods of analysis are agnostic on species, and the continuing decline in the cost of data development means that it is becoming practical to apply methods described here to almost any species on Earth.

# References

Achaz, G. 2009. "Frequency Spectrum Neutrality Tests: One for All and All for One." *Genetics* 183: 249–258.

Aguade, M., N. Miyashita, and C. H. Langley. 1989. "Reduced Variation in the Yellow-Achaete-Scute Region in Natural Populations of Drosophila Melanogaster." *Genetics* 122: 607–615.

Ahlquist, K. D., M. M. Bañuelos, A. Funk, J. Lai, S. Rong, F. A. Villanea, and K. E. Witt. 2021. "Our Tangled Family Tree: New Genomic Methods Offer Insight into the Legacy of Archaic Admixture." *Genome Biology and Evolution* 13.

Akashi, H. 1995. "Inferring Weak Selection From Patterns of Polymorphism and Divergence at 'Silent' Sites in Drosophila DNA." *Genetics* 139: 1067–1076.

Akey, J. M., G. Zhang, K. Zhang, L. Jin, and M. D. Shriver. 2002. "Interrogating a High-density SNP Map for Signatures of Natural Selection." *Genome Res* 12: 1805–1814.

Alonzo, S. H., and M. R. Servedio. 2019. "Grey Zones of Sexual Selection: Why is Finding a Modern Definition So Hard?" *Proceedings of the Royal Society B: Biological Sciences* 286: 20191325.

Altenberg, L., U. Liberman, and M. W. Feldman. 2017. "Unified Reduction Principle for the Evolution of Mutation, Migration, and Recombination." *Proceedings of the National Academy of Sciences* 114: E2392–E2400.

Antonio, M. L., Z. Gao, H. M. Moots, M. Lucci, F. Candilio, S. Sawyer, V. Oberreiter, et al. 2019. "Ancient Rome: A Genetic Crossroads of Europe and the Mediterranean." *Science* 366: 708–714.

Atkinson, Q. D., R. D. Gray, and A. J. Drummond. 2007. "mtDNA Variation Predicts Population Size in Humans and Reveals a Major Southern Asian Chapter in Human Prehistory." *Molecular Biology and Evolution* 25: 468–474.

Bagheri, H. C. 2006. "Unresolved Boundaries of Evolutionary Theory and the Question of How Inheritance Systems Evolve: 75 Years of Debate on the Evolution of Dominance." *Journal of Experimental Zoology Part B* 306: 329–359.

Bank, C., G. B. Ewing, AnnaFerrer-Admettla, M. Foll, and J. D. Jensen. 2014. "Thinking Too Positive? Revisiting Current Methods of Population Genetic Selection Inference." *Trends in Genetics* 30: 540–546.

Barrick, J. E., M. R. Kauth, C. C. Strelioff, and R. E. Lenski. 2010. "Escherichia Coli rpoB Mutants Have Increased Evolvability in Proportion to Their Fitness Defects." *Molecular Biology and Evolution* 27: 1338–1347.

Barton, N., and S. P. Otto. 2005. "Evolution of Recombination Due to Random Drift." *Genetics* 169: 2353–2370.

Becks, L., and A. F. Agrawal. 2012. "The Evolution of Sex Is Favoured During Adaptation to New Environments." *PLOS Biology* 10: e1001317.

Begun, D. J., and C. F. Aquadro. 1992. "Levels of Naturally Occurring DNA Polymorphism Correlate with Recombination Rates in *D. Melanogaster*." *Nature* 356: 519–520.

Bell, A. D., C. J. Mello, J. Nemesh, S. A. Brumbaugh, A. Wysoker, and S. A. McCarroll. 2020. "Insights Into Variation in Meiosis from 31,228 Human Sperm Genomes." *Nature* 583: 259–264.

Bell, G. 2017. "Evolutionary Rescue." *Annual Review of Ecology, Evolution, and Systematics* 48: 605–627.

Bersaglieri, T., P. C. Sabeti, N. Patterson, T. Vanderploeg, S. F. Schaffner, J. A. Drake, M. Rhodes, D. E. Reich, and J. N. Hirschhorn. 2004. "Genetic Signatures of Strong Recent Positive Selection at the Lactase Gene." *American Journal of Human Genetics* 74: 1111–1120.

Birdsell, J. A., and C. Wills. 2003. "The Evolutionary Origin and Maintenance of Sexual Recombination: A Review of Contemporary Models." In *Evolutionary Biology*, edited by Ross J. Macintyre and Michael T. Clegg, 27–138. Boston, MA: Springer US.

Birky, C. W., Jr., and J. B. Walsh. 1988. "Effects of Linkage on Rates of Molecular Evolution." *Proceedings of the National Academy of Sciences* 85: 6414–6418.

Bishop, J. A. 1972. "An Experimental Study of the Cline of Industrial Melanism in *Biston betularia* (L.) (Lepidoptera) between Urban Liverpool and Rural North Wales." *Journal of Animal Ecology* 41: 209–243.

Boerlijst, M. C., S. Bonhoeffer, and M. A. Nowak. 1996. "Viral Quasi-Species and Recombination." *Proceedings of the Royal Society B: Biological Sciences* 263: 1577–1584.

Box, G. E. P. 1979. "Robustness In the Strategy of Scientific Model Building." In *Robustness in Statistics*, edited by R. L. Launer and G. N. Wilkinson, 201–236. New York: Academic Press.

Boyko, A. R. 2011. "The Domestic Dog: Man's Best Friend In the Genomic Era." *Genome Biology* 12: 216.

Boyko, A. R., P. Quignon, L. Li, J. J. Schoenebeck, J. D. Degenhardt, K. E. Lohmueller, K. Zhao, et al. 2010. "A Simple Genetic Architecture Underlies Morphological Variation in Dogs." *PLOS Biology* 8: e1000451.

Boyko, A. R., S. H. Williamson, A. R. Indap, J. D. Degenhardt, R. D. Hernandez, K. E. Lohmueller, M. D. Adams, et al. 2008. "Assessing the Evolutionary Impact of Amino Acid Mutations in the Human Genome." *PLOS Genetics* 4: e1000083.

Broman, K. W., J. C. Murray, V. C. Sheffield, R. L. White, and J. L. Weber. 1998. "Comprehensive Human Genetic Maps: Individual and Sex-Specific Variation in Recombination." *American Journal of Human Genetics* 63: 861–869.

Browning, S. R., and B. L. Browning. 2015. "Accurate Non-parametric Estimation of Recent Effective Population Size from Segments of Identity by Descent." *American Journal of Human Genetics* 97: 404–418.

Buffalo, V. 2021. "Quantifying the Relationship Between Genetic Diversity and Population Size Suggests Natural Selection Cannot Explain Lewontin's Paradox." *eLife* 10: e67509.

Buffalo, V., S. M. Mount, and G. Coop. 2016. "A Genealogical Look at Shared Ancestry on the X Chromosome." *Genetics* 204: 57–75.

Bull, J. J., L. A. Meyers, and M. Lachmann. 2005. "Quasispecies Made Simple." *PLOS Computational Biology* 1: e61.

Bustamante, C. D., A. Fledel-Alon, S. Williamson, R. Nielsen, M. Todd Hubisz, S. Glanowski, and D. M. Tanenbaum. 2005. "Natural Selection on Protein-Coding Genes in the Human Genome." *Nature* 437: 1153–1157.

Caballero, A. 1994. "Developments in the Prediction of Effective Population Size." *Heredity* 73: 657–679.

Cabrera, V. M. 2021. "Human Molecular Evolutionary Rate, Time Dependency and Transient Polymorphism Effects Viewed Through Ancient and Modern Mitochondrial DNA Genomes." *Scientific Reports* 11: 5036.

Cann, R. L., M. Stoneking, and A. C. Wilson. 1987. "Mitochondrial DNA and Human Evolution." *Nature* 325: 31–36.

Carlson, J., A. E. Locke, M. Flickinger, M. Zawistowski, S. Levy, R. M. Myers, and M. Boehnke. 2018. "Extremely Rare Variants Reveal Patterns of Germline Mutation Rate Heterogeneity in Humans." *Nature Communications* 9: 3753.

Ceballos, F. C., P. K. Joshi, D. W. Clark, M. Ramsay, and J. F. Wilson. 2018. "Runs of Homozygosity: Windows Into Population History and Trait Architecture." *Nature Reviews Genetics* 19: 220–234.

Chang, J. T. 1999. "Recent Common Ancestors of All Present-day Individuals." *Advances in Applied Probability* 31: 1002–1026.

Charlesworth, B. 1978. "Model for Evolution of Y Chromosomes and Dosage Compensation." *Proceedings of the National Academy of Sciences* 75: 5618–5622.

Charlesworth, B. 1994. *Evolution in Age-Structured Populations.* 2nd ed. Cambridge: Cambridge University Press.

Charlesworth, B., M. T. Morgan, and D. Charlesworth. 1993. "The Effect of Deleterious Mutations on Neutral Molecular Variation." *Genetics* 134: 1289–1303.

Chuang, J. S., O. Rivoire, and S. Leibler. 2009. "Simpson's Paradox in a Synthetic Microbial System." *Science* 323: 272–275.

Civetta, A., and J. M. Ranz. 2019. "Genetic Factors Influencing Sperm Competition." *Frontiers in Genetics* 10: 820.

Clark, A. G., X. Wang, and T. Matise. 2010. "Contrasting Methods of Quantifying Fine Structure of Human Recombination." *Annual Review of Genomics and Human Genetics* 11: 45–64.

Clarke, C. A., B. S. Grant, F. M. M. Clarke, and T. Asami. 1994. "A Long Term Assessment of *Biston Betularia* (L.) in One UK Locality (Caldy Common near West Kirby, Wirral), 1959–1993, and Glimpses Elsewhere." *The Linnean* 10: 18–26.

Clarke, C. A., G. S. Mani, and G. Wynne. 2008. "Evolution in Reverse: Clean Air and the Peppered Moth." *Biological Journal of the Linnean Society* 26: 189–199.

Colgrave, N. 2002. "Sex Releases the Speed Limit on Evolution." *Nature* 420: 664–666.

Collins, F. S., E. D. Green, A. E. Guttmacher, and M. S. Guyer. 2003. "A Vision for the Future of Genomics Research." *Nature* 422: 835–847.

Comeron, J. 2017. "Background Selection as Null Hypothesis in Population Genomics: Insights and Challenges from *Drosophila* Studies." *Philosophical Transactions of the Royal Society B: Biological Sciences* 372: 20160471.

Cook, L. M. 2003. "The Rise and Fall of the Carbonaria Form of the Peppered Moth." *Quarterly Review of Biology* 78: 399–417.

Cook, L. M., B. S. Grant, I. J. Saccheri, and J. Mallet. 2012. "Selective Bird Predation on the Peppered Moth: The Last Experiment of Michael Majerus." *Biology Letters* 8: 609–612.

Coop, G. 2020. "Population and Quantitative Genetics." Accessed January 20. https://gcbias.org/population -genetics-notes/.

Coop, G., and M. Przeworski. 2007. "An Evolutionary View of Human Recombination." *Nature Reviews Genetics* 8: 23–34.

Cooper, T. F. 2007. "Recombination Speeds Adaptation by Reducing Competition between Beneficial Mutations in Populations of Escherichia Coli." *PLOS Biology* 5: e225.

Coyne, J. A., and H. A. Orr. 2004. *Speciation*. Sunderland, MA: Sinauer.

Crow, J. F. 1991. "Why is Mendelian Segregation So Exact?" *Bioessays* 13: 305–312.

Crow, J. F., and M. Kimura. 1970. *An Introduction to Population Genetics Theory*. New York: Harper & Row.

Cvijović, I., B. H. Good, and M. M. Desai. 2018. "The Effect of Strong Purifying Selection on Genetic Diversity." *Genetics* 209: 1235–1278.

Cvijović, I., B. H. Good, E. R. Jerison, and M. M. Desai. 2015. "Fate of a Mutation in a Fluctuating Environment." *Proceedings of the National Academy of Sciences* 112: E5021–E5028.

da Fonseca, R. R., A. Albrechtsen, G. E. Themudo, J. Ramos-Madrigal, J. A. Sibbesen, L. Maretty, M. L. Zepeda-Mendoza, P. F. Campos, R. Heller, and R. J. Pereira. 2016. "Next-Generation Biology: Sequencing and Data Analysis Approaches for Non-Model Organisms." *Mar Genomics* 30: 3–13.

Dance, A. 2021. "The Incredible Diversity of Viruses." *Nature* 595: 22–25.

Dapper, A. L., and B. A. Payseur. 2017. "Connecting Theory and Data to Understand Recombination Rate Evolution." *Philosophical Transactions of the Royal Society B: Biological Sciences* 372: 20160469.

Darwin, C. 1859. *On the Origin of Species by Means of Natural Selection*. London: John Murray.

Darwin, C. 1871. *The Descent of Man, and Selection in Relation to Sex*. London: John Murray.

Darwin, C., and A. R. Wallace. 1858. "On the Tendency of Species to Form Varieties; and on the Perpetuation of Varieties and Species by Natural Means of Selection." *Proceedings of the Linnean Society of London* 3: 46–50.

de Visser, J. A. G. M., J. Hermisson, G. P. Wagner, L. A. Meyers, H. Bagheri-Chaichian, J. L. Blanchard, and L. Chao. 2003. "Perspective: Evolution and Detection of Genetic Robustness." *Evolution* 57: 1959–1972.

de Visser, J. A. G. M., and J. Krug. 2014. "Empirical Fitness Landscapes and the Predictability of Evolution." *Nature Reviews Genetics* 15: 480–490.

de Visser, J. A. G. M., C. W. Zeyl, P. J. Garrish, J. L. Blanchard, and R. Lenski. 1999. "Diminishing Returns from Mutation Supply Rate in Asexual Populations." *Science* 283: 404–406.

Desai, M., D. S. Fisher, and A. W. Murray. 2007. "The Speed of Evolution and Maintenance of Variation in Asexual Populations." *Current Biology* 17: 385–394.

Desai, M. M., and D. S. Fisher. 2007. "Beneficial Mutation-Selection Balance and the Effect of Linkage on Positive Selection." *Genetics* 176: 1759–1798.

Dickerson, R. E. 1971. "The Structure of Cytochrome *c* and the Rates of Molecular Evolution." *Journal of Molecular Evolution* 1: 26–45.

Drummond, A. J., A. Rambaut, B. Shapiro, and O. G. Pybus. 2005. "Bayesian Coalescent Inference of Past Population Dynamics from Molecular Sequences." *Molecular Biology and Evolution* 22: 1185–1192.

Dumont, B. L., and B. A. Payseur. 2008. "Evolution of the Genomic Rate of Recombination in Mammals." *Evolution* 62: 276–294.

Dutheil, J. Y. 2017. "Hidden Markov Models in Population Genomics." In *Hidden Markov Models: Methods and Protocols*, edited by D. R. Westhead and M. S. Vijayabaskar, 149–164. New York: Springer New York.

Dykhuizen, D. 2005. "Species Numbers in Bacteria." *Proceedings. California Academy of Sciences* 56: 62–71.

Enard, D., P. W. Messer, and D. A. Petrov. 2014. "Genome-wide Signals of Positive Selection in Human Evolution." *Genome Research* 24: 885–895.

Escarmís, C., M. Dávila, and E. Domingo. 1999. "Multiple Molecular Pathways for Fitness Recovery of an RNA Virus Debilitated by Operation of Muller's Ratchet." *Journal of Molecular Biology* 285: 495–505.

Estes, S., and M. Lynch. 2003. "Rapid Fitness Recovery in Mutationally Degraded Lines of Caenorhabditis Elegans." *Evolution* 57: 1022–1030.

Ewens, W. J. 1979. *Mathematical Population Genetics*. Berlin, Germany: Springer-Verlag.

Eyre-Walker, A., and P. D. Keightley. 2007. "The Distribution of Fitness Effects of New Mutations." *Nature Reviews Genetics* 8: 610–618.

Eyre-Walker, A., and P. D. Keightley. 2009. "Estimating the Rate of Adaptive Molecular Evolution in the Presence of Slightly Deleterious Mutations and Population Size Change." *Molecular Biology and Evolution* 26: 2097–2108.

Felsenstein, J. 1974. "The Evolutionary Advantage of Recombination." *Genetics* 78: 737–756.

Felsenstein, J. 1992. "Estimating Effective Population Size from Samples of Sequences: Inefficiency of Pairwise and Segregating Sites as Compared to Phylogenetic Estimates." *Genetical Research* 59: 139–147.

Felsenstein, J. 2019. "Theoretical Evolutionary Genetics." Accessed October 20, 2020. https://evolution.gs.washington.edu/pgbook/pgbook.pdf.

Field, Y., E. A. Boyle, N. Telis, Z. Gao, K. J. Gaulton, D. Golan, L. Yengo, et al. 2016. "Detection of Human Adaptation During the Past 2000 Years." *Science* 354: 760–764.

Firnberg, E., J. W. Labonte, J. J. Gray, and M. Ostermeier. 2014. "A Comprehensive, High-Resolution Map of a Gene's Fitness Landscape." *Molecular Biology and Evolution* 31: 1581–1592.

Fisher, R. A. 1954. "A Fuller Theory of 'Junctions' in Inbreeding." *Heredity* 8: 187–197.

Fogle, C. A., J. L. Nagle, and M. M. Desai. 2008. "Clonal Interference, Multiple Mutations and Adaptation in Large Asexual Populations." *Genetics* 180: 2163–2173.

Fragata, I., A. Blanckaert, M. A. D. Louro, D. A. Liberles, and C. Bank. 2019. "Evolution in the Light of Fitness Landscape Theory." *Trends in Ecology & Evolution* 34: 69–82.

Frank, S. A., and M. Slatkin. 1992. "Fisher's Fundamental Theorem of Natural Selection." *Trends in Ecology & Evolution* 7: 92–95.

Frankham, R. 1995. "Effective Population Size/Adult Population Size Ratios in Wildlife: A Review." *Genetical Research* 66: 95–107.

Freedman, A. H., and R. K. Wayne. 2017. "Deciphering the Origin of Dogs: From Fossils to Genomes." *Annual Review of Animal Biosciences* 5: 281–307.

Galtier, N., and M. Rousselle. 2020. "How Much Does $N_e$ Vary Among Species?" *Genetics* 216: 559–572.

Gavrilets, S. 1997. "Evolution and Speciation on Holey Adaptive Landscapes." *Trends in Ecology & Evolution* 12: 307–312.

Genomes Project Consortium. 2012. "An Integrated Map of Genetic Variation from 1,092 Human Genomes." *Nature* 491: 56–65.

Gerrish, P. J., and R. E. Lenski. 1998. "The Fate of Competing Beneficial Mutation in an Asexual Population." *Genetica* 102/103: 127–144.

Gillespie, J. H. 1984. "Molecular Evolution Over the Mutational Landscape." *Evolution* 38: 1116–1129.

Gillespie, J. H. 2000. "Genetic Drift in an Infinite Population: The Pseudohitchhiking Model." *Genetics* 155: 909–919.

Gillespie, J. H. 2004. *Population Genetics: A Concise Guide*. Baltimore, MD: Johns Hopkins University Press.

Goddard, M. R., H. C. Godfray, and A. Burt. 2005. "Sex Increases the Efficacy of Natural Selection in Experimental Yeast Populations." *Nature* 434: 636–640.

Goodenough, U., and J. Heitman. 2014. "Origins of Eukaryotic Sexual Reproduction." *Cold Spring Harbor Perspectives in Biology* 6: a016154.

Goyal, S., D. J. Balick, E. R. Jerison, R. A. Neher, B. I. Shraiman, and M. M. Desai. 2012. "Dynamic Mutation–Selection Balance as an Evolutionary Attractor." *Genetics* 191: 1309–1319.

Grant, B. S., D. F. Owen, and C. A. Clarke. 1996. "Parallel Rise and Fall of Melanic Peppered Moths in America and Britain." *Journal of Heredity* 87: 351–357.

Gutenkunst, R. N., R. D. Hernandez, S. H. Williamson, and C. D. Bustamante. 2009. "Inferring the Joint Demographic History of Multiple Populations from Multidimensional SNP Frequency Data." *PLOS Genetics* 5: e1000695.

Hahn, M. W. 2019. *Molecular Population Genetics*. New York: Oxford University Press.

Haigh, J. 1978. "The Accumulation of Deleterious Genes in a Population—Muller's Ratchet." *Theoretical Population Biology* 14: 251–267.

Hajdinjak, M., F. Mafessoni, L. Skov, B. Vernot, A. Hübner, Q. Fu, E. Essel, et al. 2021. "Initial Upper Palaeolithic Humans in Europe had Recent Neanderthal Ancestry." *Nature* 592: 253–257.

Haldane, J. B. S. 1927. "A Mathematical Theory of Natural and Artificial Selection. Part V: Selection and Mutation." *Proceedings of the Cambridge Philosophical Society* 23: 838–844.

Halligan, D. L., and P. D. Keightley. 2009. "Spontaneous Mutation Accumulation Studies in Evolutionary Genetics." *Annual Review of Ecology and Systematics* 40: 151–172.

Hanson, G., and J. Coller. 2018. "Codon Optimality, Bias and Usage in Translation and mRNA Decay." *Nature Reviews Molecular Cell Biology* 19: 20–30.

Harris, K., and R. Nielsen. 2013. "Inferring Demographic History from a Spectrum of Shared Haplotype Lengths." *PLOS Genetics* 9: e1003521.

Harris, K., and J. K. Pritchard. 2017. "Rapid Evolution of the Human Mutation Spectrum." *eLife* 6: e24284.

Hartfield, M., and P. D. Keightley. 2012. "Current Hypotheses for the Evolution of Sex and Recombination." *Integrative Zoology* 7: 192–209.

Hartl, D. L. 2020. *A Primer of Population Genetics and Genomics*. 4th ed. Oxford: Oxford University Press.

Hartl, D. L., and A. G. Clark. 2007. *Principles of Population Genetics*. 4 ed. Sunderland, MA: Sinauer.

Hayes, B. J., P. M. Visscher, H. C. McPartlan, and M. E. Goddard. 2003. "Novel Multilocus Measure of Linkage Disequilibrium to Estimate Past Effective Population Size." *Genome Research* 13: 635–643.

Hermisson, J., and P. S. Pennings. 2017. "Soft Sweeps and Beyond: Understanding the Patterns and Probabilities of Selection Footprints Under Rapid Adaptation." *Methods in Ecology and Evolution* 8: 700–716.

Hernandez, R. D., J. L. Kelley, E. Elyashiv, S. C. Melton, A. Auton, G. McVean, G. Sella, M. Przeworski, and 1000 Genomes Project. 2011. "Classic Selective Sweeps were Rare in Recent Human Evolution." *Science* 331: 920–924.

Ho, S. Y. W., and B. Shapiro. 2011. "Skyline-Plot Methods for Estimating Demographic History from Nucleotide Sequences." *Molecular Ecology Resources* 11: 423–434.

Homburger, J. R., A. Moreno-Estrada, C. R. Gignoux, D. Nelson, E. Sanchez, P. Ortiz-Tello, B. A. Pons-Estel, et al. 2015. "Genomic Insights into the Ancestry and Demographic History of South America." *PLOS Genetics* 11: e1005602.

Hudson, R. R. 1987. "Estimate the Recombination Parameter of a Finite Population Model Without Selection." *Genetical Research Cambridge* 50: 245–250.

Hudson, R. R. 1990. "Gene Genealogies and the Coalescent Process." In *Oxford Surveys in Evolutionary Biology*, edited by D. Futuyma and J. Antonovics, 1–44. New York: Oxford University Press.

Hudson, R. R. 2015. "A New Proof of the Expected Frequency Spectrum under the Standard Neutral Model." *PLOS ONE* 10: e0118087.

Huff, C. D., D. J. Witherspoon, T. S. Simonson, J. Xing, W. S. Watkins, Y. Zhang, and T. M. Tuohy 2011. "Maximum-Likelihood Estimation of Recent Shared Ancestry (ERSA)." *Genome Research* 21: 768–774.

Iles, M. M., K. Walters, and C. Cannings. 2003. "Recombination Can Evolve in Large Finite Populations Given Selection on Sufficient Loci." *Genetics* 165: 2249–2258.

Iram, S., E. Dolson, J. Chiel, J. Pelesko, N. Krishnan, Ö. Güngör, B. Kuznets-Speck, et al. 2020. "Controlling the Speed and Trajectory of Evolution with Counterdiabatic Driving." *Nature Physics* 17, 135–142.

Iriarte, A., G. Lamolle, and H. Musto. 2021. "Codon Usage Bias: An Endless Tale." *Journal of Molecular Evolution* 89: 589–593.

Jensen, J. D. 2014. "On the Unfounded Enthusiasm for Soft Selective Sweeps." *Nature Communications* 5: 5281.

Jónsson, H., P. Sulem, B. Kehr, S. Kristmundsdottir, F. Zink, E. Hjartarson, M. T. Hardarson, et al. 2017. "Parental Influence on Human Germline De Novo Mutations in 1,548 Trios from Iceland." *Nature* 549: 519–522.

Jorde, L. B., and S. P. Wooding. 2004. "Genetic Variation, Classification and 'Race'." *Nature Genetics* 36: S28–S33.

Karczewski, K. J., L. C. Francioli, G. Tiao, B. B. Cummings, J. Alföldi, Q. Wang, R. L. Collins, et al. 2020. "The Mutational Constraint Spectrum Quantified from Variation in 141,456 Humans." *Nature* 581: 434–443.

Karlin, S. 1983. The Eleventh R.A. Fisher Memorial Lecture. *Proceedings of the Royal Society B* 219: 1216.

Karlin, S., and J. McGregor. 1974. "Towards a Theory of the Evolution of Modifier Genes." *Theoretical Population Biology* 5: 59–103.

Keinan, A., J. C. Mullikin, N. Patterson, and D. Reich. 2007. "Measurement of the Human Allele Frequency Spectrum Demonstrates Greater Genetic Drift in East Asians than in Europeans." *Nature Genetics* 39: 1251–1255.

Keinan, A., J. C. Mullikin, N. Patterson, and D. Reich. 2009. "Accelerated Genetic Drift on Chromosome X During the Human Dispersal out of Africa." *Nature Genetics* 41: 66–70.

Keinan, A., and D. Reich. 2010. "Human Population Differentiation Is Strongly Correlated with Local Recombination Rate." *PLOS Genetics* 6: e1000886.

Kettlewell, H. B. D. 1955. "Selection Experiments on Industrial Melanism in the Lepidoptera." *Heredity* 9: 323–342.

Kettlewell, H. B. D. 1956. "Further Selection Experiments on Industrial Melanism in the Lepidoptera." *Heredity* 10: 287–301.

Kimura, M. 1969. "The Number of Heterozygous Nucleotide Sites Maintained in a Finite Population Due to Steady Flux of Mutations." *Genetics* 61: 893–903.

Kimura, M. 1983. *The Neutral Theory of Molecular Evolution*. Cambridge, UK: Cambridge University Press.

Kimura, M., and J. F. Crow. 1964. "The Number of Alleles that Can be Maintained in a Finite Population." *Genetics* 49: 725–738.

Kirin, M., R. McQuillan, C. S. Franklin, H. Campbell, P. M. McKeigue, J. F. Wilson. 2010. "Genomic Runs of Homozygosity Record Population History and Consanguinity." *PLOS ONE* 5: e13996.

Kong, A., J. Barnard, D. F. Gudbjartsson, G. Thorleifsson, G. Jonsdottir, S. Sigurdardottir, and B. Richardsson. 2004. "Recombination Rate and Reproductive Success in Humans." *Nature Genetics* 36: 1203–1206.

Korol, A. B., and K. G. Iliadi. 1994. "Increased Recombination Frequencies Resulting from Directional Selection for Geotaxis in Drosophila." *Heredity* 72: 64–68.

Kulathinal, R. J., S. M. Bennett, C. L. Fitzpatrick, and M. A. F. Noor. 2008. "Fine-Scale Mapping of Recombination Rate in *Drosophila* Refines its Correlation to Diversity and Divergence." *Proceedings of the National Academy of Sciences* 105: 10051–10056.

Kumar, S. 2005. "Molecular Clocks: Four Decades of Evolution." *Nature Reviews Genetics* 6: 654–662.

LaBar, T., and C. Adami. 2017. "Evolution of Drift Robustness in Small Populations." *Nature Communications* 8: 1012.

Lang, G. I., D. P. Rice, M. J. Hickman, E. Sodergren, G. M. Weinstock, D. Botstein, and M. M. Desai. 2013. "Pervasive Genetic Hitchhiking and Clonal Interference in Forty Evolving Yeast Populations." *Nature* 500: 571–574.

Lawson, D. J., L. van Dorp, and D. Falush. 2018. "A Tutorial on How Not to Over-Interpret STRUCTURE and ADMIXTURE Bar Plots." *Nature Communications* 9: 3258.

LeClerc, J. E., B. Li, W. L. Payne, and T. A. Cebula. 1996. "High Mutation Frequencies Among Escherichia Coli and Salmonella Pathogens." *Science* 274: 1208–1211.

Leffler, E. M., K. Bullaughey, D. R. Matute, W. K. Meyer, L. Ségurel, A. Venkat, P. Andolfatto, and M. Przeworski. 2012. "Revisiting an Old Riddle: What Determines Genetic Diversity Levels within Species?" *PLOS Biology* 10: e1001388.

Leffler, E. M., Z. Gao, S. Pfeifer, L. Ségurel, A. Auton, O. Venn, R. Bowden, et al. 2013. "Multiple Instances of Ancient Balancing Selection Shared Between Humans and Chimpanzees." *Science* 399: 1578–1582.

Lessard, S. 1997. "Fisher's Fundamental Theorem of Natural Selection Revisited." *Theoretical Population Biology* 52: 119–136.

Levy, S. F., J. R. Blundell, S. Venkataram, D. A. Petrov, D. S. Fisher, and G. Sherlock. 2015. "Quantitative Evolutionary Dynamics Using High-Resolution Lineage Tracking." *Nature* 519: 181–186.

Lewontin, R. C. 1972. "The Apportionment of Human Diversity." In *Evolutionary Biology: Volume 6*, edited by T. Dobzhansky, et al., 381–398. New York: Springer US.

Lewontin, R. C. 1974. *The Genetic Basis of Evolutionary Change*. New York: Columbia University Press.

Lewontin, R. C., and J. L. Hubby. 1966. "A Molecular Approach to the Study of Genic Heterozygosity in Natural Populations II: Amount of Variation and Degree of Heterozygosity in Natural Populations of *Drosophila Pseudoobscura*." *Genetics* 54: 595–609.

Lewontin, R. C., and J. Krakauer. 1973. "Distribution of Gene Frequency as a Test of the Theory of the Selective Neutrality of Polymorphisms." *Genetics* 74: 175–195.

Li, H., and R. Durbin. 2011. "Inference of Human Population History from Individual Whole-Genome Sequences." *Nature* 475: 493–496.

Lindsey, H. A., J. Gallie, S. Taylor, and B. Kerr. 2013. "Evolutionary Rescue from Extinction is Contingent on a Lower Rate of Environmental Change." *Nature* 494: 463–467.

Loeb, L. A. 2011. "Human Cancers Express Mutator Phenotypes: Origin, Consequences and Targeting." *Nature Reviews Cancer* 11: 450–457.

Lord, B. D., R. N. Martini, and M. B. Davis. 2022. "Understanding How Genetic Ancestry May Influence Cancer Development." *Trends in Cancer* 8: 276–279.

Lynch, M. 2010. "Evolution of the Mutation Rate." *Trends in Genetics* 26: 345–352.

Lynch, M., M. S. Ackerman, J.-F. Gout, H. Long, W. Sung, W. . Thomas, and P. L. Foster. 2016. "Genetic Drift, Selection and the Evolution of the Mutation Rate." *Nature Reviews Genetics* 17: 704–714.

Lynch, M., R. Bürger, D. Butcher, and W. Gabriel. 1993. "The Mutational Meltdown in Asexual Populations." *Journal of Heredity* 84: 339–344.

Lynch, M., and J. S. Conery. 2003. "The Origins of Genome Complexity." *Science* 302: 1401–1404.

MacLeod, I. M., D. M. Larkin, H. A. Lewin, B. J. Hayes, and M. E. Goddard. 2013. "Inferring Demography from Runs of Homozygosity in Whole-genome Sequence, with Correction for Sequence Errors." *Molecular Biology and Evolution* 30: 2209–2223.

Malécot, G. B. L. 1948. *Les mathématiques de l'hérédité*. Paris: Masson et Cie.

Margoliash, E. 1963. "Primary Structure and Evolution of Cytochrome *c*." *Proceedings of the National Academy of Sciences, USA* 50: 672–679.

Marsden, C. D., D. O.-D. Vecchyo, D. P. O'Brien, J. F. Taylor, O. Ramirez, C. Vilà, T. Marques-Bonet, R. D. Schnabel, R. K. Wayne, and K. E. Lohmueller. 2016. "Bottlenecks and Selective Sweeps During Domestication Have Increased Deleterious Genetic Variation in Dogs." *Proceedings of the National Academy of Sciences* 113: 152–157.

Maruyama, T., and M. Kimura. 1974. "A Note on the Speed of Gene Frequency Changes in Reverse Directions in a Finite Population." *Evolution* 28: 161–163.

Maynard Smith, J. 1970. "Natural Selection and the Concept of a Protein Space." *Nature* 225: 563–565.

Maynard Smith, J., and J. Haigh. 1974. "The Hitch-hiking Effect of a Favourable Gene." *Genetical Research* 23: 23–35.

McDonald, J. H., and M. Kreitman. 1991. "Adaptive Protein Evolution at the *Adh* Locus in *Drosophila*." *Nature* 351: 652–654.

McDonald, M. J., D. P. Rice, and M. M. Desai. 2016. "Sex Speeds Adaptation by Altering the Dynamics of Molecular Evolution." *Nature* 531: 233–236.

Mendel, J. G. 1866. "Versuche über Pflanzenhybriden." *Verhandlungen des naturforschenden Vereines in Brünn* Bd. IV für das Jahr 1865: 3–47.

Menozzi, P., A. Piazza, and L. Cavalli-Sforza. 1978. "Synthetic Maps of Human Gene Frequencies in Europeans." *Science* 201: 786–792.

Messer, P. W. 2009. "Measuring the Rates of Spontaneous Mutation From Deep and Large-Scale Polymorphism Data." *Genetics* 182: 1219–1232.

Messer, P. W., and D. A. Petrov. 2013. "Population Genomics of Rapid Adaptation by Soft Selective Sweeps." *Trends in Ecology & Evolution* 28: 659–669.

Meyer, D., and G. Thomson. 2001. "How Selection Shapes Variation of the Human Major Histocompatibility Complex: A Review." *Annals of Human Genetics* 65: 1–26.

Meyer, M., M. Kircher, M.-T. Gansauge, H. Li, F. Racimo, S. Mallick, and J. G. Schraiber, et al. 2012. "A High-Coverage Genome Sequence from an Archaic Denisovan Individual." *Science (New York, N.Y.)* 338: 222–226.

"Microbiology by numbers." 2011. *Nature Reviews Microbiology* 9: 628–628.

Miralles, R., P. J. Gerrish, A. Moya, and S. F. Elena. 1999. "Clonal Interference and the Evolution of RNA Viruses." *Science* 285: 1745–1747.

Mooney, J. A., A. Yohannes, and K. E. Lohmueller. 2021. "The Impact of Identity by Descent on Fitness and Disease in Dogs." *Proceedings of the National Academy of Sciences* 118: e2019116118.

Moorjani, P., N. Patterson, J. N. Hirschhorn, A. Keinan, L. Hao, G. Atzmon, E. Burns, H. Ostrer, A. L. Price, and D. Reich. 2011. "The History of African Gene Flow into Southern Europeans, Levantines, and Jews." *PLOS Genetics* 7: e1001373.

Moreno-Estrada, A., C. R. Gignoux, J. C. Fernández-López, F. Zakharia, M. Sikora, A. V. Contreras, V. Acuña-Alonzo, et al. 2014. "The Genetics of Mexico Recapitulates Native American Substructure and Affects Biomedical Traits." *Science* 344, 6189: 1280–1285.

Mountain, J. L., and U. Ramakrishnan. 2005. "Impact of Human Population History on Distributions of Individual-Level Genetic Distance." *Human Genomics* 2: 4.

Muller, H. J. 1928. "The Measurement of Gene Mutation Rate in *Drosophila*, Its High Variability, and Its Dependence upon Temperature." *Genetics* 13: 279–357.

Mustonen, V., and M. Lässig. 2009. "From Fitness Landscapes to Seascapes: Non-Equilibrium Dynamics of Selection and Adaptation." *Trends in Genetics* 25: 111–119.

Myers, S., L. Bottolo, C. Freeman, G. McVean, and P. Donnelly. 2005. "A Fine-Scale Map of Recombination Rates and Hotspots Across the Human Genome." *Science* 310: 321–324.

Neher, R. A. 2013. "Genetic Draft, Selective Interference, and Population Genetics of Rapid Adaptation." *Annual Review of Ecology, Evolution, and Systematics* 44: 195–215.

Neher, R. A., B. I. Shraiman, and D. S. Fisher. 2010. "Rate of Adaptation in Large Sexual Populations." *Genetics* 184: 467–481.

Nei, M., and W. H. Li. 1973. "Linkage Disequilibrium in Subdivided Populations." *Genetics* 75: 213–219.

Nielsen, R., C. Bustamante, A. G. Clark, S. Glanowski, T. B. Sackton, M. J. Hubisz, A. Fledel-Alon, et al. 2005. "A Scan for Positively Selected Genes in the Genomes of Humans and Chimpanzees." *PLOS Biology* 3: e170.

Nielsen, R., and M. Slatkin. 2013. *An Introduction to Population Genetics*. Sunderland, MA: Sinauer Associates.

NIH. 2021a. "The Future of Health Begins with You." Accessed May, 6, 2022. https://allofus.nih.gov/.

NIH. 2021b. "National Human Genome Research Institute DNA Sequencing Costs: Data." Accessed April 6. https://www.genome.gov/about-genomics/fact-sheets/DNA-Sequencing-Costs-Data.

Novembre, J. 2016. "Pritchard, Stephens, and Donnelly on Population Structure." *Genetics* 204: 391–393.

Novembre, J. 2022. "The Background and Legacy of Lewontin's Apportionment of Human Genetic Diversity." *Philosophical Transactions of the Royal Society B: Biological Sciences* 377: 20200406.

Novembre, J., T. Johnson, K. Bryc, Z. Kutalik, A. R. Boyko, A. Auton, A. Indap, et al. 2008. "Genes Mirror Geography within Europe." *Nature* 456: 98–101.

Nowak, M. A. 2006. *Evolutionary Dynamics: Exploring the Equations of Life*. Harvard University Press.

Oakeshott, J. G., J. B. Gibson, P. R. Anderson, W. R. Knibb, D. G. Anderson, and G. K. Chambers. 1982. "Alcohol Dehydrogenase and Glycerol-3-Phosphate Dehydrogenase Clines in Drosophila melanogaster on Different Continents." *Evolution* 36: 86–96.

Ohta, T. 1973. "Slightly Deleterious Mutant Substitutions in Evolution." *Nature* 246: 96–98.

Okasha, S. 2008. "Fisher's Fundamental Theorem of Natural Selection—A Philosophical Analysis." *The British Journal for the Philosophy of Science* 59: 319–351.

Ortega-Del V. D., K. E. Lohmueller, and J. Novembre. 2022. "Haplotype-Based Inference of the Distribution of Fitness Effects." *Genetics* 220.

Otto, S. P. 2009. "The Evolutionary Enigma of Sex." *The American Naturalist* 174: S1–S14.

Otto, S. P. 2013. "Evolution of Modifier Genes and Biological Systems." In *Princeton Guide to Evolution*, edited by D. A. Baum and D. J. Futuyma, 253–260. Princeton, NJ: Princeton University Press.

Otto, S. P. 2020. "Selective Interference and the Evolution of Sex." *Journal of Heredity* 112: 9–18.

Otto, S. P., and A. C. Gerstein. 2008. "The Evolution of Haploidy and Diploidy." *Current Biology* 18: R1121–R1124.

Otto, S. P., and T. Lenormand. 2002. "Resolving the Paradox of Sex and Recombination." *Nature Reviews Genetics* 3: 252–261.

Pagani, L., D. J. Lawson, E. Jagoda, A. Mörseburg, A. Eriksson, M. Mitt, and F. Clemente, et al. 2016. "Genomic Analyses Inform on Migration Events During the Peopling of Eurasia." *Nature* 538: 238–242.

Palamara, P. F., T. Lencz, A. Darvasi, and I. Pe'er. 2012. "Length Distributions of Identity by Descent Reveal Fine-scale Demographic History." *American journal of human genetics* 91: 809–822.

Palkopoulou, E., S. Mallick, P. Skoglund, J. Enk, N. Rohland, H. Li, and A. Omrak, et al. 2015. "Complete Genomes Reveal Signatures of Demographic and Genetic Declines in the Woolly Mammoth." *Current Biology* 25: 1395–1400.

Park, S.-C., and J. Krug. 2013. "Rate of Adaptation in Sexuals and Asexuals: A Solvable Model of the Fisher–Muller Effect." *Genetics* 195: 941–955.

Park, S.-C., D. Simon, and J. Krug. 2010. "The Speed of Evolution in Large Asexual Populations." *Journal of Statistical Physics* 138: 381–410.

Patterson, N., M. Isakov, T. Booth, L. Büster, C.-E. Fischer, I. Olalde, and H. Ringbauer, et al. 2022. "Large-scale Migration into Britain during the Middle to Late Bronze Age." *Nature* 601: 588–594.

Patterson, N., P. Moorjani, Y. Luo, S. Mallick, N. Rohland, Y. Zhan, T. Genschoreck, T. Webster, and D. Reich. 2012. "Ancient Admixture in Human History." *Genetics* 192: 1065–1093.

Peñalba, J. V., and J. B. W. Wolf. 2020. "From Molecules to Populations: Appreciating and Estimating Recombination Rate Variation." *Nature Reviews Genetics* 21: 476–492.

Penn, D. J., and W. K. Potts. 1999. "The Evolution of Mating Preferences and Major Histocompatibility Complex Genes." *The American Naturalist* 153: 145–164.

Pennings, P. S., S. Kryazhimskiy, and J. Wakeley. 2014. "Loss and Recovery of Genetic Diversity in Adapting Populations of HIV." *PLoS Genet* 10: e1004000.

Peter, B. M. 2016. "Admixture, Population Structure, and F-Statistics." *Genetics* 202: 1485–1501.

Peter, B. M. 2022. "A Geometric Relationship of $F_2$, $F_3$ and $F_4$-Statistics with Principal Component Analysis." *Philosophical Transactions of the Royal Society B: Biological Sciences* 377: 20200413.

Poelwijk, F. J., V. Krishna, and R. Ranganathan. 2016. "The Context-Dependence of Mutations: A Linkage of Formalisms." *PLOS Computational Biology* 12: e1004771.

Poelwijk, F. J., M. Socolich, and R. Ranganathan. 2019. "Learning the Pattern of Epistasis Linking Genotype and Phenotype in a Protein." *Nature Communications* 10 (1): 4213.

Poon, A., and L. Chao. 2005. "The Rate of Compensatory Mutation in the DNA Bacteriophage fX174." *Genetics* 170: 989–999.

Poon, A., B. H. Davis, and L. Chao. 2005. "The Coupon Collector and the Supressor Mutation: Estimating the Number of Compensatory Mutations by Maximum Likelihood." *Genetics* 170: 1323–1332.

Powell, J. E., P. M. Visscher, and M. E. Goddard. 2010. "Reconciling the Analysis of IBD and IBS in Complex Trait Studies." *Nature Reviews Genetics* 11: 800–805.

Pritchard, J. K., M. Stephens, and P. Donnelly. 2000. "Inference of Population Structure Using Multilocus Genotype Data." *Genetics* 155: 945–959.

Prüfer, K., C. de Filippo, S. Grote, F. Mafessoni, P. Korlević, M. Hajdinjak, B. Vernot, et al. 2017. "A High-coverage Neandertal Genome from Vindija Cave in Croatia." *Science* 358: 655–658.

Prüfer, K., F. Racimo, N. Patterson, F. Jay, S. Sankararaman, S. Sawyer, A. Heinze, et al. 2014. "The Complete Genome Sequence of a Neanderthal from the Altai Mountains." *Nature* 505: 43–49.

Pybus, O. G., A. Rambaut, and P. H. Harvey. 2000. "An Integrated Framework for the Inference of Viral Population History from Reconstructed Genealogies." *Genetics* 155: 1429–1437.

Ralph, P., and G. Coop. 2013. "The Geography of Recent Genetic Ancestry across Europe." *PLOS Biology* 11: e1001555.

Ramachandran, S., O. Deshpande, C. C. Roseman, N. A. Rosenberg, M. W. Feldman, and L. Luca Cavalli-Sforza. 2005. "Support from the Relationship of Genetic and Geographic Distance in Human Populations for a Serial Founder Effect Originating in Africa." *Proceedings of the National Academy of Sciences of the United States of America* 102: 15942–15947.

Rand, D. M., and L. M. Kann. 1996. "Excess Amino Acid Polymorphism in Mitochondrial DNA: Contrasts Among Genes from Drosophila, Mice and Humans." *Molecular Biology and Evolution* 13: 735–748.

Raynes, Y., and P. D. Sniegowski. 2014. "Experimental Evolution and the Dynamics of Genomic Mutation Rate Modifiers." *Heredity* 113: 375–380.

Reich, D. 2018. *Who We Are and How We Got Here: Ancient DNA and the New Science of the Human Past.* New York: Pantheon Press.

Reich, D., K. Thangaraj, N. Patterson, A. L. Price, and L. Singh. 2009. "Reconstructing Indian Population History." *Nature* 461: 489–494.

Reich, D. E., R. E. Green, M. Kircher, J. Krause, N. Patterson, E. Y. Durand, B. Viola, et al. 2010. "Genetic History of an Archaic Hominin Group from Denisova Cave in Siberia." *Nature* 468: 1053–1060.

Ritz, K. R., M. A. F. Noor, and Nadia D. Singh. 2017. "Variation in Recombination Rate: Adaptive or Not?" *Trends in Genetics* 33 (5): 364–374.

Rogers, A. R., and H. Harpending. 1992. "Population Growth Makes Waves in the Distribution of Pairwise Genetic Differences." *Molecular Biology and Evolution* 9: 552–569.

Rosenberg, N. A., J. K. Pritchard, J. L. Weber, H. M. Cann, K. K. Kidd, L. A. Zhivotovsky, and M. W. Feldman. 2002. "Genetic Structure of Human Populations." *Science* 298: 2381–2385.

Rutherford, A. 2017. *A Brief History of Everyone Who Ever Lived.* New York: The Experiment, LLC.

Sabeti, P. C., D. E. Reich, J. M. Higgins, H. Z. P. Levine, D. J. Richter, S. F. Schaffner, S. B. Gabriel, et al. 2002. "Detecting Recent Positive Selection in the Human Genome from Haplotype Structure." *Nature* 419: 832–837.

Sabeti, P. C., P. Varilly, B. Fry, J. Lohmueller, E. Hostetter, C. Cotsapas, X. Xie, et al. 2007. "Genome-wide Detection and Characterization of Positive Selection in Human Populations." *Nature* 449: 913–918.

Sams, A. J., and A. R. Boyko. 2019. "Fine-Scale Resolution of Runs of Homozygosity Reveal Patterns of Inbreeding and Substantial Overlap with Recessive Disease Genotypes in Domestic Dogs." *G3 (Bethesda)* 9: 117–123.

Samuk, K., B. Manzano-Winkler, K. R. Ritz, and M. A. F. Noor. 2020. "Natural Selection Shapes Variation in Genome-wide Recombination Rate in *Drosophila pseudoobscura*." *Current Biology* 30: 1517–1528.e6.

Sankararaman, S., S. Mallick, M. Dannemann, K. Prüfer, J. Kelso, S. Pääbo, N. Patterson, and D. Reich. 2014. "The Genomic Landscape of Neanderthal Ancestry in Present-day Humans." *Nature* 507: 354–357.

Sankararaman, S., S. Mallick, N. Patterson, and D. Reich. 2016. "The Combined Landscape of Denisovan and Neanderthal Ancestry in Present-Day Humans." *Current Biology* 26: 1241–1247.

Sattath, S., E. Elyashiv, O. Kolodny, Y. Rinott, and G. Sella. 2011. "Pervasive Adaptive Protein Evolution Apparent in Diversity Patterns around Amino Acid Substitutions in Drosophila Simulans." *PLOS Genetics* 7: e1001302.

Saupe, T., F. Montinaro, C. Scaggion, N. Carrara, T. Kivisild, E. D'Atanasio, Ruoyun Hui, et al. 2021. "Ancient Genomes Reveal Structural Shifts After the Arrival of Steppe-related Ancestry in the Italian Peninsula." *Current Biology* 31: 2576–2591.e12.

Schiffels, S., and R. Durbin. 2014. "Inferring Human Population Size and Separation History from Multiple Genome Sequences." *Nature Genetics* 46: 919–925.

Shemirani, R., G. M. Belbin, C. L. Avery, E. E. Kenny, C. R. Gignoux, and J. L. Ambite. 2021. "Rapid Detection of Identity-by-descent Tracts for Mega-scale Datasets." *Nature Communications* 12: 3546.

Shen, P., F. Wang, P. A. Underhill, C. Franco, W.-H. Yang, A. Roxas, R. Sung, et al. 2000. "Population Genetic Implications from Sequence Variation in Four Y Chromosome Genes." *Proceedings of the National Academy of Sciences* 97: 7354–7359.

Silander, O. K., O. Tenaillon, and L. Chao. 2007. "Understanding the Evolutionary Fate of Finite Populations: The Dynamics of Mutational Effects." *PLOS Biology* 5: e94.

Simon, J.-C., F. Delmotte, C. Rispe, and T. Crease. 2003. "Phylogenetic Relationships Between Parthenogens and Their Sexual Relatives: The Possible Routes to Parthenogenesis in Animals." *Biological Journal of the Linnean Society* 79: 151–163.

Singer, G. A. C., and D. A. Hickey. 2000. "Nucleotide Bias Causes a Genomewide Bias in the Amino Acid Composition of Proteins." *Molecular Biology and Evolution* 17: 1581–1588.

Singhal, S., S. M. Gomez, and C. L. Burch. 2019. "Recombination Drives the Evolution of Mutational Robustness." *Current Opinion in Systems Biology* 13: 142–149.

Soni, V., and A. Eyre-Walker. 2022. "Factors That Affect the Rates of Adaptive and Nonadaptive Evolution at the Gene Level in Humans and Chimpanzees." *Genome Biology and Evolution* 14.

Stajich, J. E., and M. W. Hahn. 2004. "Disentangling the Effects of Demography and Selection in Human History." *Molecular Biology and Evolution* 22: 63–73.

Stapley, J., P. G. D. Feulner, S. E. Johnston, A. W. Santure, and C. M. Smadja. 2017. "Variation in Recombination Frequency and Distribution Across Eukaryotes: Patterns and Processes." *Philosophical Transactions of the Royal Society B: Biological Sciences* 372: 20160455.

Stumpf, M. P. H., and G. A. T. McVean. 2003. "Estimating Recombination Rates from Population-genetic Data." *Nature Reviews Genetics* 4: 959–968.

Suárez, P., J. Valcárcel, and J. Ortín. 1992. "Heterogeneity of the Mutation Rates of Influenza A Viruses: Isolation of Mutator Mutants." *Journal of Virology* 66: 2491–2494.

Sved, J. A. 1971. "Linkage Disequilibrium and Homozygosity of Chromosome Segments in Finite Populations." *Theoretical Population Biology* 2: 125–141.

Sweetlove, L. 2011. "Number of Species on Earth Tagged at 8.7 Million." *Nature.*

Tajima, F. 1983. "Evolutionary Relationship of DNA Sequences in Finite Populations." *Genetics* 105: 437–460.

Tenesa, A., P. Navarro, B. J. Hayes, D. L. Duffy, G. M. Clarke, M. E. Goddard, and P. M. Visscher. 2007. "Recent Human Effective Population Size Estimated from Linkage Disequilibrium." *Genome research* 17: 520–526.

Thompson, E. A. 2013. "Identity by Descent: Variation in Meiosis, Across Genomes, and in Populations." *Genetics* 194: 301–326.

Thompson, E. A. 2018. "From 1949 to 2018: R. A. Fisher's Theory of Junctions." *Journal of Animal Breeding and Genetics* 135: 335–336.

Tishkoff, S. A., F. A. Reed, A. Ranciaro, B. F. Voight, C. C. Babbitt, J. S. Silverman, K. Powell, et al. 2007. "Convergent Adaptation of Human Lactase Persistence in Africa and Europe." *Nature Genetics* 39: 31–40.

Van den Bergh, B., J. E. Michiels, T. Wenseleers, E. M. Windels, P. Vanden Boer, D. Kestemont, L. De Meester, et al. 2016. "Frequency of Antibiotic Application Drives Rapid Evolutionary Adaptation of Escherichia Coli Persistence." *Nature Microbiology* 1: 16020.

van't Hof, A. E., N. Edmonds, M. Dalíková, F. Marec, and I. J. Saccheri. 2011. "Industrial Melanism in British Peppered Moths Has a Singular and Recent Mutational Origin." *Science* 332: 958–960.

Venter, J. C., M. D. Adams, E. W. Myers, P. W. Li, R. J. Mural, G. G. Sutton, H. O. Smith, et al. 2001. "The Sequence of the Human Genome." *Science* 291: 1304–1351.

Vigue, C. L., and F. M. Johnson. 1973. "Isozyme Variability in Species of the Genus Drosophila. VI. Frequency-property-environment Relationships of Allelic Alcohol Dehydrogenases in *D. melanogaster.*" *Biochemical Genetics* 9: 213–227.

Voight, B. F., S. Kudaravalli, X. Wen, and J. K. Pritchard. 2006. "A Map of Recent Positive Selection in the Human Genome." *PLOS Biology* 4: e72.

Wagner, G. P., and P. Krall. 1993. "What is the Difference Between Models of Error Thresholds and Muller's Ratchet?" *Journal of Mathematical Biology* 32: 33–44.

Wakeley, J. 2009. *Coalescent Theory: An Introduction.* Greenwood Village, Colorado: Roberts and Company.

Wakeley, J., and J. Hey. 1997. "Estimating Ancestral Population Parameters." *Genetics* 145: 847–855.

Walsh, B., and M. Lynch. 2018. *Evolution and Selection of Quantitative Traits.* 1st ed. Oxford: Oxford University Press.

Wang, E. T., G. Kodama, P. Baldi, and R. K. Moyzis. 2006. "Global Landscape of Recent Inferred Darwinian Selection for *Homo sapiens.*" *Proceedings of the National Academy of Sciences* 103: 135–140.

Wang, J. 2005. "Estimation of Effective Population Sizes from Data on Genetic Markers." *Philosophical Transactions of the Royal Society B: Biological Sciences* 360: 1395–409.

Wang, J., E. Santiago, and A. Caballero. 2016. "Prediction and Estimation of Effective Population Size." *Heredity* 117: 193–206.

Watterson, G. A. 1975. "On the Number of Segregating Sites in Genetical Models without Recombination." *Theoretical Population Biology* 7: 256–276.

Weinreich, D. M. 2010. "Predicting Molecular Evolutionary Trajectories in Principle and in Practice." In *Encyclopedia of Life Sciences (ELS).* Chichester, UK: John Wiley & Sons.

Weinreich, D. M., and L. Chao. 2005. "Rapid Evolutionary Escape by Large Populations from Local Fitness Peaks is Likely in Nature." *Evolution* 59: 1175–1182.

Weinreich, D. M., Nigel F. Delaney, Mark A. Depristo, and Daniel L. Hartl. 2006. "Darwinian Evolution Can Follow Only Very Few Mutational Paths to Fitter Proteins." *Science* 312: 111–114.

Weinreich, D. M., Y. Lan, C. Scott Wylie, and Robert B. Heckendorn. 2013. "Should Evolutionary Geneticists Worry about High Order Epistasis?" *Current Opinion in Development and Genetics* 23: 700–707.

Weinreich, D. M., S. Sindi, and R. A. Watson. 2013. "Finding the Boundary between Evolutionary Basins of Attraction, and Implications for Wright's Fitness Landscape Analogy." *Journal of Statistical Mechanics*: P01001.

Weinreich, D. M., R. A. Watson, and L. Chao. 2005. "Perspective: Sign Epistasis and Genetic Constraint on Evolutionary Trajectories." *Evolution* 59: 1165–1174.

Weisman, A. 1887. "On the Signification of the Polar Globules." *Nature* 36: 607–609.

Weissman, D. B., and N. H. Barton. 2012. "Limits to the Rate of Adaptive Substitution in Sexual Populations." *PLOS Genetics* 8: e1002740.

Weissman, D. B., M. W. Feldman, and D. S. Fisher. 2010. "The Rate of Fitness-Valley Crossing in Sexual Populations." *Genetics* 186: 1389–1410.

Weissman, D. B., and O. Hallatschek. 2014. "The Rate of Adaptation in Large Sexual Populations with Linear Chromosomes." *Genetics* 196: 1167–1183.

Wilke, C. 2005. "Quasispecies Theory in the Context of Population Genetics." *BMC Evolutionary Biology* 5: 44.

Woodward, Billy, J. Baker, M. P. Kinney, A. Ingram, and T. Anderson. 2023. "Who Saved the Most Lives in History?!". Accessed January 21, 2023. https://www.scienceheroes.com/.

Wright, S. 1932. "The Roles of Mutation, Inbreeding, Crossbreeding and Selection in Evolution." In *Proceedings of the Sixth International Congress of Genetics,* edited by D. F. Jones, 356–366. Menasha, WI: Brooklyn Botanic Garden.

Wright, S. 1943. "Isolation by Distance." *Genetics* 28: 114–138.

Wright, S. 1949. "The Genetical Structure of Populations." *Annals of Eugenics* 15: 323–354.

Yang, Z. 2014. *Molecular Evolution: A Statistical Approach.* New York: Oxford University Press.

Zuckerkandl, E., and L. Pauling. 1962. "Molecular Disease, Evolution, and Genetic Heterogeneity." In *Horizons in Biochemistry,* edited by M. Kasha and B. Pullman, 189. New York: Academic Press.

# Index

Note: Entries in *italics* denote a definition, found on the underlined page. Page numbers in *italics* denote a figure. Page numbers in **bold** denote a table.

Absolute Wrightian fitness. *See Wrightian fitness*
*Absorbing barrier,* 57, *66,* 75, 79–80, 88
*Admixture,* 198
  detection with Tajima's *D* statistic, 199
  estimation from haplotype structure, 203–204
*Age-structured population,* 7
Alcohol dehydrogenase (*Adh*) gene, 42–43
*Allele,* 21
  beneficial (*see* Beneficial allele)
  deleterious (*see* Deleterious allele)
  fate of new at moment of appearance, 59–62, 80, 84, *85*
  infinite alleles model, 94–97, *96,* 186
  *modifier,* 169–174
  mutation, nomenclature compared, 132
  *parental,* 92
Allele frequencies
  cline (*see Cline*)
  deterministic predictions, 7, 21–25, 35–36, 38, 40–43
  evolution under mutation alone, 21–23
  mathematical connection to genotype frequencies, 30–34
  mathematical connection to haplotype frequencies, 117–119
  stochastic predictions, 70–73, 88–91, *90*
  variance in nature, 182
Allele frequency clines, 42–43
*Allozygote,* 46
*Alphabet,* 138
*Ancestral nucleotide,* 104, 106
*Ancestral recombination graph (ARG),* 148–151, *149*
*Ancient DNA,* 176
*Anisogamous,* 30, 170
*Antimutator,* 173
Apportionment of human diversity, 49–50
Approximation, 14
  diffusion (*see Diffusion approximations*)
  infinite alleles approximation (*see Infinite alleles model*)

infinite sites approximation (*see Infinite sites model*)
  strong selection/weak mutation (SSWM) 141, 143, 151–152, 162, *163,* 165
  Taylor, 60–62, *61,* 73, 85
*ARG. See Ancestral recombination graph*
Artificial selection in dogs, 177
*Asexual organisms,* 28, 29
  effective population size estimation in, 185–190
  expected coalescence times under recurrent selective sweeps, 151–153
  Muller's ratchet and, 166–167
*Assortative mating,* 44–45, **45,** 200
*Asymptotic approach,* 15
Autosomes, 204–205
*Autozygote,* 46

*Background selection,* 159–160, 169, 173, 211–212, 214
*Back mutation,* 23
  beneficial allele production by, 26
  error catastrophe halted by, 135–136
  Muller's ratchet halted by, 168
  rate, 23
*Backward diffusion approximation,* 68, 72–73
  probability of fixation for a selectively neutral allele, 79–82
  probability of fixation for selected allele, 82–86
*Balancing selection,* 39, 212–213. *See also Overdominance*
*Bateman–Mukai* method, 178–180
Beneficial allele. *See also* Mutation, beneficial
  back, reversion or compensatory mutation, via, 26, 136
  driver (*see Driver mutation*)
  establishment, 53–54, 59–62, 83–84
  fixation, 53, 59, 62–63, 85
  genetic draft, 151–159
  fixation by random genetic drift, 63
  stochastic loss, 59, 62
  time to fixation, 87–88

β-globin locus, 38, 205
Biased component of probability mass movement
    over time. *See* Translation of stochastic
    probability distributions over time
*Biased random walk,* 57
*Binomially distribution,* 74
Biological species concept, 40
Biparental organisms
    life cycle, *29*
    reproduction, 2, 21
*Birth-death process,* 79
*Biston Betularia* (peppered moth), *9,* 9–10, 14, 43,
    177
Blood groups, human, 34
Box, George, 2, 14
*Branching process,* 55, 59, 84–85, *85,* 143

Cann, Rebecca, 185
*Carrying capacity,* 7, 9
*Census size,* 108, 112, 182
Centimorgans (cM), 120. *See also Genetic map*
    *distance*
Chaperones, molecular, 173–174
*Chapman–Kolmogorov equation (P),* 62–73, *66*
    diffusion approximations of, 67–73
    representing biology in, 73–79
*Chromosome,* 120
    degeneration of Y, 168
    free recombination between, 120–121
    linkage groups, 120
    as mosaic of MRCAs, 198
    segregation during meiosis, 116, 195–196
    sex, 168, 204–205
*Cline,* 42–43
Clonal interference model. *See Fisher–Muller effect*
*Clonal reproduction,* 2–3, 28–29, *29,* 170
*Closed-form expression,* 21, 67, 78
*Coalesce,* 97
Coalescent effective population size. *See Effective*
    *population size*
*Coalescent event,* 97
    population expansion, effect of, 187
    probability density through time, 189, 194
*Coalescent theory,* 97–108, 187, 189
    probability of incomplete lineage sorting (ILS),
        106–108
Coalescent times
    background selection, with, 159–160
    expectation under Wright–Fisher model, 98–101
    expectation with migration, 111–112
    genetic draft, with 151–159
    Hamming distances and, 187
    immediately after selective sweep with
        recombination, 154
    multilocus, 151–160, *153, 156*
    pairwise sequential Markovian coalescent,
        estimation with, 194
    population size and structure, influence on,
        187–188, *188*
    single-locus, 98–102
    soft selective sweep, after, 215–216, *215*
*Codon,* 183
*Codon usage bias,* 183–184
Coefficient of linkage disequilibrium (*D*), 118

*Combinatorially complete datasets,* 143
*Compensatory mutation* or *second site mutation,* 136
*Compound parameter,* 8
Conditionally beneficial mutation, 136
Constant of integration, 5, 89, 91
Covariance, 122–123
Critical frequency, of beneficial allele establishment,
    84
Crossover events, 198

Darwin, Charles, 3, 15, 30, 46, 205
*Degrees of freedom,* 118–119, 124, 139
Deleterious allele. *See also* Mutation, deleterious
    background selection, 159–160, 211–212
    compensatory mutation created by, 136
    error catastrophe, 133–136
    Muller's ratchet, 166–168
    mutation/selection equilibrium, 23–25, 128–132
    nearly neutral mutations, 86
    probability of fixation, 86
    selection coefficient, 96
    time to fixation, 87
    time to loss, 87
Deleterious mutation rate, 128–129, 133–134, *134,*
    169, 173, 178
Deletions, 23
*Deme,* 40, 42, 47–48, 50
    isolation between recently subdivided, 106,
        202–203
    reciprocal monophyly between, 106
*Demography,* 108, 211
    sex-specific, 204–205
Denisovans, 176, 204
*Density-dependent selection,* 18
*Dependent variables,* 2
*Derivatives,* 4
*Derived nucleotide,* 104, 106
*Deterministic models,* 2
DFE. *See Distribution of fitness effects*
*Diffusion* of stochastic probability distributions over
    time, 58, *69,* 69–70
*Diffusion approximations*
    backward (*see Backward diffusion approximation*)
    of Chapman–Kolmogorov equation, 67–73
    forward (*see Forward diffusion approximation*)
Diffusion equations, representing biology in, 73–79
Diffusion theory, computing times to fixation and
    loss in, 87
*Dioecious,* 30
*Diploid organisms,* 28–30
    asexual, 29
    natural selection in, 35–39, *38*
    nonrandom mating in, 43–50
    phase of life cycle, 28–30, *29,* 35, 115
    population genetics of reproduction, 30–34
*Direct selection. See Modifier theory*
*Disassortative mating,* 44–45
*Distribution of fitness effects* (*DFE*), 10, 209, 214
Divergence, 49, 81, *81,* 93, 158–159, 206–207, 209, 211
*Diversifying selection,* 40, 42–43, 49, 199, 213–214
Dogs, population genetics of, 177
Dominance, 36, 116, 127
    *fully dominant,* 37
    *fully recessive,* 37

genic, _37_
modifiers of, 174
*overdominance*, _37_–39, 82
*partially dominant*, _37_
*partially recessive*, _37_
*underdominance*, _37_–39
*Dominance coefficient* (*h*), _36_, 76, 117, 124, 127
  estimating from mutation accumulation lines, 179
  in nature, 36
Draft. *See Genetic draft*
Drift. *See Random genetic drift*
*Drift-barrier hypothesis*, _86_
Drift load. *See Genetic load*
*Driver mutation*, _151_–159
  in nature, 211–216
*Drosophila melanogaster*, 42–43, 103, 120, 158
*Drosophila pseudoobscura*, 172
*Drosophila simulans*, 212
Drug resistance, 214–216

*Effective population size* (*N*e), _108_–112
  *coalescent*, _110_–112, 153
  *eigenvalue*, _110_
  genetic draft and, 153
  *inbreeding*, _110_
  population subdivision and, 111–112
  unequal sex ratio and, 110
  *variance*, _110_–111, 182
Effective population size estimation in nature
  direct observation of reproductive variance, via, 182
  distribution of pairwise Hamming distance, via, 185–188, *187*
  equilibrium, assuming, 182–184
  linkage disequilibrium, from, 184–185
  nonequilibrium, in asexuals, 185–190
  nonequilibrium, in sexuals, mutation-based, 190–195
  nonequilibrium, in sexuals, recombination-based, 195–198
  pairwise sequential Markovian coalescent, with, 193–195
  site frequency spectrum (SFS), from, 190–191
  skyline plots, with, 188–190
  Tajima's *D* statistic, from, 191–193
Eigen, Manfred, 26
Eigenvalue effective population size. *See Effective population size*
*Epistasis*, _116_
  on fitness landscape, 139–147
  linkage disequilibrium, influence on, 125–127, *125–127*
  magnitude, 141, *142*
  negative, 125–127, *125–127*, 171 (*see also Hill–Robinson effect*)
  positive, 125–127, *125–127*
  selection, and, 123–128 (*see also Hill–Robinson effect*)
  sex, evolution of, and, 170–171
  sign, 141–144, *142, 145*
  Wrightian (ε), 124
*Equilibrium*, _22_–23
  migration/drift (*see Migration/drift equilibrium*)
  migration/mutation/drift, 203

migration/selection, 41–42
mutation/drift (*see Mutation/drift equilibrium*)
mutation/selection, 23–25, _24_, *25*, 128–132, *131* (*see also Error Threshold; Haldane–Muller principle*)
polymorphism, under mutation alone, 21–23
polymorphism, under selection alone, 38–39
recombination/drift, 150–151, 181, 184–185
*Error catastrophe*, _25_, 25–27
  multilocus, 129, 133–136
*Error threshold*, _25_
  multilocus, 133–136, *134*
  one locus, *25*, 25–27
*Escherichia coli*, 7, 10, *144*
Established population, size of, 57
*Establishment*, _53_–54
  beneficial allele, of 53–54, 59–62, *83*, 151–153, 162–164, *163, 164, 215*
  population, of 54–59, *55, 56*
*Eukaryotes*, _23_
Evolution 1–2
  drug resistance in *E. coli*, *144*
  drug resistance in HIV, 214–216
  human evolutionary history, 176
  of mutation rate, 173
  by natural selection, 7–21, *8, 9, 12*
  nearly neutral model of molecular, 86, 96
  neutral theory of molecular, 80–82, *81*, 205, 209–210
  parallel, 9, 214
  of sex, 170–172
*Evolutionary rescue*, _28_
*Exome*, _175_–176
*Expected value*, _52_. See also *Stochastic models*
*Exponential distribution*, _93_
Exponential growth of a population, 3–7, *4*
*External branches*, _191_
Extinction, 26–28
  mutational meltdown, by 167–168
  probability of in new population, 54–56

*Facultatively sexual organism*, _116_
*Female*, _30_
*Fertility selection*, _35_
First derivatives. *See Derivatives*
Fisher, R. A., 15, 198
*Fisher–Muller effect*, _162_–166, *163, 164*
  *clonal interference model*, _164_, *164*
  evidence in laboratory populations of microbes, 164–165
  *multiple mutations model*, _163_–164, *164*
Fisher's fundamental theorem of natural selection, 15–18, 126, 138, 171
Fisher's theory of junctions, 198
Fitness
  absolute Wrightian (*W*), 19, 51
  genic, 37
  heterozygote, 36–37
  Malthusian (*r*), 7–8, 26–27, 79, 124
  relative Wrightian (*w*), 19, 35, 60, 123–124
*Fitness function*, _130_, 133–134
  fitness landscape, 138
  multilocus error threshold, 133–135, *134*
  *rotationally symmetric*, _130_, 137–138, 159–160

*Fitness landscape,* 138–144, *145*
　crossing fitness valleys on, 144–147, *146*
　epistasis, 139–147
　multilocus adaptation on, 140–143
　peaks (*see Fitness peaks*)
　valley (*see Fitness valley*)
Fitness peaks
　global, 139–147, *142, 145*
　local, 139, 144, *145,* 146–147
　number of, 140–143
*Fitness seascape,* 138
*Fitness valley,* 139
　crossing of, 144–147, *146*
　symmetric, 144, *145*
Fitness variance. *See also Fisher's fundamental*
　　*theorem of natural selection*
　modifier alleles act on, 171
　in a mutation-accumulation experiment, 178–179
　in a population, 15–18, 126–127, 160, 168, 171
　population bottleneck and, 178–179
　recombination, influence on, 126–127, 171
*Fixation,* 11
　beneficial allele or mutation, 7–12, 53, 59, 62–63,
　　85, 151, 211
　deleterious allele or mutation, 62, 145
　probability for selected allele, 77–79, 82–87
　probability for selectively neutral allele,
　　79–80
　sequential and simultaneous, in crossing fitness
　　valleys 145, 147
　time for, 12–15, 87–88, 102
　using backward diffusion approximation to study,
　　79–88
*Fixation index (F),* 43–45
　inbreeding coefficient ($F_I$), 46, 195–196
　$F_{IS}$, 50
　$F_{IT}$, 50
　$F_{ST}$, 49–50, *94,* 200–201, 213
　$F_t$, 92–93
Folded site frequency spectrum. *See Site frequency*
　　*spectrum*
*Forward diffusion approximation,* 68, 72–73
　to study internal equilibrium allele frequencies
　　events, 88–91
*Free recombination,* 120–121, 148
*Frequency-dependent selection,* 18

Galton-Watson process, 97
*Gametes,* 28–30
*Gene,* 22, 81, *81,* 205–206
*Genealogy,* 97–98, *98*
　external branches, 191
　marginal, 148–150, *149*
　modeling under Wright–Fisher model, 98–102
　population size and structure influence on
　　coalescent, 187–188, *188*
　population subdivision, and (*see Incomplete lineage*
　　*sorting*)
　with recombination (*see Ancestral recombination*
　　*graph*)
Gene flow, 39–43, 91, 106–108, 201–203, *202*
*Genetic code,* 183
*Genetic draft,* 151–159, 152
　effective population size under, 153
Genetic drift. *See Random genetic drift*

*Genetic load,* 24
　drift load, 25, 162
　inbreeding load, 25
　lag load, 25
　migration load, 25
　mutation load, 24, 25, 26, 129
　recombination load, 25
　segregation load, 25, 82
　substitution load, 25, 81–82
*Genetic map distance (r),* 119–120, *156,* 180, *204*
*Genetic map length (R),* 120–121, 165, 195–196, *196*
*Genetic mapping,* 120
Genetic parameters, measurement of, 177–181
　mutation rate ($\mu$), 178–180
　recombination rate (r), 180–181
Genetic polymorphism. *See* Polymorphism
*Genetic recombination,* 31, 115–116, 119–121
　and background selection, 159–160
　evolution of, 171
　free, 120–121, 148
　and genealogies, 148–151, *149*
　and genetic draft, 154–159
　and the Hill–Robertson effect, 160–166, 168
　linkage disequilibrium, influence on, 121, 126–127,
　　*127*
　in noneukaryotic organisms, 116
　obligate and facultative, 116
　parallels to migration, formal, 150
　reproduction-independent, 116
　and selection, 126–128, *127* (*see also Genetic draft;*
　　*Hill–Robertson effect*)
　widespread in nature, 116, 170
Genetic variation
　created by mutation, 21
　created by recombination, 116, **122,** 148–150, *149*
　eliminated by genetic drift, 91–94
　eliminated by natural selection, 11, 103, 151–160
　maintained by migration, 41–43, 91
　maintained by mutation, 22, 88–90, *90*
　required for natural selection (*see Fisher's*
　　*fundamental theorem of natural selection*)
　segregating, 11
*Genic selection,* 36–37, 38
Genome sequence data 1, 102, 132, 175–177, 180,
　　196–198, 203, 205, 211–214
*Genotype,* 30
Genotype frequencies
　one locus, 30–34, **32**
　multilocus, 115–117
　nonrandom mating, and, 43–45, **45**
　Random genetic drift, impact on (*see*
　　*Heterozygosity*)
　Wahlund effect, impact on 47–50
*Geometric distribution,* 92–93, 99, 186–187, 189
*Global fitness peak. See Fitness peaks*
*Group or optimality selection model,* 171–172

Haldane, J. B. S., 25
*Haldane–Muller principle,* 25, 128–129, 165, 173
*Hamming distance,* 137–138
　equilibrium distribution of pairwise, 185–188, *187,*
　　190
　demography, effect of, 188
　in mitochondrial DNA, 185–187, *187*
　recombination, effect of, 181

*Haploid organisms,* 28–30
  phase of life cycle, 28–30, *29,* 35, 115–117
  selection in haploid phase, 29–30, 123–124
*Haplotype,* 117
  Hamming distance between, 137
  sequence space, *137,* 137–141, *145*
Haplotype frequencies, 121, **122**
  mathematical connection to allele frequencies,
    117–119
  multilocus Wahlund effect, impact on 121–123
  at mutation/selection equilibrium, 128–133, *131*
  recombination, influence on, 121
Haplotype structure in a population
  admixture estimation from, 203–204
  detecting natural selection from, 213–214
  population size estimation, from 195–198
Hard selection. *See Selection*
Hard selective sweep. *See Selective sweep*
Hardy, H. G., 32
*Hardy–Weinberg frequencies* 31–32
  derivation 32–34
  deviations due to nonrandom mating, 44–47
  deviations due to nonrandom mating and
    population subdivision, 50.
  deviations due to population subdivision, 47–50, *48*
  deviations due to random genetic drift, 91–94, *94*
*Harmonic mean,* 108–109
HCV (hepatitis C virus), 190, 194
*Heterozygosity,* 32, 34. *See also Fixation index;*
    *Hardy–Weinberg frequencies*
  background selection, and 159–160
  biased effect induced by stochasticity, 93–94, *94*
  effective population size estimates from, 182–184
  genetic draft, and 151–159, *153, 156*
  Lewontin's paradox and (*see Lewontin's paradox*)
  migration/drift equilibrium of, under infinite
    alleles model, 97
  mutation/drift equilibrium of, under infinite alleles
    model, 94–96, *96,* 186
  mutation/drift equilibrium of, under infinite sites
    model, 102–103
  natural selection detection from, 211–212
  per locus (*H*) (*see Per locus heterozygosity*)
  per site (*π*) (*see Per site heterozygosity*)
*Heterozygote,* 30
  fitness, 36–37 (*see also Dominance*)
  loss under perfect assortative mating, 44–45
  loss under random genetic drift 91–94, *94*
Hidden Markov model (HMM), 194
*Hill–Robertson effect,* 85, 160–162, **161**
  Fisher–Muller effect, 162–166, *164, 164,* 165
  Muller's ratchet, 162, 166–168
*Hitchhiking or passenger mutations,* 151–159,
    *156*
  antimutator hitchhiking, 173
  mutator hitchhiking, 169
HIV (human immunodeficiency virus)
  drug resistance, 214–216
  genome, 137
  mutation rate, 103
  population expansion, 189–190
*Homolog,* 30–31
*Homozygote,* 30
Human evolutionary history, 176
Human genome project, 175

Human genome sequences
  ancient DNA, 176
  mutation rate estimation from, 180
  number published, 175–176

*Identity by descent* (*IBD*), 45–47
  coalescence and, 97
  demography and, 195–198, *196,* 203–204
*ILS. See Incomplete lineage sorting*
*Inbreeding,* 45–47
  pedigree collapse, 198
  random mating and, 91–95
*Inbreeding coefficient. See Fixation index*
Inbreeding effective population size. *See Effective*
    *population size*
Inbreeding load. *See Genetic load*
*Incomplete lineage sorting* (*ILS*), 106–108, *107*
*Independent variables,* 2
*Indicator function* (*I*), 18, 94, *94*
*Indirect selection. See Modifier theory*
Inductive proof, 104–105
*Infinite alleles model,* 94–96, *96,* 99, 186
*Infinite sites model,* 102–106, 153, 186–188, 192–193,
    203, 206–208
Insertion mutations, 23
Integral operator (∫), 52, 102
*Internal equilibrium,* 38, 88
*Intrinsic rate of growth,* 3
*Introgression,* 203–204
*Island models of population structure,* 40

Joint site frequency spectrum. *See Site frequency*
    *spectrum*
Jukes–Cantor model, 206–208
Junctions, Fisher's theory of, 198

Karlin, Sam, 2
Kimura, Motoo, 80–82

Lactase, 214
Lag load. *See Genetic load*
Landsteiner, Karl, 34
Law of large numbers, 53–54, 56–57, 62, 84, 88
Lewontin, Richard, 49–50
Lewontin's *D',* 119
Lewontin's paradox, 96, *96,* 109, 152–153, 158
*Linkage disequilibrium,* 118–119, *119*
  coefficient (*D*), 118 (*see also Lewontin's D'; and*
    *Pearson's correlation coefficient*)
  decay of, in estimating admixture dates, 203–204, *204*
  epistasis, influence of, 125–127, *125–127*
  evolution of sex, and, 170–171
  Hill–Robertson effect, and 160–168
  multilocus Wahlund effect and, 121–123
*Linkage groups,* 120
*Locus* (plural: *loci*), 21
*Logarithm,* 6
  base, 6
  log order, 6
  semi-log plot, 5, 6
*Logistic model,* 7

*Magnitude epistasis,* 141, *142*
Major histocompatibility complex (MHC) locus,
    38, 44

*Male*, 30, 170
*Malthusian fitness or parameter (r)*, 3, 5, 7–8, 19, 27, 79, 124
*Malthusian selection coefficient (s)*, 8, 20
*Marginal genealogies*, 148–150, *149*
*Mathematical models*, 1–2
  deterministic, 2
  solving, 1–2, 13
  stochastic, 2, 52
*Mating*, 29
  assortative, 44–45, **45**, 200
  disassortative, 44–45
  nonrandom (*see Nonrandom mating*)
  random (*see Random mating*)
Matrix multiplication, 65, *66*
McDonald/Kreitman test, 209–211, **210**
Mean
  computing, 16–17, 52
  Chapman–Kolmogorov transition matrix, of (*M*), 71–72, 75–76, 78
Mean fitness, 15–17, 19, 24, *25*, 35, 125, 129
*Meiosis*, 28–33, *29*, **32**, 35, 48, 115–117, 119–120, **122**, 154, 169, 195
*Meiotic drive*, 31, 33, 34
*Mendelian segregation*, 31, **32**, **45**, 117
*Mendel's first law of segregation*, 31, 46
*Mendel's second law of independent assortment*, 120
*Metapopulation*, 40, 47, 49, 121–123, 199, 201–202, *202*
MHC (major histocompatibility complex) locus, 38, 44
Migration, 39–43. *See also Gene flow*
  estimating from $F_{ST}$, 200–201
  estimating from haplotype structure, 203–205
  estimating from joint site frequency spectrum (jSFS), 201–203, *202*
  out-of-Africa by humans, 176, 185 191, 194, 203, 205
  parallels to mutation, formal, 41–42, 91, 97, 150
  parallels to recombination, formal, 150
  sex-biased effects on, 204–205
  time to coalescence, and, 111–112
*Migration/drift equilibrium*, 91
  allele frequencies under diffusion approximation at, 91
  heterozygosity under Wright–Fisher model, 97, 100
  joint site frequency spectrum (jSFS) at, 203
Migration load. *See Genetic load*
Migration/mutation/drift equilibrium, 203
*Migration/selection equilibrium*, 41–42
*Missense mutation. See Nonsynonymous mutation*
Mitochondria, 185
Mitochondrial DNA (mtDNA), 185–188, *187*
*Mitochondrial Eve*, 185
*Mitosis*, 29
MN blood groups, 34
Models, 2. *See also specific models*
*Modifier alleles (modifiers)*, 169–174
*Modifier theory*, 168–174, *169*
  evolution of mutation rate, 173
  evolution of sex, 170–172
  short- and long-term fitness consequences, 171–172
Molecular chaperones, 173–174
*Molecular clock hypothesis*, 80–81

genetic draft and, 159
quantifying natural selection with, 205–209
rate constancy in units of time, 86
selective sweeps and, 159
*Monoecious*, 30
*Monomorphic*, 11
*Moran model of random genetic drift*, 77–79
  probability of fixation of selected allele, under, 78–79, *85*
  Wright–Fisher model compared, 78–79
Morgan, Thomas Hunt, 120
*Most recent common ancestor (MRCA)*, 97–98, *98*. *See also Incomplete lineage sorting*
  expected number of generations since under Wright–Fisher model, 101–102
  inferential time horizon, due to, 190–191
  reciprocal monophyly, 106–107
  sexually reproducing individuals, many of, 198
  shared by two loci, probability of 150
*MSMC. See Multiple sequential Markovian coalescent*
Muller, H. J., 178
*Muller's ratchet*, 162, 166–168
  degeneration of Y chromosomes, 168
  mutational robustness and, 174
Multilocus coalescent theory, 148–151
*Multilocus genotype*, 115
Multiple mutations model. *See Fisher–Muller effect*
*Multiple sequential Markovian coalescent (MSMC)*, 194–195
Mutagenic drugs, 27
Mutation, 21–28
  allele, nomenclature compared, 132
  ancestral, 104
  back or reversion, 23, 26, 136, 141, 206
  beneficial, 26, 140–144, 209–211 (*see also* Beneficial allele; *Driver mutation*)
  compensatory or conditionally beneficial or second site, 136
  deleterious, 23–26, *25*, 128–136, *131*, *134*, 166–168, 178–179 (*see also* Deleterious allele)
  derived, 104
  infinite alleles model of (*see Infinite alleles model*)
  infinite sites model of (*see Infinite sites model*)
  kinds, 23 (*see also Nonsense mutation; Nonsynonymous mutation; Point mutation; and Synonymous mutation*)
  maintenance of genetic variation, role in, 21, 94–96
  nearly neutral, 86
  neutral, 79–82
  parallels to migration, formal, 41–42, 91, 97, 150
  polarization of, 106
  rates (*see* Mutation rate)
  recurrent beneficial, 151–154, *153*, 157–159
  recurrent deleterious, 23–26, 128–136, 159–160, 166–168
  two-way, 21–23, 88–90
*Mutation-accumulation (MA) lines*, 178–180
*Mutation/drift equilibrium*. 89. *See also Infinite alleles model; Infinite sites model*
  allele frequency probability density at, 88–91, *90*
  allelic heterozygosity at, 94–97, *96*, 99
  effective population size estimates, assuming, 182–183

number of variable sites in a sample at, 103
  pairwise Hamming distance at, 187, *187*
  per-site heterozygosity at, 102–103
  site frequency spectrum (SFS) at, 104–106
Mutation load. *See Genetic load*
Mutation rate ($\mu$, $v$), 21. *See also* Mutation
  empirical measurement of, 23, 178–180
  evolution of, 173
  neutral, 80–81, 96, 184–185, 205–206
  per allele, 95
  per nucleotide, 102
  population size, inverse relationship with, 86
  symmetric, 22
*Mutation/selection equilibrium*, 23–25, <u>24</u>, *25*,
    128–132, *131*. *See also Error Threshold;*
    *Haldane–Muller principle*
*Mutational bias*, <u>23</u>, 183–184
*Mutational meltdown*, 167–<u>168</u>
*Mutational robustness*, 173–<u>174</u>
*Mutational saturation*, <u>207</u>, *207*
*Mutational trajectories*, <u>141</u>–144, *142, 144*
*Mutator*, <u>169</u>, 173

*Natural selection*, <u>3</u>–21. *See also Selection*
  biased influence on allele frequency, 62
  in *Biston betularia*, 9, *9*
  Darwin's model of, in diploids, 35–39, *38*
  Darwin's model of, in haploids, 7–12
  detecting in nature, 205–217, **210**
  Fisher's fundamental theorem of, 15–18, 39, 126,
    171
  linkage disequilibrium, and, 124–125 (*see also*
    *Hill–Robertson effect*)
  Moran model, in, 78–79
  nonoverlapping generations, assuming, 18–21, 27,
    123–127
  overlapping generations, assuming, 7–12
  probability of fixation for selected allele, 79, 82–86,
    *83, 85*
  purging of recurrent deleterious mutations, 23–25,
    *25*, 127–132, *131* (*see also Error threshold;*
    *Muller's ratchet*)
  and recombination, 126–127, *127*
  time course of, *8, 12, 38, 125, 127*
  Wright–Fisher model, in, 76
Neanderthals, 176, 203–204
*Nearly neutral alleles*, <u>86</u>
Nearly neutral model of molecular evolution, 86, 96
Negative (purifying) selection. *See Selection,*
    *negative*
*Neutrality Index* (NI), <u>210</u>–211
Neutral mutation rate. *See Mutation rate, neutral*
Neutral theory of molecular evolution, 80–82, 86,
    205–210
Next-generation sequencing, 177
*Nonoverlapping generations model*, <u>18</u>
  deterministic, 18–21, 30–39, 123–136
  in stochastic, 59–67, *66*, 73–77
Nonrandom mating, 40
  in diploids, 43–50
  population structure and, 47
  within subdivided population, 50
*Nonrecombinant gamete*, <u>116</u>
*Nonsense mutation*, <u>183</u>
*Nonsynonymous mutations*, <u>183</u>, 208–212, **210**

*Obligately sexual organism*, <u>116</u>
Ohta, Tomoko, 86
*One-step process*, <u>79</u>
*On the Origin of Species* (Darwin), 3, 15, 46
*Optimality or group selection model*, <u>171</u>–172
*Orthologs*, <u>205</u>–210
*Outlier loci*, <u>213</u>
Out-of-Africa migration, 176, 191, 194, 203, 205
*Overdominance*, <u>37</u>–39, 82
*Overlapping generations model* <u>18</u>
  deterministic 3–18, 20–21
  stochastic, 77–79

Pairwise coalescent times. *See Coalescent times*
Pairwise Hamming distances
  population size estimation, using, 185–188, *187*
  recombination rate estimation, using, 181
*Pairwise sequential Markovian coalescent* (PSMC),
    193–195, *194*
*Parallel evolution*, 9, <u>214</u>
*Paralogs*, <u>206</u>
*Parameters*, <u>2</u>
  compound, <u>8</u>, 24, 42, 83, 84, 87, 89, 95, 97, 108,
    145, 155, 177
*Parental allele*, <u>92</u>
*Parthenogenic*, <u>170</u>
Partial derivative. *See Derivatives*
*Passenger mutations. See Hitchhiking or passenger*
    *mutations*
PCA. *See Principal component analysis*
Pearson's correlation coefficient ($r_p$), 119
  population size estimation using, 184
  recombination rate estimation using, 181
*Pedigree*, <u>46</u>, 178
*Pedigree collapse*, <u>198</u>
Percolation theory, 140
Per locus heterozygosity ($H$), 91–97, *94, 96*, 99,
    186–187
Per site heterozygosity ($\pi$), 103, 153–154, *156*, 157,
    186, 192–193
*Phenotype*, <u>16</u>
*Physical distance*, <u>155</u>–158, *156*
Physical recombination rate, 155–158, 160, 165, 180,
    211–212
*Point mutation*, <u>23</u>, 137, 180
Poisson distribution, 59–60, 128–129, 132, 166
*Polarize*, of alleles or mutations, <u>106</u>
*Polymorphic*, <u>11</u>
Polymorphism. *See also Single-nucleotide*
    *polymorphism*
  alcohol dehydrogenase (*Adh*) gene, in, 42
  maintained by migration, 41–43, *42*, 91
  maintained by mutation, 21–22, 88–90, *90*
  maintained by balancing selection, 39–40, 213
*Population bottleneck*, <u>96</u>, 211
  effective population size, influence on, 109
  fitness variance and, 178–179
  heterozygosity reduction, 96
  mitochondrial DNA, during organismal
    reproduction, 185
  mutation accumulation lines, during, 178–179
  out-of-Africa migration, 176, 191, 194
  selective sweep as genealogical, 211–213
  site frequency spectrum, impact on, 191
  Tajima's $D$ statistic, impact on, 192–193

Population expansion, 187–188, *188,* 190–193, 199
Population size
  carrying capacity (*K*), 7
  census size, 108–109, 111–112, 152–153, *153,* 182
  decline and Muller's ratchet, 167–168
  deterministic theory, exponential growth of, 3–7, *5*
  effective (*see Effective population size*)
  estimation in nature, 181–198, *187, 188, 196*
  stochastic theory, during and after establishment
    of, 54–59
  the Wright–Fisher model and, 75
Population structure and subdivision
  coalescence, influence on, 111–112
  detecting, 198–205, *202, 204*
  $F_{ST}$, 49–50
  incomplete lineage sorting, 106–108, *107*
  models of, 40
  Wright's shifting balance theory, 144
Positive selection. *See Selection*
*Principal component analysis (PCA),* 199–200
*Probability density function,* 52
*Probability mass function,* 16–17, 52, 64, 72
Probability of coalescence, 98–101, 109, 155
Probability of fixation
  for alleles as function of starting frequency, 83, *83*
  population size, influence on, 86
  for selected allele, 77–79, 82–86, *85*
  for selectively neutral allele, 79–80
Proofs by induction, 104
*PSMC. See Pairwise sequential Markovian
    coalescent*
Purifying selection. *See Selection*

*Quasispecies,* 26
*Quod erat demonstrandum* (Q.E.D.), 17

Race, human. *See Apportionment of human
    diversity*
*Random genetic drift,* 53, 62–63. *See also* Moran
    model of random genetic drift; Wright–Fisher
    model of random genetic drift
  beneficial alleles can be fixed by, 63
  heterozygosity and, 91
Random mating, 31–32, **32**
  inbreeding under Wright–Fisher model, in spite of
    93–94, *94*
Ratio of allelic frequencies under natural selection, 12
*Realized value,* 52
*Recessive. See Dominance*
*Reciprocal monophyly,* 106–107
*Reciprocal sign epistasis, 142, 143, 145*
*Recombinant gamete,* 116
Recombination. *See Genetic recombination*
Recombination/drift equilibrium, 150
  effective population size estimates, assuming,
    184–185
  recombination rate estimates, assuming 180–181
Recombination load. *See Genetic load*
Recombination rate (*r*), 119–120
  heterozygosity correlated with, 158–160
  hotspots, 155
  measuring, 180–181
  physical, 155, 157–158, 160, 166, 180
  and selection, 126–128, *127* (*see also Genetic draft*;
    *Hill–Robertson effect*)

*Recurrence equation,* 54, 59, 65, 67, 92, 95, 97, 150,
    186, 207
*Reduction Principle,* 172–173
Redundancy, of genetic code, 183
Relative Wrightian fitness. *See Wrightian fitness*
Reproduction
  clonal, 2, 11, 21, 28–29, *29,* 116, 170
  diploid, 28–34, *29*
  overlapping and nonoverlapping generation
    models of, 18
  parthenogenic, evolutionary advantage of
    170
  sexual (*see Sexual reproduction*)
Reproductive isolation, 40
Reproductive variance
  among carriers of a beneficial allele, 59
  direct observation of, 182
  in the Wright–Fisher model, 75
*Reversion,* 23. *See also Back mutation*
*Root* (algebraic), 24
*Rotationally symmetric* fitness function, 130,
    137–138, 159, 166, 168
*Runs of homozygosity (ROH),* 195–197, *196,* 203,
    205, 211, 213–214

*Sampling with replacement,* 73–75, 92
Scaling relationships, 84, 86. *See also Parameters,
    compound*
Second derivatives. *See Derivatives*
*Second-site mutation* or *compensatory mutation,*
    136
*Segregating variation,* 11
Segregation
  chromosome, 116, 169, 195–196
  *Mendel's first law of segregation,* 31, 46
Segregation load. *See Genetic load*
Selected allele
  probability of fixation for, 79, 82–86, *83, 85*
  time to fixation for new, 87–88
Selection. *See also Natural selection*
  artificial, 177
  background, 159–160, 169, 173, 211–212, 214
  balancing, 39, 212–213
  density-dependent, 18
  direct (*see Modifier theory*)
  diversifying, 40, 42–43, 49, 199, 213–214
  fertility, 35
  frequency-dependent, 18
  genic, 37
  group, 171–172
  in haploid phase, 29–30
  indirect (*see Modifier theory*)
  *hard,* 27, *28*
  hard and soft selective sweeps, 214–217, *215*
  indirect (*see Modifier theory*)
  long- and short-term, conflict between, 171–172
  *migration/selection equilibrium,* 41–42
  mutation/selection equilibrium (*see Mutation/
    selection equilibrium*)
  *negative* or *purifying,* 11, 86, 130–135, 160, 166,
    205, 209–211
  *positive,* 11, 151, 162, 209, 212, 214
  *sexual,* 30
  *soft,* 27, *28,* 167
  *viability,* 35

Selection coefficient (*s*), 8, 10, *39, 42–43*
  Malthusian, 8
  Wrightian, 20
Selective fixation. *See Fixation*
*Selective interference,* 160
*Selectively accessible trajectories,* 141–144, *144*
Selectively neutral allele or mutation
  neutral theory of molecular evolution, 81–82
  passenger or hitchhiking mutations, 151, 154–155, *156*
  probability of fixation for, 79–80, 86
*Selective sweep,* 151–159, *153,* 163–166, *163, 164,* 211
  driver mutation (*see Driver mutation*)
  *hard,* 214–217, *215*
  *soft* (*see Soft selective sweep*)
*Selfing* (self-fertilization), 46
Semi-log plot. *See Logarithms*
*Sequence space,* 136–137, *137, 145,* 199
Sex, evolution of, 170–172
Sex chromosomes, 120, 168, 204–205
*Sexes,* 30
Sex ratio, unequal, 110, 182
Sex-specific demographic history, 204–205
*Sexual dimorphisms,* 30
*Sexual organism,* 28, *29*
  number of genealogical ancestors, 198
*Sexual reproduction,* 2, 28–30, *29*
  evolutionary persistence of, 170
  *facultatively sexual organisms,* 116
  multilocus, 115–121
  *obligately sexual organisms,* 116
  one-locus, 28–34
  pace of adaptation, influence of, 126–127, *126, 127,* 162–166, *163, 164*
*Sexual selection,* 30
SFS. *See Site frequency spectrum*
*Shifting balance theory,* 144
Sickle-cell anemia, 38, 205
*Sign epistasis,* 141–144, *142*
  reciprocal, *142,* 143–145, *145*
*Single-nucleotide polymorphism (SNP),* 180, 182, 184–185
  joint site frequency spectrum (jSFS), 201–203, *202*
  lactase persistence causal for, 214
  McDonald/Kreitman test, 210, **210**
  population-specific frequencies of, 200
  principal component analysis of, 199
  selectively neutral, 182, 208–209
  site frequency spectrum (SFS), 190–191, 212–213
  Tajima's *D* statistic 191–193, 199
*Single-sperm genotyping,* 180–181
*Site frequency spectrum (SFS),* 104–106, *105, 106*
  detecting historical population size changes, 190–191
  detecting natural selection from, 212–213
  *folded site frequency spectrum,* 106, *106*
  *joint site frequency spectrum (jSFS),* 201–203, *202*
Skyline plots, 188–190, *189*
SNP. *See Single-nucleotide polymorphism*
Soft selection. *See Selection*
*Soft selective sweep,* 216–217, *215*
  *de novo* or *multiple origin soft selective sweep,* 216
  *single-origin* or *standing variation soft selective sweep,* 216
*Solving,* a mathematical model, 1–2, 13
Speciation, 40

Sperm competition, 29–30
*SSWM. See Strong selection/weak mutation*
*State variable,* 52. *See also Stochastic models*
  expected and realized values of, 52
Statistical inference, 16, 148, 175, 179–181, 185, 189
*Stepping stone model,* 40
*Stochastic matrix,* 64–65, 68, 74–78
*Stochastic models,* 2, 52. *See also State variable*
  state space, 52
  state transition, 63–64 (*see also Stochastic matrix*)
  stochastic variable (*see State variable*)
Stochastic probability distribution, 52
  diffusion over time of, 58
  translation over time of, 58
*Stochastic tunneling,* 146
*String,* 138
*Strong selection/weak mutation (SSWM),* 141, 143, 151–152, 162, *163,* 165
STRUCTURE, 200
Substitution load. *See Genetic load*
*Summary statistic,* 16
*Symmetric mutation rate,* 22, 90–91, *90*
*Syngamy,* 28–30, *29,* 32–33, 35, 115, 117, 127
*Synonymous codon,* 183
*Synonymous mutation,* 183–184, 208–211, **210**

Tajima's *D* statistic, 191–193, 199
Taylor approximation, 60–62, *61,* 73, 85, 92, 129, 133, 207
Taylor expansion, 60–62, *61*
TEM-1 β-lactamase, 144, *144*
Time to coalescence. *See Coalescent times*
Time to fixation. *See Fixation*
Time to loss, 87
Transition matrix (*T*), 64–69, *66, 69,* 72–74
  for Moran model, 78
  for Wright–Fisher model, 74
Transition probability, 63–65
*Translation of stochastic probability distributions over time,* 57–58, *58, 69,* 69–72
Translocations, 40
*Two-fold cost of anisogamous sexual reproduction,* 170

Unbiased component of probability mass movement over time. *See Diffusion of stochastic probability distribution over time*
*Underdominance,* 37–39
Uniparental life cycle, *29*

*Viability selection,* 35
Variables
  *dependent,* 2
  *independent,* 2
  *state,* 52
Variance
  allele frequencies, 182
  computing of, 16–17, 52
  Chapman–Kolmogorov transition matrix, of (*V*), 71–72, 75–76, 78
  covariance, 122–123
  fitness (*see Fitness variance*)
  partitioning of, with *F* statistics, 49
  reproductive, direct observation of, 182
  in reproductive success, natural selection and, 211

Variance effective population size. *See Effective population size*
*Velocity of adaptation (V),* 162–166, *163,* 171–172

*Wahlund effect,* 47
  heterozygosity lost to random genetic drift, offset by, 93–94, *94*
  one-locus, 47–50, *48*
  two-locus, 121–123
Waiting time
  selective sweeps, expected for, 151–152
  beneficial mutations, expected for, 141, 164
  transit a fitness valley, expected to, 145–147, *146*
Wallace, Alfred Russel, 3
Watson–Crick pairings, 136
Weinberg, Wilhelm, 32
*Well-mixed population,* 39, 42, 106, 108, 153, 198, 201–203, *202*
*Wild type,* 8
Wright, Sewall, 19, 44, 138–139, 144–145
  shifting balance theory, 144
*Wright–Fisher model of random genetic drift,* 73–75
  heterozygosity under (*see Heterozygosity*)
  incorporating biased processes into, 76–77
  modeling genealogies under, 98–102
  Moran model compared, 78–79, *85*
  probability of fixation under, 79–80, 82–86
*Wrightian epistasis coefficient (ε),* 124–127
*Wrightian fitness*
  absolute (W), 19, 35, 51
  relative (w), 19, 35, 60, 123
*Wrightian selection coefficient (s),* 20, 36, 76, 123

X chromosome, 168, 204–205

Y chromosome, 168, 188, 190

*Zygote,* 28, 30–31, 33, 115–116